# WITHDRAWN
## UTSA Libraries

# YEAST TECHNOLOGY

# YEAST TECHNOLOGY

## Second Edition

**Gerald Reed**
*Universal Foods Corporation*
*Milwaukee, Wisconsin*

**Tilak W. Nagodawithana**
*Universal Foods Corporation*
*Milwaukee, Wisconsin*

An **avi** Book
Published by Van Nostrand Reinhold
New York

To our wives, Helen (in memoriam) and Swarna

An AVI Book
(AVI is an imprint of Van Nostrand Reinhold)

Copyright © 1991 by Van Nostrand Reinhold
Library of Congress Catalog Card Number 90-33144
ISBN 0-442-31892-8

All rights reserved. No part of this work covered by the copyright hereon may be reproduced or used in any form by any means — graphic, electronic, or mechanical, including photocopying, recording, taping, or information storage and retrieval systems — without written permission of the publisher.

Printed in the United States of America

Van Nostrand Reinhold
115 Fifth Avenue
New York, New York 10003

Van Nostrand Reinhold International Company Limited
11 New Fetter Lane
London EC4P 4EE, England

Van Nostrand Reinhold
480 La Trobe Street
Melbourne, Victoria 3000, Australia

Nelson Canada
1120 Birchmount Road
Scarborough, Ontario M1K 5G4, Canada

16 15 14 13 12 11 10 9 8 7 6 5 4 3 2 1

**Library of Congress Cataloging-in-Publication Data**
Reed, Gerald.
   Yeast technology/Gerald Reed, Tilak W. Nagodawithana. — 2nd ed.
    p.   cm.
   "An AVI book."
   Includes bibliographical references.
   ISBN 0-442-31892-8
   1. Yeast.  I. Nagodawithana, Tilak W.  II. Title.
TP580.R43   1991
664'.68 — dc20                                   90-33144

# CONTENTS

Foreword  vii
Preface  ix

|   | Introduction | 1 |

Preservation, 2  Nutrition, 3

| 1 | General Classification of Yeast | 7 |

Classification, 10  Impact of Recent Taxonomic Revision on Industry, 30  Yeasts of Economic Importance, 31  Culture Collection, 34

| 2 | Yeast Genetics | 37 |

Life Cycle, 38  Techniques Applicable to Yeast Strain Development, 42  Baker's Yeast, 56  Brewer's Yeast, 72

| 3 | Brewer's Yeast | 89 |

General Characteristics of Yeast, 91  Yeast Characteristics Important for Brewing, 92  Brewing Process, 105  Brewery Contaminants, 121  Biochemistry of Brewing, 126  Recent Developments, 138

| 4 | Wine Yeasts | 151 |

Ecology, 151  Terminology, 151  Natural Yeasts and Their Occurrence on Grapes and in Musts, 154  Natural Fermentations, 156  Description of Species, 157  Selected Pure Culture Yeasts and Active Dry Wine Yeasts, 159  Preparation of Starter Cultures, 162  Biochemistry of Wine Fermentation, 165  The Killer Factor, 184  By-products of the Alcoholic Fermentation and Aroma Compounds, 185  Microbiological Reduction of Acids, 193  Genetic Manipulation of Wine Yeasts, 198  Microbial Spoilage

vi  CONTENTS

of Wines, 199   *Bótrytis cineria*, 201   Biogenic Amines and Ethyl Carbamate, 202   Technology of Wine Making, 204   Further Readings, 215

## 5  Distiller's Yeasts — 225

Whisky, 227   Distillates from Sugar-Containing Raw Materials, 238   Characteristics and Strains of Distiller's Yeasts, 242   Flavor Compounds, 248   Fermentation Alcohol as Fuel, 252   Further Readings, 256

## 6  Baker's Yeast Production — 261

Manufacturing Process Outline, 262   Strains, 263   Principles of Aerobic Growth, 264   Raw Materials, 271   Environmental Parameters, 280   Practice of Aerobic Growth, 284   Further Readings, 309

## 7  Use of Yeast in Baking — 315

Function of Yeast in Baking, 316   Bread Flavor, 334   White Pan Bread Technology, 336   Active Dry Yeast, 359   Further Readings, 362

## 8  Yeast-Derived Products — 369

Flavor Products and Flavor Enhancers, 370   Nutritional Yeast, 385   Colorants Derived from Yeast, 390   Yeast-Derived Enzymes, 393   Products of Pharmaceutical and Cosmetic Value, 399

## 9  Food and Feed Yeast — 413

Composition, 415   Use of Yeast as a Major Protein Source, 424   Production of Biomass, 426   Further Readings, 437

## 10  Use of Yeasts in the Dairy Industry — 441

Cheese, 441   Lactose-free Milk, 441   Acidophilus-Yeast Milk, 442   Kefir, 442   Koumiss, 444   Desugaring of Eggs, 444

Index   446

# FOREWORD

Yeasts are the active agents responsible for three of our most important foods—bread, wine, and beer—and for the almost universally used mind/personality-altering drug, ethanol. Anthropologists have suggested that it was the production of ethanol that motivated primitive people to settle down and become farmers. The Earth is thought to be about 4.5 billion years old. Fossil microorganisms have been found in Earth rock 3.3 to 3.5 billion years old. Microbes have been on Earth for that length of time carrying out their principal task of recycling organic matter as they still do today. Yeasts have most likely been on Earth for at least 2 billion years before humans arrived, and they play a key role in the conversion of sugars to alcohol and carbon dioxide.

Early humans had no concept of either microorganisms or fermentation, yet the earliest historical records indicate that by 6000 B.C. they knew how to make bread, beer, and wine. Earliest humans were foragers who collected and ate leaves, tubers, fruits, berries, nuts, and cereal seeds most of the day much as apes do today in the wild. Crushed fruits readily undergo natural fermentation by indigenous yeasts, and moist seeds germinate and develop amylases that produce fermentable sugars. Honey, the first concentrated sweet known to humans, also spontaneously ferments to alcohol if it is by chance diluted with rainwater.

Thus, yeasts and other microbes have had a long history of 2 to 3.5 billion years. Humans inherited the microbes along with their foods when they arrived on Earth probably about a million years ago. It is only about 300 years since Leeuwenhoek discovered "animalcules" and a little more than 100 years since Pasteur showed the connection between microbes and fermentation. Since then, knowledge of microbes has accumulated at an ever-increasing rate, and the present book covering only one group of microbes—the yeasts—demonstrates the large amount of information currently available on yeast science and yeast technology.

This book is an authoritative, comprehensive condensation of yeast knowledge. It contains the basic scientific knowledge of yeasts and also their practical use in the production of bread, wine, beer, and distilled beverages. There are very good chapters on yeast taxonomy, yeast genetics, food and feed yeasts, and yeast-derived products. It is truly a comprehensive reference on the subject that will be valuable to microbiologists, food scientists, and others interested in yeast food/beverage interactions. It covers a wide range of yeast technology including nutritional aspects, enzymes and nucleotides, and the production of flavor-enhancing compounds — all subjects of great interest to the food industry and to the consumer. The authors are to be congratulated on a fine addition to the scientific literature.

Keith H. Steinkraus
*Cornell University*

# PREFACE

This book is a broad introduction to the technology of yeast as it is used in the food and beverage industries. The main features of commercial processes are explained and illustrated, but the book is not meant to be an operating manual. Rather, it will guide the reader through the principles underlying commercial operations.

Every technology is based on art, science, or a blend of art and science – on what we commonly call know-how. In yeast technology the transition from an art to a science has proceeded since publication of the first edition in 1973, due to progress in engineering, yeast genetics, computer applications, and improvements in analytical techniques. In some instances, scientific investigations have drastically affected traditional processes; in others, they have merely provided a rationale for what has been done traditionally.

The first two chapters provide an introduction into the scientific basis of commercial processes. They deal with the classification, biology, and genetics of yeasts. The following chapters describe the various industries that are based on the use of yeast, principally baking, brewing, wine making, the production of distilled beverages, the propagation of food and feed yeasts, and the production of major yeast-derived products such as yeast extracts and enzymes.

In addition, the monograph stresses the various scientific and technological factors common to several industries for which the use of yeast is basic. It concentrates essential technological information in one handy volume and provides references for further detailed study. The first edition of the book has been particularly well received by food technologists and by chemists, physicists, and engineers interested in the technology of yeast. We hope that the second edition, which has been completely rewritten, will be equally well received.

The invaluable help of Mrs. Betty Blue in the preparation of the manuscript is gratefully acknowledged.

# INTRODUCTION

A few definitions are in order. The word *ferment* means to seethe with agitation or excitement. It is derived from the Latin *fermentum* (yeast) or *fermentare* (to cause to rise). In its broadest sense fermentation includes all microbial processes used industrially, as well as fermentations carried out in the home. It also includes microbial spoilage, as when we speak of fermented honey or wine turned to vinegar. Food fermentations are the production of food and beverages. These are the main subjects of this text. The production of such food ingredients as organic acids, amino acids, and enzymes will be treated only in outline.

Food fermentations can be usefully divided into alcoholic fermentations carried out by yeasts, acid fermentations carried out by bacteria, mixed alcoholic/acid fermentations, and fungal (mold) fermentations. It should be remembered that this division is not clear-cut. For instance, the production of white bread, rolls, and buns is an alcoholic fermentation. But sourdough bread and soda crackers are produced by alcoholic/acid fermentations. Table 1 lists the major food fermentations based on this classification.

All food fermentations started as natural (spontaneous) fermentations, that is, fermentation of any raw material by its indigenous microflora. Examples are the spontaneous alcoholic fermentation of crushed grapes or the spontaneous "souring" of milk. Such processes can be carried out successively by mixing a retained portion of the fermented food with fresh raw material. This food fermentation was the only kind carried out until the end of the nineteenth century. Even today several industrial processes

## 2  INTRODUCTION

Table 1.  Major Food Fermentations

| | |
|---|---|
| Alcoholic fermentations | White bread, beer, wine, whiskey |
| Acid fermentations | Cheese, yogurt, vinegar, sausage, sauerkraut, olives, pickles |
| Mixed alcoholic/ acid fermentations | Kefir, soda crackers, sourdough bread |
| Mold (fungal) fermentations | Blue cheese, sake, soy sauce, and many other oriental fermented foods |

are spontaneous fermentations: all processes that produce sauerkraut, pickles, and fermented olives; some wine fermentations; and the production of soda crackers and San Francisco sourdough bread. In the home the production of yogurt by inoculation with a spoonful of store-bought yogurt is a typical example.

The initiation of spontaneous fermentations is a good indication that the practice of fermenting foods must have arisen in prehistoric times. Very primitive cultures already practiced simple biotechnological steps involving enzymes: for instance, the practice of chewing grain and then spitting it into a pot to convert some of the starch to fermentable sugars by salivary amylase; or the practice of dehairing animal hides by soaking them in urine, where urease produces ammonia.

Records left by the earliest civilizations, such as the Babylonian or the Egyptian, clearly show their knowledge of brewing and baking. The history of these arts will not be discussed here, and the reader is referred to the excellent books by Jacob (1944) and Darby, Grivetti, and Ghallioungui (1977).

## PRESERVATION

This aspect, which is common to almost all food fermentations, is often neglected, possibly because it is taken for granted. The preservative action is due to the production of ethanol or to the lowering of the pH by acid formation and, in many instances, to both these reactions. Obviously, this case is true for alcoholic beverages, depending, of course, largely on the ethanol concentration. Beer keeps reasonably well; wine can be preserved well for many years (unless exposed to the oxygen of the air); and whiskey keeps indefinitely. Acid production is responsible for the preservation of fermented vegetables. Sauerkraut, pickles, and fermented olives can be kept through a long winter, but cabbages, cucumbers, and olives are perishable. In the dairy industry, acid fermentation permits the longtime storage of cheese, though milk is highly perishable. The preservative effect of fermentation is particularly important in Third World countries. Kozaki (1978) said quite simply, "The people of Southeast Asia have an old custom of food storage by fermentation methods. For example, rice wine, sugar cane wine, pickles, fish sauce, fermented noodles, oncom, and tempe."

## NUTRITION

Some fermentations lead to enrichment with vitamins. This situation has been fairly well established for apple cider and beer, although there is not much literature on this subject. Thiamin, riboflavin, and pantothenic acid may be enriched in beer by microbial action. In African maize beer (beer made from corn) a partial substitution of maize beer for whole meal maize resulted in doubling the daily riboflavin and niacin intake (see also Reed 1981).

Fermented foods are highly significant contributors to our diet. In the United States, the annual consumption of fermented foods per capita in 1986 was as follows: 35 pounds of cheese (including cottage and process-type cheese); about 60 pounds of bread and rolls; 3.5 pounds of soda crackers and pretzels; 3.2 pounds of yeast-raised sweet rolls; 6 pounds of olives, pickles, and sauerkraut. The average person drank 23.8 gallons of beer, 2.3 gallons of wine, and 1.8 gallons of distilled alcoholic beverages. Usage of vinegar was 0.82 gallons per person (about 70% for use in processed foods).

The contribution of various nutrients to our total diet is considerable. Taking only the most important categories of fermented foods, namely cheese, beer, and bread and rolls, we find the average contribution as a percentage of the RDA (recommended daily allowance of the Food and Drug Administration) to be as follows: Fermented foods contribute 20% of calories, 37% of protein, 50% of calcium, 45% of phosphorus, 13% of iron, 11% of vitamin A, 17% of vitamin $B_1$, 25% of vitamin $B_2$, and 16% of niacin. The contribution of vitamin C is negligible.

Table 2 shows the calculation for each of the major categories of fermented foods. Not surprisingly, cheese is a major contributor to protein nutrition. Bread and rolls contribute significantly to the intake of iron and vitamins $B_1$, $B_2$, and niacin, largely because of the fortification of bread with vitamins and iron. It is important to realize that these are average values based on the total consumption in the United States. They may vary greatly among individuals, age groups, and geographic areas, and they will certainly vary seasonally.

Table 2. Nutritional Contribution of Fermented Foods

|  |  | g | cal | Prot. g | Ca mg | P mg | Fe mg | A IU | $B_1$ mg | $B_2$ mg | Niacin mg |
|---|---|---|---|---|---|---|---|---|---|---|---|
| RDA |  |  | 2400 | 65 (45) | 800 | 800 | 18 | 5000 | 1.4 | 1.7 | 20 |
| Cheese (cheddar) | a | 24 | 96 | 6.0 | 180 | 115 | 0.2 | 310 | 0.01 | 0.11 |  |
|  | b | 48.5 | 173 | 10.9 | 325 | 208 | 0.36 | 561 | 0.02 | 0.2 |  |
|  | c |  | 7.2% | 24.2% | 40% | 26% | 2% | 11% | 1.3% | 12% |  |
| Bread and rolls | a | 907 | 2494 | 81.6 | 871 | 925 | 22.7 |  | 2.45 | 1.81 | 21.8 |
|  | b | 78 | 215 | 7.0 | 75 | 80 | 2 |  | 0.21 | 0.16 | 1.9 |
|  | c |  | 9% | 11% | 9% | 10% | 11% |  | 15% | 9.4% | 9.5% |
| Beer | a | 246 g | 103 | 0.74 | 12.3 | 74 |  |  | 0.01 | 0.07 | 1.4 |
|  | b | 240 ml | 101 | 0.72 | 12 | 72 |  |  | 0.01 | 0.07 | 1.4 |
|  | c |  | 4.2% | 1.1% | 1.5% | 9% |  |  | 0.7% | 4.1% | 7% |
| Total percentages (rounded off) |  |  | 20% | 37% | 50% | 45% | 13% | 11% | 17% | 25% | 16% |

[a] Values for a given weight.
[b] Per capita daily intake.
[c] Percentage of RDA contributed.

Food fermentations involve a loss of energy (calories). This subject is rarely if ever mentioned in the literature. There is obviously a loss of nutrients and calories when cheese is made from cow's milk. Cheese whey, which contains whey proteins, minerals (Ca, P), and lactose, is partially used in processed foods, but most of it is fed to animals or discarded. But there is another type of loss due to the metabolism of the microorganisms. In most fermentations it is difficult to calculate, but for alcoholic fermentations it can be approximated quite well. The conversion of a given weight of fermentable sugar (glucose) to ethanol leads to a loss of about 13% of the calories, based on the conversion of 1 mol of glucose to 2 mol of ethanol. This loss applies to alcoholic beverages. During the production of yeast-fermented baked goods all or most of the alcohol is driven off in the oven. Here the loss of *total* calories of the raw materials (flour and sugar) may amount to 5-10%.

## REFERENCES

Darby, W., L. Grivetti, and P. Ghalliugui, 1977. *Food: Gift of Osiris*. Academic Press, London.

Jacob, H. E. 1944. *Six Thousand Years of Bread*. Doubleday, Doran & Co., Garden City, N.J.

Kozaki, M. 1978. Lactic acid bacterial flora of fermented foods in Southeast Asia. *Ann. Reports, Int. Center Coop. Res. and Dev. in Microbial Eng.* **1:**363.

Reed, G. 1981. Use of microbial cultures: Yeast products. *Food Technol.* **35**(1):89-94.

CHAPTER

# 1

# GENERAL CLASSIFICATION OF YEASTS

The term *systematics* applies to the study of the classification of life forms and is in a general sense synonymous with taxonomy. For centuries, taxonomy referred to describing various forms of life and their component parts. Despite the usefulness of such a system, relationships between life forms were little understood. Following the rise of Darwinism, systematic biology began to develop and replace what was simply a nomenclatural science. Today many biologists have gone into the molecular field attempting to understand the minute biochemical changes at a molecular level that led to evolution. Their contributions toward a clarification of the mechanism of evolution were most useful.

In the field of mycology, it has been the objective of every systematic mycologist to become interested in the differences as well as the similarities to predict relatedness between various forms of fungi. Basic units that appeared similar were grouped together, to give rise to a series of groups that in turn were arranged systematically to bring out the relatedness between the different groups, perhaps from an evolutionary point of view.

Taxonomy (from the Greek *taxis* meaning arrangements and *nomos* meaning law), however, was one specific branch of systematics that included classification, identification, and nomenclature. Classification provided a logical arrangement of these life forms based on their relationships. An operationally useful classification, once established, was found useful for identification of unknown organisms by comparison with characteristics of already identified and classified organisms and for the placement of new discoveries within the hierarchy. Nomenclature (from the Latin *nomen*, meaning name and *clare*, to call) is the naming of taxonomic groups of organisms as it applies to variety, species, genus, family, or order.

A phylogenetic classification represents relationships between organisms on the basis of their probable origin. This classification is possible if sufficient fossil data of primitive and intermediate forms can be recovered intact to reconstruct the evolutionary ladder. Although this approach could be applicable to animal and plant kingdoms, it has not been practical for most microorganisms, particularly the yeasts. The small size and the delicate nature of the yeast cells account for the lack of yeast fossils. The alternative to the phylogenetic system of classification is the phenetic system based on the comparison of easily recognized characteristics of existing organisms. By comparison, this approach represents an artificial arrangement based on the deliberate selection of a specified number of criteria that meet the desired objective.

Following van Leeuwenhoek's first description of yeast in beer in 1680, little additional information about yeast appeared until the 1830s, due to the lack of good-quality microscopes, too few persevering naturalists, and the absence of an organized system of naming and arranging the life forms. The latter need was finally met in the middle of the eighteenth century by Carolus Linnaeus (Carl von Linne, 1707-1778), the Swedish botanist who assembled the first recognized system of naming and arranging plants and animals. Although Linnaeus was unfamiliar with the microbial world, he placed the then-described fungi that included the mushrooms under the 24th class cryptogamia. Submicroscopic life forms including the yeast received superficial attention mainly because many scientists at the time accepted the theory of spontaneous generations.

The binomial system that is widely used today consisting of Latin generic and specific names was first introduced by Linnaeus. In pre-Linnaean times, names given to many different species were often long and cumbersome, often describing the animal or the plant. Linnaean system reduced the number of words used to designate a species to two words: one for the genus and the other for the species. This system was soon accepted and the tenth edition of Linnaeus' *Systema Natura* that appeared in 1758 was considered the starting point of taxonomic nomenclature (Bailey 1933).

Yeast taxonomy began in 1837 with Meyer's assigning of the genus *Saccharomyces* (from the Greek *sakehar*, meaning sugar and *mykes*, meaning fungus) to yeast. Half a century later, the first concept of naming species to yeast emerged based on the pioneering studies by Emil Christian Hansen at Carlsberg Laboratories on the pure culture techniques. Hansen differentiated and classified different yeast cultures he isolated in pure form on the basis of cellular and ascosporal morphology, sugar assimilation and fermentation characteristics, and tolerance of yeast to various temperatures. As a result of this discovery of new strains, the list of taxa lengthened and there was an immediate need to extend the number of phenotypic characters to be examined for more accurate differentiation. Yeast taxonomy has undergone a slow but steady evolution since then, and the taxonomic system we know today is not perfect and may thus be considered a stage in the evolution toward a perfect classification.

The present taxonomic system for yeasts used by almost all mycologists and prepared through the cooperative efforts of a number of taxonomists throughout the world was published by Kreger-van Rij (1984). The first edition of *The Yeasts: A Taxonomic Study*, by Lodder and Kreger-van Rij (1952), listed 26 genera and 164 species, which in the next 18 years increased to 39 genera and 349 species as described in the second edition by Lodder (1970). Since then, several new species have been discovered and recent developments in taxonomy have required the use of new criteria for the description of yeasts taxa for establishing an improved system of classification. The third edition, which presents the most recent classification of the yeasts by Kreger-van Rij (1984), describes 60 genera and 500 species. It is the most recent stage in the evolution toward an ideal classification. Following is a chronological summary of events since the publication of *Species Plantarum* in 1753 by Linnaeus describing the pioneering work of systematic botany up to the present yeast taxonomy: (Bailey 1933).

*Prescience Period*

| | |
|---|---|
| ca. 3000 B.C. | Bakery scenes in carvings among the ruins of ancient Memphis. |
| 1383 A.D. | Brewery founded in Munich. |
| 1680 | van Leeuwenhoek observes and sketches yeast. |
| 1753-58 | Linnaeus proposes binomial nomenclature and introduces systematics of plants and animals. |
| 1810 | Gay-Lussac formulates equation of alcoholic fermentation. |

*Era of Emerging Zymology and Microbiology*

| | |
|---|---|
| 1822-32 | Persoon and Fries advance classification of fungi; yeasts regarded as fungi; Persoon names them *Mycoderma*. |
| 1825-39 | Budding of yeasts observed independently by Schwann, Kützing, de la Tour; Schwann describes spores. |
| 1838 | Meyer names beer yeast *Saccharomyces cerevisiae*. |
| 1857-63 | Pasteur proves microbial origin of fermentation and explains its anaerobic nature. |

| | |
|---|---|
| 1866 | deBary observes life cycle of yeasts; he recognizes it as *Ascomycetes*. |
| 1870 | Reess observes yeast ascospores and their germination; he redefines the yeast genus *Saccharomyces*. |
| 1881-83 | Hansen begins study of pure cultures and single-cell isolation of brewer's yeast. |
| 1888 | Marx advocates pure yeast cultures in wine making. |
| 1896 | Hansen publishes comprehensive scientific system of classifying yeasts. |

*Modern Milestones*

| | |
|---|---|
| 1920-28 | Guilliermond expands Hansen's methods of yeast classification. |
| 1931-52 | Stelling-Dekker, Lodder, Diddens, and Kreger-van Rij bring unity and order to yeast taxonomy. |
| 1934 | Winge and Satava discover alternation of haplophase and diplophase in *Saccharomyces*. |
| 1943 | Lindegren finds heterothallism in *Saccharomyces*. |
| 1944 | Electron microscope perfected. |
| 1951 | Wickerham advances yeast characterization with new techniques. |
| 1952 | Publication of the first edition on yeast taxonomy by Lodder and Kreger-van Rij. |
| 1970 | Lodder and an international team update classification of yeast. |
| 1978 | DNA-DNA homology experiments by Price et al. for species delimitation. |
| 1984 | Kreger-van Rij and an international team update 1970 yeast classification. |

## CLASSIFICATION

Although most unicellular forms of yeast appeared almost identical under primitive microscopes, early investigators recognized their genetic and physiological variability through their experiences with different types of alcoholic fermentations. With the advent of the pure culture technique, there was a dramatic improvement in the quality of alcoholic beverages as demonstrated in the brewing industry. These developments coupled with the emphasis on the need for strain identification subsequently led to the introduction of a variety of test procedures to distinguish and characterize different yeast types. From the surveys that followed, there was sufficient evidence to indicate that reasonably sharp differences do exist that would separate a yeast population into distinguishable assemblages. Although the intention of early researchers was to catalog data on yeast characteristics for future reference, it became clear that this information could also be utilized successfully for the classification of yeasts following the criteria that were previously applied to bacterial taxonomy.

Genetic variation within a yeast population was so minor that it became a matter of scientific tact to decide what degree of genotypic diversity should justify the splitting of the assemblage into two or more basic taxonomic units. Such basic units can thus be referred to as *species*. It is difficult to provide a precise definition for species. Nevertheless, a species consists of a collection of strains that show minor differences with respect to certain physiological characteristics, but they are capable of successful interbreeding. These strains within a species share a high degree of phenotypic similarity while maintaining appreciable dissimilarity from other assemblages of the same general kind. The generic name, likewise, indicates a degree of relationship. Several species that may be included within a genus are more closely related to one another than they are to species in another closely related genus. A combination of characters distinguishes the species of one genus from species in other genera. Delimitation of a genus is arbitrary and depends on the choice of characters. Likewise, various degrees of similarity and differences are also indicated by families, orders, and higher taxa.

The most widely used classification of yeast found in *The Yeasts: A Taxonomic Study*, edited by Kreger-van Rij (1984), has made use of new critera in its preparation, making it somewhat different from the previous classification edited by Lodder (1970). Although earlier attempts were made to show the importance of base composition of nuclear DNA (reported as mol% G+C) to predict the relatedness among strains, a difference in base composition as small as 2 mol% G+C has resulted in significant loss in DNA homology. Subsequent research has made it possible to show the importance of a high degree of DNA-DNA complementarity for successful interfertility. Based on these data, the presently accepted taxonomic system considers a high DNA base sequence homology, that is, greater than *Ca* 80%, with the accompanying interfertility to be the critical feature for the delimitation of species. Likewise, characters like the ultrastructure of the cell wall and the chemical composition of the coenzyme Q have been used in the generic differentiation.

Several biochemical properties other than fermentation and assimilation studies have also been suggested for use in classification with limited success. Campbell (1970), employing computer numerical analysis, correlated 48 characteristics of 28 species of *Saccharomyces* with 10 serological groups. The largest of these groups included the industrial yeasts. In a similar manner, Poncet (1967) applied Adansonian principles (Sneath

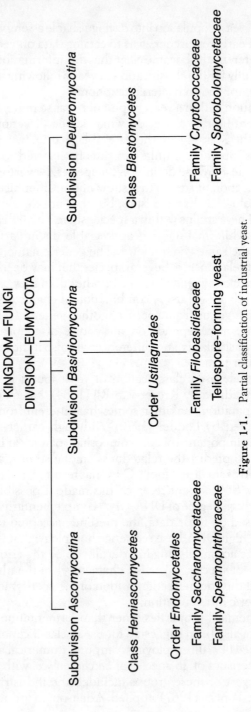

**Figure 1-1.** Partial classification of industrial yeast.

1964) to numerical classification of the genus *Pichia*. In more recent studies (Chen 1981; Kock et al. 1985) yeast lipids have been considered of taxonomic importance, despite the variations known to occur under different cultural conditions. Another aid to classification, proton magnetic resonance spectra of yeast polymers, has been proposed and evaluated by Gorin and Spencer (1970).

All the member organisms belonging to the group fungi are divided into four classes: *Phycomycetes* (*Zygomycetes*), *Ascomycetes*, *Basidiomycetes*, and *Deuteromycetes* (*Fungi Imperfecti*) (Fig. 1-1). This classification was primarily based on the vegetative growth characteristics and on the nature of the spores, if formed. Most yeasts in nature belong to the subdivision *Ascomycetes* (sporogenous) and *Fungi Imperfecti* (asporogenous). A few belonged to the *Basidiomycetes*. Although *Phycomycetes* could show the yeast morphology under certain cultural conditions, their normal existence is in the filamentous form. Hence, this group of fungi will not be classified under yeasts.

## Ascomycetes

The ability to form sexual spores is the primary criterion for the placement of these yeasts under the subdivision *Ascomycetes*. The sexual spores produced by the ascosporogenous yeasts exhibit such significant differences in size, shape, and other important morphological characteristics that they have become important criteria for the identification and characterization of this spore-bearing group of yeasts (Fig. 1-2 and Table 1-1). These ascosporogenous yeasts are placed in the class *Hemiascomycetes* under two families, namely, *Saccharomycetaceae* and *Spermophthoraceae* of the order *Endomycetales* (Fig. 1-3).

There are three genera classified under the family *Spermophthoraceae*, namely, *Coccidiascus*, *Metschnikowia*, and *Nematospora*. These genera have needle- or spindle-shaped ascospores (Fig. 1-2h). These are also characterized by the presence of asci significantly larger than the vegetative cells.

The second family, *Saccharomycetaceae*, is made up of four subfamilies: *Schizosaccharomycetoideae*, *Nadsonioideae*, *Lipomycetoideae*, and *Saccharomycetoideae*. These subfamilies are distinguished from each other on the basis of vegetative and sexual reproduction characteristics.

### Subfamily Schizosaccharomycetoideae

This subfamily is distinguished from the other three subfamilies by the way it reproduces vegetatively. A single genus, *Schizosaccharomyces*, repro-

# 14  GENERAL CLASSIFICATION OF YEASTS

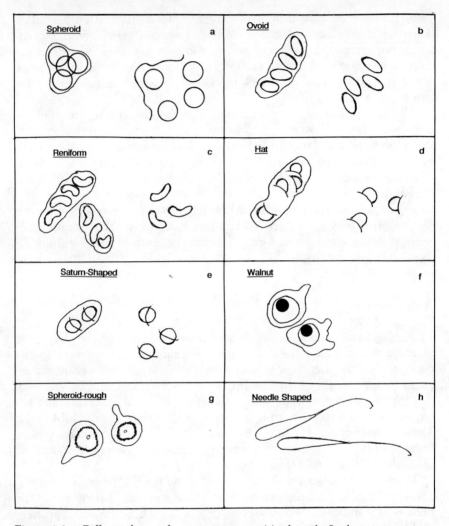

**Figure 1-2.** Different shapes of ascospores in yeast: (a) spheroid—*Saccharomyces cerevisiae*; (b) ovoid—*Schizosaccharomyces pombe*; (c) reniform—*Kluyveromyces marxianus*; (d) hat—*Hansanula anomala*; (e) saturn—*Hansanula saturnus*; (f) walnut—*Schwannomyces occidentalis*; (g) rough spheroid—*Debaromyces hansenii*; (h) needle-shaped—*Metschnikowia bicaspidata*.

Table 1-1. Characteristics of Ascosporogenous Yeasts

| Genus | Number of Species | Type of Budding | Spore Shape | Mol% of G+C | Coenzyme Q |
|---|---|---|---|---|---|
| Schizosaccharomyces | 4 | Fission | Round, oval, lunate | 40-42 | 9 or 10 |
| Hansaniospora | 6 | Polar | Hat | 28.8-40.5 | 6 |
| Nadsonia | 3 | Polar | Spheroid, rough | 39.5 | 6 |
| Saccharomycodes | 1 | Polar | Oval | 38.3 | 6 |
| Wickerhamia | 1 | Polar | Cap | 37.6 | 9 |
| Lipomyces | 5 | Multilateral | Spheroid | 47.6 | 9 |
| Coccidiascus | 1 | — | Spindle-shaped | | |
| Nematospora | 1 | Multilateral | Whiplike | 40.2 | 5 or 6 |
| Metschnikowia | 6 | Multilateral | Needle-shaped | 42.2-49.0 | 9 |
| Ambrosiozyma | 3 | Multilateral | Hat | 35.1-40.2 | 7 |
| Arthroascus | 1 | — | Hat | 31.5 | 8 |
| Arxiozyma | 1 | Multilateral | Spheroid, rough | 31.8-34.9 | 6 |
| Citeromyces | 1 | Multilateral | Spheroid or hat | 44.4-45.1 | 8 |
| Clavispora | 1 | Multilateral | Conical | 43.4-46.3 | 8 |
| Cyniclomyces | 1 | Multilateral | Oval | 34.5 | |
| Debaryomyces | 11 | Multilateral | Spheroid, rough | 38-40 | 9 |
| Dekkera | 2 | Multilateral | Hat | 38.2-41.1 | 9 |
| Guillermondella | 1 | Multilateral | Oval or lunate | 30.2 | 8 |
| Hansenula | 27 | Multilateral | Hat | 31.8-50.2 | 7 or 8 |
| Issatchenkia | 4 | Multilateral | Spheroid, rough or smooth | 32.4-41.1 | 7 |
| Kluyveromyces | 11 | Multilateral | Kidney, oval, or spheroid | 33.4-46.7 | 6 |
| Lodderomyces | 1 | Multilateral | Oval, smooth | 39.5-39.8 | 9 |
| Pachysolen | 1 | Multilateral | Hat, saturn | 42.4 | 8 |
| Pachytichospora | 1 | Multilateral | Spheroid, smooth | 32.0-34.4 | |
| Pichia | 56 | Multilateral | Hat | 28.3-51.9 | 7, 8, or 9 |
| Saccharomyces | 7 | Multilateral | Oval or spheroid | 31.8-42.0 | 9 |
| Saccharomycopsis | 6 | Multilateral | Hat | 43.4 | 8 |
| Schwanniomyces | 1 | Multilateral | Walnut, rough | 34.6-35.4 | 9 |
| Sporopachydermis | 3 | Multilateral | Oval, smooth | 43.2-49.6 | |
| Stephanvascus | 1 | — | Hat | | |
| Torulaspora | 3 | Multilateral | Spheroid, rough | 42.2-47.5 | 6 |
| Wickerhamiella | 1 | Multilateral | Oval, rough | 48.6 | |
| Wingea | 1 | Multilateral | Oval, smooth | 42.7 | 9 |
| Yarrowia | 1 | Multilateral | Oval, hat, rough Saturn or walnut | 49.5-50.2 | 9 |
| Zygosaccharomyces | 8 | Multilateral | Round, smooth | 34.3-44.6 | 6 |

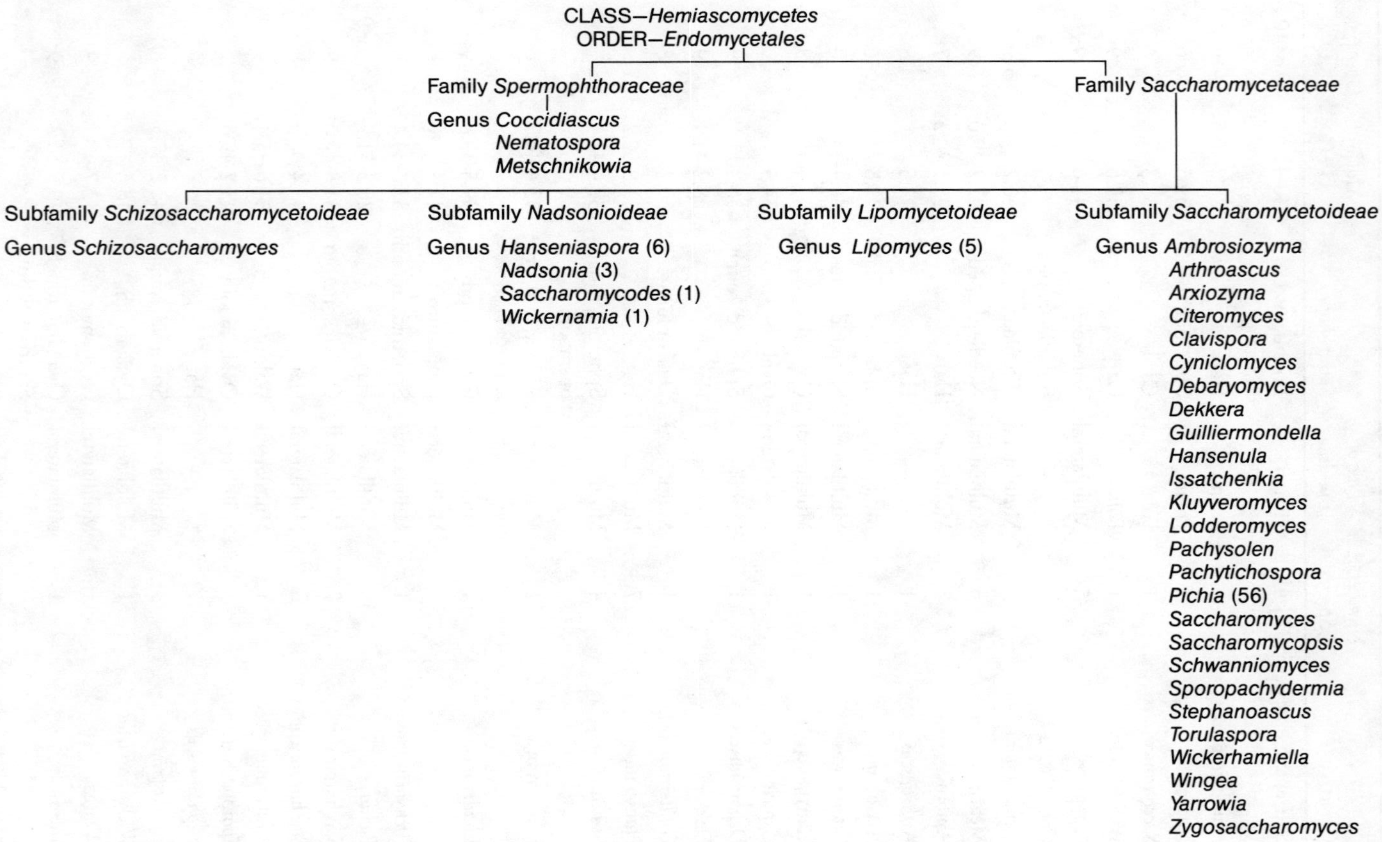

CLASS—*Hemiascomycetes*
ORDER—*Endomycetales*

```
├── Family Spermophthoraceae
│     Genus Coccidiascus
│           Nematospora
│           Metschnikowia
│
└── Family Saccharomycetaceae
      ├── Subfamily Schizosaccharomycetoideae
      │     Genus Schizosaccharomyces
      │
      ├── Subfamily Nadsonioideae
      │     Genus Hanseniaspora (6)
      │           Nadsonia (3)
      │           Saccharomycodes (1)
      │           Wickernamia (1)
      │
      ├── Subfamily Lipomycetoideae
      │     Genus Lipomyces (5)
      │
      └── Subfamily Saccharomycetoideae
            Genus Ambrosiozyma
                  Arthroascus
                  Arxiozyma
                  Citeromyces
                  Clavispora
                  Cyniclomyces
                  Debaryomyces
                  Dekkera
                  Guilliermondella
                  Hansenula
                  Issatchenkia
                  Kluyveromyces
                  Lodderomyces
                  Pachysolen
                  Pachytichospora
                  Pichia (56)
                  Saccharomyces
                  Saccharomycopsis
                  Schwanniomyces
                  Sporopachydermia
                  Stephanoascus
                  Torulaspora
                  Wickerhamiella
                  Wingea
                  Yarrowia
                  Zygosaccharomyces
```

**Figure 1-3.** Classification of the order *Endomycetales*.

duces vegetatively either by fission, which is the splitting of a single vegetative cell subsequent to the formation of a wall across the middle of the elongated cell (Fig. 1-4b), or the true mycelium may split into small segments called arthrospores (Fig. 1-4f). This mode of cell division is distinct from all other yeasts that reproduce vegetatively by budding. An ascus contains four to eight ascospores that are round, oval, or reniform

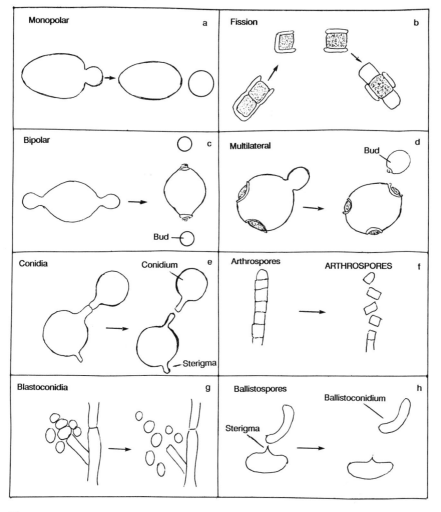

**Figure 1-4.** Conidiation in yeast: (a) monopolar budding—*Melassesia*; (b) fission—*Schizosaccharomyces*; (c) bipolar budding—*Saccharomycoda, Hanseniospora*, etc.; (d) multilateral budding—*Saccharomyces, Candida, Hansenula, Pichia*, etc.; (e) conidia on stalks—*Sterigmatomyces*; (f) arthrospores from hyphae—*Trichosporon*; (g) blastoconidia on hyphae—*Candida albicans*; (h) ballistospores on sterigmata—*Bullera, Sporobolomyces*.

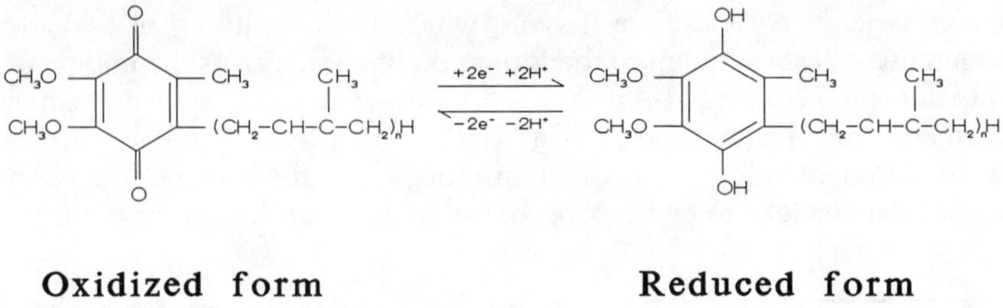

## 18 GENERAL CLASSIFICATION OF YEASTS

$$\underset{\text{Oxidized form}}{\text{CH}_3\text{O}\underset{\text{O}}{\overset{\text{O}}{\bigcirc}}\text{(CH}_2\text{-CH-C-CH}_2)_n\text{H}} \quad \underset{-2e^- \; -2H^+}{\overset{+2e^- \; +2H^+}{\rightleftarrows}} \quad \underset{\text{Reduced form}}{\text{CH}_3\text{O}\underset{\text{OH}}{\overset{\text{OH}}{\bigcirc}}\text{(CH}_2\text{-CH-C-CH}_2)_n\text{H}}$$

**Figure 1-5.** Coenzyme Q: the oxidized and reduced states.

(Fig. 1-2), and are generally released from the ascus when fully matured. Species belonging to this genus show vigorous fermentation and have either Co-Q-9 or Co-Q-10 systems (Yamada, Arimoto, and Kondo 1973) where the number signifies the number of isoprene units/molecule of the enzyme, coenzyme-Q, associated with the electron transport chain (Table 1-1 and Fig. 1-5). These yeasts do not produce high concentrations of ethanol. *Schizosaccharomyces pombe* ferments malic acid to ethanol and $CO_2$ and has been used commercially in champagne fermentations. Fermentations with *Schizosaccharomyces pombe* gave wines with reduced titratable acidity because of its ability to carry out malolactic fermentations, but the resulting wines were thought to be inferior in quality.

*Subfamily Nadsonioideae*

Members of the subfamily *Nadsonioideae* are characterized by their ability to reproduce by bipolar budding (Fig. 1-4c). The vegetative cells have multiple scars at the two poles of lemon-shaped cells. The four genera described in this chapter are differentiated by the shape of the spores, the method by which the ascus is formed, and the structure of the spore wall.

In the genus *Nadsonia*, which includes three species, conjugation between mother cell and bud results in the formation of a zygote that transforms into an ascus either in the mother cell, the bud, or in a new bud later developed at the opposite pole of the mother cell. Each ascus after meiosis or reduction division bears one or two ascospores that are often spherical, brown, and warty. *Nadsonia* species appear to be present in exudates of certain broad-leafed trees. These yeasts appear brown on isolation plates due to the production of brown spores.

The six species in the genus *Hanseniaspora* are differentiated by the shape of their respective ascospores as described by Kreger-van Rij (1977). All *Hanseniaspora* species require inositol for growth and are known to ferment sugars with low tolerance to ethanol. H. guilliermondii and H. apiculata are commonly present in grapes and are generally responsible for the initiation of spontaneous wine fermentations.

The third genus of the Nadsonioideae is Saccharomycodes with a single species, Saccharomycodes ludwigii. Each ascus has four ascospores that conjugate in pairs while still in the ascus to give rise to the diploid vegetative cells. This species is highly resistant to sulfur dioxide and remains active in grape musts with $SO_2$ concentrations as high as 500 ppm. Thus, it may occur as a spoilage organism in highly sulfited wines. S. ludwigii is capable of producing to 8-11% ethanol by volume.

The fourth genus of Nadsonioideae is Wickerhamia with a single species, W. florescens, isolated from faeces of wild squirrels. It produces cap-shaped ascopores generally with one spore per ascus. Although the other three genera belonging to Nadsonioideae have the Co-Q system 6, W. florescens has been shown to contain CO-Q-9 (Table 1-1) (Yamada et al. 1976).

*Subfamily Lipomycetoideae*

This subfamily includes one genus, Lipomyces, with at least five species identified by Phaff and Kurtzman (1984). These species have been isolated from soils in many parts of the world. The method of ascus formation varies from species to species. Each ascus may contain 2 to 30 or more amber ascospores that are smooth with irregular folds. These ascospores may or may not be released.

*Subfamily Saccharomycetoideae*

Several yeast genera of considerable industrial importance are grouped under the subfamily Saccharomycetoideae. These genera generally show multilateral budding with true mycelia occasionally splitting into arthrospores. At least there are two genera, Pichia and Hansenula, having few species that can produce true hyphae.

In the most recent classification of yeasts by Kreger-van Rij (1984), four additional genera have been included to bring the total number of genera to 26. Saccharomyces telluris was found unique among other Saccharomyces species in that it was the only species that produced worty spores. It was later reclassified in a new genus, Arxiozyma, by van der Walt and Yarrow (1984). Likewise, Zygosaccharomyces and Torulaspora, which were previously assigned to the genus Saccharomyces, have recently been separated from this genus, thereby adding two new genera, Zygosaccharomyces and Torulospora, to the subfamily Saccharomycetoideae (Barnett, Payne, and Yarrow

1979). These two new genera are distinguished by their mode of conjugation and sporulation characteristics. Barnett, Payne, and Yarrow (1979) have included three species, *T. delbrueckii*, *T. globosa*, and *T. pretoriensis*, in the genus *Torulospora*. Five *Saccharomyces* species, *S. rosei*, *S. fermentati*, *S. inconspicuus*, *S. saitoanus*, and *S. vafer*, were merged as a single species, *Torulospora delbrueckii* (Kreger-van Rij 1984).

*Saccharomycopsis lipolytica*, the perfect state of *Candida lipolytica*, was found to differ from other *Saccharomycopsis* species with respect to a number of characteristics. For example, *S. lipolytica* has a Co-Q of 9, which is different from all other *Saccharomycopsis* species that have a Co-Q value of 8 (Yamada et al. 1976). *S. lipolytica* was also unique in that it has septa with a central connection either open or closed (Kreger-van Rij and Veenhuis 1976). All other *Saccharomycopsis* species have septa with plasmodesmata. In the most recent classification, this species has been renamed *Yarrowia lipolytica* under a new genus *Yarrowia* (van der Walt and von Arx 1980).

The genus *Saccharomyces* as defined by Lodder and Kreger-van Rij (1952) included several species, some of which were unrelated to *Saccharomyces cerevisiae* with respect to certain important criteria. One of these groups of species included *S. lactis*, *S. fragilis*, and *S. marxianus*, which unlike the true *Saccharomyces* species, formed fragile asci that readily ruptured and released spores into the environment at maturity. These genera produced ascospores that were lunate in contrast to the round ascospores produced by the *Saccharomyces* species. In view of these fundamental property differences, there was sufficient justification to place *S. fragilis*, *S. marxianus*, and *S. lactis* under the genus *Kluyveromyces* by Lodder (1970). Accordingly, in the new classification *K. marxianus*, *K. fragilis*, and *K. lactis* represented *S. marxianus*, *S. fragilis*, and *S. lactis* (Table 1-2).

Studies on mass matings of auxotrophic mutant strains (Johannsen and van der Walt, 1978) have led to a reduction of the number of *Kluyveromyces* species, which were incorrectly grouped according to the previous classification by Lodder (1970). According to the classification by Kreger-van Rij (1984), *K. marxianus* has been delimited on the basis of interfertility and is comprised of seven syngamous varieties (Johannsen 1980). Hence, *K. marxianus* and *K. fragilis* have been merged as single species *K. marxianus* var. *marxianus* (van der Walt 1970). Similarly, *K. lactis* was renamed *K. marxianus* var. *lactis* (van der Walt 1970) (Table 1-2). *Candida pseudotropicalis*, which is now referred to as *C. kefyr* according to the most recent taxonomic revision (Kreger-van Rij 1984), is considered the imperfect form of *K. marxianus*. *Kluyveromyces* strains have the unique ability to assimilate and ferment lactose. Thus, such strains are used commercially for the production of biomass or alcohol from whey.

Lodder's updated classification in 1970 united several species pre-

Table 1-2. Historical Nomenclature of Yeasts of Industrial Importance (1952-1984)

| 1952 Classification (Lodder and Kreger-van Rij 1952) | 1970 Classification (Lodder 1970) | 1984 Classification (Kreger-van Rij 1984) |
|---|---|---|
| Saccharomyces bayanus<br>S. oviformis<br>S. pastorianus<br>S. cheriensis<br>S. beticus | S. bayanus | |
| S. cerevisiae<br>S. cerevisiae var. ellipsoideus<br>S. willianus<br>S. vini | S. cerevisiae | S. cerevisiae |
| S. carlsbergensis<br>S. logos<br>S. uvarum | S. uvarum | |
| S. chevalieri<br>S. italicus | S. chevalieri<br>S. italicus<br>S. acetic<br>S. diastaticus | |
| S. acidifaciens<br>S. bailii<br>S. elegans | S. bailii | Zygosaccharomyces bailii |
| S. fragilis<br>S. marxianus | K. fragilis<br>K. marxianus | K. marxianus var. marxianus |
| S. lactis | K. lactis | K. marxianus var. lactis |

viously maintained as separate taxa by Stelling-Dekker (1931) and Lodder and Kreger-van Rij (1952). Among these are a number of physiologically similar *Saccharomyces* species that these authors considered different on the basis of their size, shape, and fermentation characteristics. Accordingly, *S. cerevisiae* and its *ellipsoideus* variety, *S. willianus*, and *S. vini* were merged as a single species. *S. uvarum*, *S. carlsbergensis*, and *S. logos* were united into one species, *S. uvarum*. Many other yeast species of similar physiological and minor morphological differences were united in the reclassification by Lodder (1970). For example, *S. beticus*, *S. oviformis*, *S. pastorianus*, *S. cheriensis*, and the wine and flor sherry yeast *S. bayanus* were merged into one species, *S. bayanus*. In uniting these species, the oldest legitimate epithet in all cases was retained in accordance with article 57 of the International Code of Botanical Nomenclature adopted in 1964 by the 10th Botanical Congress (Lanjouw et al. 1966).

Since 1970, there were several attempts for species delimitation by use of DNA-DNA homology studies and by comparison of nuclear DNA base composition (G+C) values. Hence, among homothallic and asporogeneous yeasts, an important criterion commonly considered was the extent to which the nuclear DNA of the two strains reassociated under optimal conditions.

The production of viable spores from a zygote subsequent to pairing and fusion of two opposite mating types of heterothallic strains was an important criterion for the inclusion of the two mating strains into a single species. Generally, a low DNA-DNA homology correlated with lack of interfertility among yeast strains, thus providing a basis for delimitation of biological species.

In the most updated classification of yeast by Kreger-van Rij (1984) the genus *Saccharomyces* has seven species. The most important basis for species differentiation as discussed previously was interfertility within the strains of the same species. Conversely, a lack of affinity for sexual union for one species with strains of other species is expected. *Saccharomyces uvarum* showed relatively high DNA-DNA homology with both *S. cerevisiae* and *S. bayanus*. Similar DNA-DNA homologies were also evident with *S. chevalieri*, *S. italicus*, *S. aceti*, and *S. diastaticus*. There was also adequate evidence to indicate the presence of interfertility between these species, resulting in the merging of these different species into one single species, *S. cerevisiae* (Kreger-van Rij 1984) (Table 1-2).

The genus *Pachysolen* with a single species *P. tanophilus* has the ability to utilize nitrate as the sole source of nitrogen, and is capable of fermenting D-xylose. Because of its high fermentation efficiency at 37°C, this species has been examined extensively for alcoholic fermentations where the substrate is D-xylose from corn-derived waste. This genus is characterized by the presence of asci that consist of cells with long necks. The cell apex has four hat-shaped ascospores. The acus wall dehisces at maturity, releasing the spores. *Schwanniomyces* is another genus that is of industrial importance because of its ability to assimilate soluble starch. Vegetative cells are haploid and a conjugation between mother cell and a bud precedes ascus formation. *Schwanniomyces occidentalis* is considered as the type species of the genus. Considering the starch-utilizing ability of these strains, there is considerable interest within the brewing community to transform brewing strains with temperature-sensitive glucoamylase genes from *Schwanniomyces castellii* for light beer production.

The genus *Pichia* includes 56 species, few of which occur most frequently in grape musts. *Pichia membranefaciens* is capable of surviving high alcohol concentrations, thereby causing wine spoilage. Although *Pichia* is nitrate-negative, it shows similarity in may respects to the nitrate-positive genus *Hansenula*. Most *Pichia* species have ascospores that are hat- or saturn-shaped and form evanescent asci. Certain species that had certain atypical characteristics such as smooth or rough round spores originally classified as *Pichia* have now been transferred to the genera *Issatchenkia* and *Debaryomyces* (Table 1-1).

The ability of the members of genus *Pichia* to grow on methanol as the sole source of carbon and energy is important from an industrial standpoint. This characteristic is also found in a number of yeast species of three other genera, namely, *Candida*, *Hansenula*, and *Torulaspora*. The methanol-utilizing pathway is thought to be similar in these yeasts and begins with the oxidation of methanol to formaldehyde, a reaction catalyzed by alcohol oxidase. The reaction results in the simultaneous reduction of oxygen to hydrogen peroxide. Therefore, yeasts sequester alcohol oxidase in a subcellular organelle termed *peroxisome* in order to avoid hydrogen peroxide toxicity.

There have been several attempts to use *Pichia pastoris* for the production of single-cell proteins from methanol. A major finding from these studies was that this organism is very amenable to genetic manipulation. Its genetic expression system has now been used extensively for the production of a number of recombinant proteins.

## Fungi Imperfecti

The imperfect or anamorphic yeasts lack the sexual stages of reproduction as represented by either the formation of the ascus or the basidium. Thus, these are referred to as asporogenous yeasts. Fungi imperfecti are related to either ascomycetous or basidiomycetous yeasts with respect to their cell wall structure, cell wall composition, conidiation, and color reactions with certain diazonium dyes. The absence of the sexual cycle in *Fungi Imperfecti* eliminates an important parameter frequently utilized for yeast classification. The present classification of the imperfect yeast is based predominantly on morphological characteristics and in a few instances on physiological characters such as fermentation properties and assimilation of nitrate or inositol.

The absence of the sexual cycle characteristic of *Fungi Imperfecti* may have resulted from the inability to identify the opposite mating type of their heterothallic haploid yeasts or may possibly be due to the lack of understanding of the sporulation process particularly with regard to the conditions that are required for sporulation. There are several imperfect yeasts that have morphological, physiological, and biochemical characteristics closely resembling their perfect counterparts with a high degree of DNA homology. Yet their evolutionary relationship is not well defined. The disappearance of the opposite mating type of a heterothallic sporing yeast may, for instance, account for the existence of only the imperfect nonsporing species with their spores growing indefinitely as haploid cells in the absence of the opposite mating type.

## 24 GENERAL CLASSIFICATION OF YEASTS

Two families are recognized among imperfect yeasts: family *Sporobolomycetaceae*, which includes 2 genera, and family *Cryptococcaceae*, with 16 genera. The species of the genera *Bullera* and *Sporobolomyces*, belonging to the family *Sporobolomycetaceae*, are characterized by the production of ballistospores. Both genera are basidiomycetous in nature.

Family *Cryptococcaceae* includes species classified in eight ascomycetous genera, six basidiomycetous genera, and two genera of mixed affinity (Fig. 1-6). Some generic characteristics within the group *Fungi Imperfecti* are presented in Table 1-3. *Brettanomyces*, an ascomycetous genus, is characterized by the production of acetic acid from glucose and is commonly isolated from grape musts and lambic beer. Two of the nine *Brettanomyces* species have their perfect states in the genus *Dekkera*. The other ascomycetous genus of industrial importance is *Kloeckera*, which includes six species that have their perfect counterparts in six *Hanseniaspora* species. *Kloeckera apiculata* is characterized by its low tolerance to ethanol and in general occurs abundantly in grape musts at the early phase of the fermentation. It forms large amounts of volatile acids and esters.

Two basidiomycetous yeasts, *Phaffia* and *Rhodotorula*, are of importance because of the considerable interest that exists worldwide to develop food colorants from natural sources. Different species of *Rhodotorula* have red or yellow color pigments due to the presence of $\beta$-carotene. A recently described genus *Phaffia* has a pink or red coloration due to its ability to produce a carotenoid pigment known as astaxanthin. The interest in astaxanthin is primarily due to its presence in the animal kingdom. It is conspicuously displayed in the plumage of birds like flamingos, in the exoskeleton of marine invertebrates like lobsters, crabs, and shrimps, or among fish such as trout and salmon where the astaxanthin is responsible for the pink color of the flesh. Johnson, Conklin, and Lewis (1977) have found that a preparation of *P. rhodozyma* is a potentially important source of astaxanthin to restore the red color in the flesh of pen-reared salmon.

Species of both genera, *Phaffia* and *Rhodotorula*, are characterized by their inability to assimilate inositol. The single species of the genus *Phaffia*, *P. rhodozyma*, has the ability to ferment glucose; however, this feature is not typical of the species of *Rhodotorula*.

*Candida* is the largest yeast genus, comprising approximately 200 species that include all the yeast species that cannot be assigned to other imperfect genera. These are predominantly ascomycetous yeasts with about 25 yeasts that are basidiomycetous in nature. *Candida* includes species with and without hyphae, mainly due to the merging of the genus *Torulopsis*, which is generally lacking a pseudomycelium, into the genus *Candida*, known for the presence of the pseudomycelium (Yarrow and Meyer 1978).

Yeasts belonging to the genus *Candida* are found in many divergent

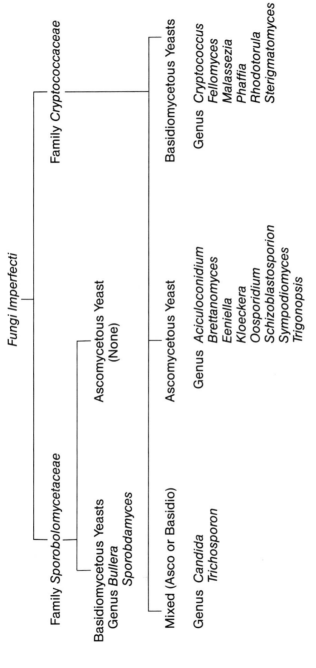

**Figure 1-6.** Classification of *Fungi Imperfecti*.

Table 1-3. Characteristics of the Fungi Imperfecti

| Genus | Number of Species | Type of Budding | Arthrospore (A) Ballistoconidia (B) | Mol% (G+C) | Coenzyme Q | True Mycelium |
|---|---|---|---|---|---|---|
| Bullera | 6 | Polar | B | 54.5 | | V |
| Sporobolomyces | 6 | Polar | B | 53.5-63 | 10 | V |
| Aciculoconidium | 1 | Multilateral | | 55.0 | | + |
| Brettanomyces | 7 | Multilateral | | 36.6-43.4 | 9 | V |
| Eeniella | 1 | Polar | | 40.2-41.4 | 9 | – |
| Kloeckera | 1 | Polar | | | | – |
| Oosporidium | 1 | Multilateral | | | | + |
| Schizoblastosporion | 1 | Polar | | 41.1-42.2 | | – |
| Sympodiomyces | 1 | – | | | | – |
| Trigonopsis | 1 | Multilateral | | 46.1 | | – |
| Cryptococcus | 24 | Polar | | 48.6-64.0 | 8, 9, 10 | – |
| Malassezia | 2 | Polar | | 62.9 | | V |
| Phaffia | 1 | Polar | | 48.3 | | – |
| Rhodotorula | 18 | Polar | | 50.2-70 | 9, 10 | V |
| Sterigmatomyces | 6 | Stengima | | 51.5 | | V |
| Candida | 196 | Multilateral | | 29.3-61 | 6, 7, 9 | V |
| Trichosporon | 8 | Polar, fission | A | 52.5-64 | | + |

*Note:* V indicates variable; + indicates present; – indicates absent.

environments. Some species are beneficial; others may cause disease and sometimes death to humans. *Candidosis* is a term generally given to infections caused by yeasts belonging to the genus *Candida*. The principal pathogen of the genus is the thrush fungus *Candida albicans*. The disease states range from superficial infections of the buccal cavity, aural canal, or vagina to deep-seated infections involving the viscera, which are often lethal. *Candida albicans* is often associated with human vulvovaginitis. The (G+C) content of *C. albicans* may vary from 34.3 to 35.6 mol%.

*Candida krusei* is of importance in the baker's yeast industry because of its occasional presence as a contaminant in bakers' yeast propagations. It is sometimes referred to as *white monilia* and is identified by the presence of clusters and chains of blastospores occurring along the pseudohyphae when grown on Dalmau plate culture on corn meal agar.

Though no known yeasts ferment L-sugars and pentose sugars, many *Candida* species grow aerobically on D-xylose or cellobiose in addition to growing on glucose, sucrose, and maltose. *Candida utilis* is especially versatile in sugar assimilation characteristics and is one of the yeasts propagated commercially in a variety of substrates including wood sugar media for feed or food uses. A number of *Candida* species grow well on various fractions of petroleum (Bell 1971). Normal alkanes and alkenes are assimilable substrates for *C. utilis*, *C. tropicalis*, and *C. lipolytica*. *C. utilis* has a Co-Q of 7 and a G+C content of 45.8 mol%.

*Candida kefyr*, which was previously referred to as *Candida pseudotropicalis*, has the ability to assimilate and ferment lactose. This species is thus commonly used for the production of ethanol or biomass from whey. This organism has a Co-Q of 6 (Yamada and Kondo 1972) and a G+C content of 41.3 mol%.

### Basidiosporogenous Yeasts

The haploid stages in the life cycle of various heterothallic *Basidiomycetes* are referred to as basidiomycetous yeasts. They are formed as a result of the germination of the basidiospore. Those basidiomycetous yeasts where the perfect state has yet to be identified are presently being classified as *Fungi Imperfecti*, as described in the previous section. Basidiomycetous yeasts undergo a characteristic color reaction with Diazonium blue B, producing a dark red color not apparent among ascomycetous yeasts with the single exception of *Sporopachydermia quercuum*. The basidiomycetous yeasts also show a bud formation, cell wall structure, and nuclear behavior during mitosis characteristically different from that of ascomycetous yeasts. The basidiosporogenous yeasts whose perfect states have been identified are classified in three groups: teliospore-producing yeasts, yeasts belonging to

# GENERAL CLASSIFICATION OF YEASTS

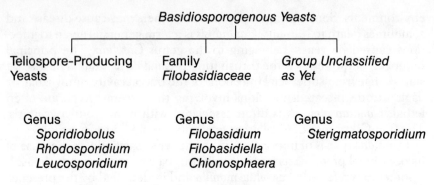

Figure 1-7. Classification of basidiosporogenous yeasts.

Table 1-4. Characteristics of Basidiosporogenous Yeasts

| Genus | Number of Species | Type of Budding | Ballistospore | Teliospores (Sexual) | Mol% (G+C) | Coenzyme Q |
|---|---|---|---|---|---|---|
| *Sporidiobolus* | 4 | Polar | + | + | 50-65 | 10 |
| *Rhodosporidium* | 9 | – | | + | 50.5-68.5 | 9, 10 |
| *Leucosporidium* | 6 | Polar | – | + | 54.1-56.1 | 8, 9 |
| *Filobasidium* | 3 | Polar | – | – | 49.8-51.5 | 9, 10 |
| *Filobasidiella* | 1 | Polar | – | – | 49-57.2 | 10 |

Note: + indicates present; − indicates absent.

the family *Filobasidiaceae*, and an unclassified group that includes a single genus *Sterigmatosporidium* (Fig. 1-7). Some of the important characteristics of basidiosporogenous yeasts are presented in Table 1-4.

## Teliospore-Producing Yeasts

The teliospore-producing group of yeasts is characterized by the presence of thick-walled, lipid-containing teliospores of various shapes on their hyphae. In general, the sexual reproductive structure of basidiomycetous yeasts is called a basidium, comprising two parts, the probasidium (or teliospore) where karyogamy takes place and the metabasidium (or promycelium) where meiosis usually occurs. There is considerable variation in the morphology of the metabasidium. The teliospores are attached to the metabasidium either directly or by short stalks.

The teliospore-producing yeasts include three genera: *Sporidiobolus*, *Rhodosporidium*, and *Leucosporidium*, each of which is characterized by the presence of two types of life cycles as described by Fell and Statzell Tallman (1984). In the first sexual mechanism, which is homothallic in nature, a single yeast cell gives rise to a uninucleate mycelium without clamp connections at the septa. The teliospores are then produced from the hyphae, which germinate to produce the metabasidia that in turn give rise to basidiospores. In the second type, compatible mating types of heterothallic yeasts conjugate to produce a dikaryotic mycelium with clamp connections. Karyogamy takes place in the teliospores and their germination produces basidia and haploid basidiospores either laterally or terminally.

The genus *Sporidiobolus* is easily distinguished by the presence of ballistospores that are basically ejected into the environment. In addition, vegetative reproduction can take place by polar budding and formation of pseudomycelia and true mycelia. This genus is capable of producing carotenoid pigments that can provide shades of pink or red on solid media. The four species accepted in this genus do not ferment sugars and are urease-positive.

The genus *Rhodosporidium* has nine species containing carotenoid pigments that give the cultures red, yellow, or orange colors. When in the yeast phase, this genus reproduces vegetatively by budding. The mycelial phase becomes conspicuous at the time of sexual reproduction. Terminal and intercalary teliospores are generally formed on the mycelium. Ballistospores are not observed at any of the phases of this genus. None of the species belonging to the genus *Rhodosporidium* ferment sugars, and they are all urease-positive.

Genus *Leucosporidium* includes six species that are white to cream-colored on solid media. Vegetative reproduction is mainly by budding. The mycelial phase becomes conspicuous at the time of sexual reproduction. Terminal or intercalary teliospores are formed on the mycelium. Ballistospores are not formed either in the yeast or mycelial phase. The species included in this genus do not ferment sugars and are urease-positive.

*Family Filobasidiaceae*

Three genera, *Filobasidium*, *Filobasidiella*, and *Chinosphaera*, are described in the second group of basidiomycetous yeasts belonging to the family *Filbasidiaceae*. These genera are characterized by the formation of slender nonseptate basidia with terminal sessile basidiospores. Teliospores are not produced by any of the genera belonging to this family.

The species accepted in the genus *Filobasidium* undergo vegetative reproduction by budding, and their sexual reproduction is heterothallic. Sessile basidiospores exist terminally on the apex of a metabasidium arranged like the petals of a flower. Of the three species, *Filobasidium capsuligenum* has fermentation ability. This species has occasionally been detected in wine cellars.

The genus *Filobasidiella* includes the species *Filobasidiella neoformans*, which has two varieties, *Filobasidiella neoformans* var. *neoformans* and *Filobasidiella neoformans* var. *bacillispora*. Both varieties are pathogenic to humans. Although asexual reproduction takes place by budding, the sexual phase is primarily heterothallic despite the presence of a few homothallic strains whose genetic control is incompletely understood. *Cryptococcus neoformans* has been regarded as the imperfect form of the genus.

The genus *Chinosphaera*, which includes one species, *Chinosphaera apobasidialis*, has a haploid phase consisting of uninucleate budding yeast cells with a white mucoid colony. Although the genus is not fully characterized, an important feature of the only species is its ability to produce a fruiting body when grown in association with a species of *Cladosporium* (Kwon-Chung 1984).

*Sterigmatosporidium*

The third group of basidiomycetous yeasts includes one species, *Sterigmatosporidium polymorphum*, which has been described by Kraepelin and Schulze (1982) as the sexual state of the imperfect genus *Sterigmatomyces*. This genus also is not fully characterized. Despite the similarities of yeast in this new taxon and the species accepted in *Sterigmatomyces*, their DNA complementarity has been relatively low (Kurtzman et al. 1984).

## IMPACT OF RECENT TAXONOMIC REVISION ON INDUSTRY

The species most commonly exploited among yeasts belong to 3 of the 60 genera described in the recent taxonomic revision. These include *Saccharomyces*, *Candida*, and *Kluyveromyces*. Although all 3 genera are of industrial importance, the genus *Saccharomyces* may be regarded as the most extensively utilized group of yeast for the benefit of humankind.

In the most recent classification of yeasts by Kreger-van Rij (1984), 7 species are described under the genus *Saccharomyces*, which is a significant reduction from the 41 species described in the earlier classification by Lodder (1970). Recent studies on yeast hybridization have demonstrated a high degree of interfertility between yeast strains that were previously classified as belonging to distinct species of the genus *Saccharomyces*. Since the updated yeast taxonomy reflects genetic or natural relationships with delimiting of yeast on the basis of interfertility, allocation of two or more inbreeding groups of yeasts to a single species *S. cerevisiae* by Kreger-van Rij (1984) was understandable.

According to the recent classification, the industrial properties of the ale yeast *S. cerevisiae* and the lager yeast *S. uvarum* were not considered to be sufficiently different to place them into separate species. Hence they were merged into a single species according to the new definition of *S. cerevisiae*. Despite this change in the nomenclature, the brewer must be aware of the distinguishing characteristics of the two yeasts and should treat them as two industrially distinct yeasts in order to achieve the desired brewing objectives. This unfortunate situation could have been avoided had some consideration been given to properties of industrial importance in the recent classification of yeasts. To the brewer, the differences between lager, ale, and nonbrewing strains of *S. cerevisiae* are sufficient to justify separate groupings however similar they may seem in the tests for classical taxonomy. Distinct differences between these strains were also demonstrated by Pedersen (1985) in fingerprint analysis of the genetic materials of these yeasts.

Merging of different species according to the recent taxonomic revision has also resulted in the grouping of formerly distinct species of yeast contaminants generally found in a variety of alcoholic fermentations in the same species as the industrial strain *S. cerevisiae*. Although the classical tests used in the most updated classification fail to differentiate contaminant from the industrially useful strain, the process microbiologist must continue to rely on previous classification schemes in order to recognize differences between such strains to ensure adequate process control. In this respect, the standardized nomenclature has offered the industrial microbiologist little support, making him or her rely more and more on previously accepted classifications.

## YEASTS OF ECONOMIC IMPORTANCE

There are nearly 50,000 fungal species thus far identified, 500 of which are classified as yeasts, yet there is only a minor fraction of these yeasts that can be considered to have commercial importance. Perhaps the best known of these are the strains of *Saccharomyces cerevisiae* used in a variety of industrial operations.

These strains of *S. cerevisiae* are employed in three main industrial processes. The first includes the production of alcoholic beverages such as wine, beer, sake, potable spirits, and to a large extent, industrial alcohol. Second, although excess brewer's or distiller's yeasts were used during the Middle Ages for the baking industry, this source became limited as the baking industry expanded. A new industry has now been set up for the production of yeast specifically to meet the growing needs of the baking industry. The third class of industrial processes is more recent and employs yeast for the production of biomass, extracts, autolysates, and flavor compounds. The yeast used in such processes can be either primary grown or spent yeast from the brewing and distilling industries.

It is only relatively recently that a fourth category has been emerging to the forefront involving the production of recombinant proteins through the genetic expression of recombinant yeast systems. In a natural habitat, yeasts are adapted to grow on sugar-based media like overripe fruits or tree exudates, and such yeasts are in a genetic balance with the environment. Even though the industrial strains selected from such populations performed satisfactorily under commercial conditions at the turn of the century, subsequent developments have placed greater demands on these traditional strains to perform even better under environments often unfavorable to the yeast. These conditions provided the major impetus for extensive strain improvement programs. The ensuing years have produced abundant experimental work that included critical evaluation of a large number of different cultures followed by the selection of the cultures that performed best with respect to a given character. Following these studies, several strain selections with improved characteristics were made that were subsequently introduced to a variety of commercial processes. Such selections of industrial importance were incapable of competing with the wild yeast flora under natural environmental conditions. Hence, it was the task of the industrial microbiologist to maintain and employ such cultures under noncompetitive conditions to achieve optimal performance under commercial conditions.

Though much of the early research on strain development was conducted using conventional techniques, these newly developed strains were again judged inadequate because they did not continue to meet the high standards expected in a world of aggressive marketing. However, this area is being explored extensively by the application of modern genetic techniques including protoplast fusion and rDNA technology. Although some progress has been made through these efforts, clearly there is still much to be learned of the genetic makeup of commercial strains to take full advantage of this technology. Despite the progress, an unfortunate consequence could be that, in the future, such recombinant strains may impose a serious threat to our already established classification based for the most part on evolution and natural selection.

## Beneficial Yeasts

Although *Saccharomyces* is by far the most widely used genus among yeasts at the commercial level, two other genera, *Candida* and *Kluyveromyces*, have shown substantial economic importance in the production of foods, flavors, feed, and alcohol from substrates like whey. Besides, there are yeasts capable of utilizing unusual substrates that are of importance in waste disposal. For example, hydrocarbons are utilized rapidly by *Candida lipolytica*. Other species that can utilize hydrocarbons include *C. tropicalis, C. guilliermondii,* and *C. intermedia.* Many of these hydrocarbon-utilizing yeasts aid in the decomposition of petroleum waste in marine environments. Marine oil slicks are commonly attacked by psychotrophic *Rhodotorula.* In waste disposal and activated sludge systems, *C. tropicalis* and *Trichosporon cutaneum* degrade benzoic acid and many of its derivatives.

The role of different species of yeast in spontaneous-type wine fermentations has been thoroughly studied within the last 40 years as described in Chapter 4. In general, different species are dominant during various stages of fermentation, depending on the ethanol concentration of the fermenting must. For example, a fermentation is initiated by yeasts with low tolerance to ethanol, like *Kloeckera apiculata, Hanseniaspora guilliermondii,* and *Candida* species. These are superceded by more alcohol-tolerant *Torulaspora delbrueckii* and *S. cerevisiae,* and finally by the species showing the greatest tolerance to ethanol like *S. oviformis, S. chevalieri,* and *S. italicus.* All three species are now grouped under *S. cerevisiae* according to the new classification. Besides these yeasts, several other yeasts like *Pichia membranefaciens, Hansenula anomala, S. ludvigii, Schizosaccharomyces pombe,* and *Torulopsis bacillaris* can occur at various stages of the spontaneous fermentation to provide the unique flavor characteristics of such wines. The application of modern biotechnology to these wines and other similar traditional products can, however, be clearly recognized by the beneficial effect of starters on food or beverage processing.

Modern fermentations performed at an industrial scale demand complete control of processes. By the application of suitable starter cultures, as in the case of beer, wine, cheese, and sausage making, previous empirical processes have become reproducible, thus making it possible to achieve uniform product quality.

In the production of sour doughs, many bakers add a commercially prepared bacterial culture containing a mixture of *Lactobacillus brevis* and *L. plantarum.* Although these bacteria can produce sufficient $CO_2$, leavening results mainly from the activity of yeast. These include *S. exiguus* and *S. inusitatis,* of which the former was considered primarily responsible for leavening of the dough.

## Spoilage-Causing Yeasts

### Contaminants

In their universal occurrence in raw materials, yeast can be a nuisance and may cause spoilage. Brewer's and baker's yeast producers have commonly used the term *wild yeast*, meaning any yeast that contaminates the pure culture employed and promoted in a given process. Wild yeasts encountered in a brewing operation often are strains of *S. cerevisiae, S. uvarum, S. bayanus,* and *S. diastaticus.* Off-flavor and turbidity are the common defects caused by wild yeasts. Spoiled beer has been attributed to the growth of *Dekkera intermedia.*

Baker's yeast production is concerned for contaminations with oxidative species of *Candida* and *Torulopsis*. In wine making, the unwanted opportunistic yeasts are primarily certain species of *Candida, Kloeckera, Brattonomyces, Hansenula,* and *Pichia. S. bailii* occurs frequently in spoiled mayonnaise and salad dressings.

*Pathogens*

Relatively few yeasts cause disease in humans and other warm-blooded animals. The most important species are found in two genera: *Candida* and *Cryptococcus. Candida albicans,* responsible for causing candidiasis in humans, is the most commonly occurring pathogenic yeast. It affects the oral mucous membranes, usually with a mild infection, and may also be the primary agent of bronchitis, pulmonary infections, and dermatitis. An average individual can also harbor the organism in the mouth, intestinal tract, or vagina.

*Cryptococcus reformans,* the most virulent yeast, is the etiological agent of cryptococcosis, a primary pulmonary infection resembling tuberculosis. Systemic infections originally in the lungs may spread throughout the body, ultimately invading the meninges and brain.

## CULTURE COLLECTION

Authentic cultures of yeasts maintained in several general and special collections have been established in the United States, the Netherlands, England, Japan, and Canada. The largest depository of industrially important yeasts and other microorganisms has been assembled by the Agricultural Research Service (ARS) at the U.S. Department of Agriculture's Northern Regional Research Laboratory (NRRL), Peoria, Illinois. Known as the ARS Culture Collection, it contains more than 9,500 species and strains of yeasts preserved by lyophilization and other methods. The ARS Culture Collection owes part of its growth to donations from investigators worldwide, and to exchanges with other collection centers, especially the American Type Culture Collection (ATC) (Rockville, Maryland), the Centraalbureau voor Schimmelcultures (Delft, Netherlands), the British

National Collection of Yeast Cultures (Nutfield, Surrey, England), the Commonwealth Mycological Institute, (Kew, Surrey, England), the Institute for Fermentation (Osaka, Japan), the Japanese Type Culture Collection (Nagao Institute, Tokyo), and the Prairie Regional Research Laboratory (Saskatoon, Saskatchewan, Canada). Only the American Type Culture Collection publishes a catalog of microorganisms in its depository. The names and addresses of 329 collections in 52 countries and the names of microbial species held in these collections are listed by the World Federation of Culture Collections.

## REFERENCES

Bailey, L. H. 1933. *How Plants Get Their Names*. MacMillan, New York. (Reprinted in 1963 by Dover Publications, New York.)
Barnett, J. A., R. W. Payne, and D. Yarrow. *A Guide to Identifying and Classifying Yeasts*. Cambridge University Press, New York.
Bell, G. H. 1971. Action of monocarboxylic acids on *Candida tropicalis* growing on hydrocarbon substrates. *Antonie van Leeuwenhoek* **37**:385-400.
Campbell, I. 1970. Comparison of serological and physiological classification of genus *Saccharomyces*. *J. Gen. Microbiol.* **63**:189-198.
Chen, E. C. H. 1981. Fatty acid profiles of some cultured and wild yeasts in brewery. *J. Am. Soc. Brew. Chem.* **39**:117-124.
Fell, J. W., and A. Statzell Tallman. 1984. Discussion of the genera belonging to the basidiosporogenous yeasts—Genus *Leucosporidium*. In *The Yeasts: A Taxonomic Study*, N. J. W. Kreger-van Rij (ed.), Elsevier Biomedical Press, Amsterdam, pp. 496-540.
Gorin, P. A. J., and J. F. T. Spencer. 1970. Proton magnetic resonance spectroscopy—an aid in identification and chemotaxonomy of yeasts. *Adv. Appl. Microbiol.* **13**:25-89.
Johannsen, E. 1980. Hybridization studies within the genus *Kluyveromyces*, van der Walt—emend. van der Walt, *Antonie van Leeuwenhoek* **46**:177-189.
Johannsen, E., and J. P. van der Walt. 1978. Interfertility as basis for the delimitation of *Kluyveromyces marxianus*. *Arch. Microbiol.* **118**:45-48.
Johnson, E. A., D. E. Conklin, and M. J. Lewis. 1977. The yeast *Phaffia rhodozyma* as a dietary pigment source for salmonids and crustaceans. *J. Fisheries Res. Board of Canada* **34**:2417-2421.
Kock, J. L. F., P. J. Botes, S. C. Erasmus, and P. M. Lategan. 1985. A rapid method to differentiate between four species of *Endomycetaceae*. *J. Gen. Microbiol.* **131**:3393-3396.
Kraepelin, G., and U. Schulze. 1982. Sterigmatosporidium gen. n., a new heterothallic basidiomycetous yeast, the perfect state of a new species of *Sterigmatomyces* Fell. *Antonie van Leeuwenhoek* **48**:471-483.
Kreger-van Rij, N. J. W. 1977. Ultrastructure of *Hanseniaspora* ascospores. *Antonie van Leeuwenhoek* **43**:225-233.
Kreger-van Rij, N. J. W. (ed.) 1984. *The Yeasts: A Taxonomic Study*. Elsevier Biomedical Press, Amsterdam, Holland.
Kreger-van Rij, N. J. W., and M. Veenhuis. 1976. Conjugation in the yeast *Guilliermondella selenospora* Nadson et Krassilnikov. *Can. J. Microbiol.* **22**:960-966.

Kurtzman, C. P., M. J. Smiley, C. J. Johnson, and M. J. Hoffman. 1984. Deoxyribonucleic acid relatedness among species of the genus *Sterigmatomyces*. *Int. J. Syst. Bacteriol.* **30**:503-522.

Kwon-Chung, K. J. 1984. Discussion of the genera belonging to the basidiosporogenous yeasts. In *The Yeasts: A Taxonomic Study*, N. J. W. Kreger-van Rij (ed.). Elsevier Biomedical Press, Amsterdam, pp. 467-482.

Lanjouw, J. et al. 1966. *International Code of Botanical Nomenclature*, J. Lanjouw (ed.), Kemink en Zoon, Utrecht, Holland.

Lodder, J. (ed.) 1970. *The Yeasts: A Taxonomic Study*, 2nd ed. North Holland Publishing Co., Amsterdam, Holland.

Lodder, J., and N. J. W. Kreger-van Rij. 1952. *The Yeasts: A Taxonomic Study*. North Holland Publishing Co., Amsterdam, Holland.

Pedersen, M. B. 1985. Brewing yeast identification by DNA fingerprinting. In *Proceedings of the 20th Congress of the European Brewery Convention.* Helsinki, pp. 203-210.

Phaff, H. J., and C. P. Kurtzman. 1984. Discussion of the genera belonging to the ascosporogenous yeasts — Genus *Lipomyces*. In *The Yeasts: A Taxonomic Study*, N. W. J. Kreger-van Rij (ed.). Amsterdam, Elsevier Biomedical Press, Amsterdam, Holland, pp. 252-262.

Poncet, S. 1967. A numerical classification of the genus *Pichia hansen* by a factor analysis method. *Antonie van Leeuwenhoek.* **33**:345-358.

Sneath, P. H. A. 1964. New approaches to bacterial taxonomy: Use of computers. *Ann. Rev. Microbiol.* **18**:335-346.

Stelling-Dekker, N. M. 1931. Die sporogenen Hefen. *Vern. kon. Ned Akad. Wetensch., Afd. Natuurk., Sect. II* **28**:1-547.

van der Walt, J. P. 1970. The genus *Kluyveromyces* van der Walt emend van der Walt. In *The Yeasts: A Taxonomic Study*, J. Lodder (ed.). North Holland Publishing Co., Amsterdam, Holland.

van der Walt, J. P., and J. A. von Arx. 1980. The yeast genus *Yarrowia*. gen. nov. *Antonie van Leeuwenhoek* **46**:517-521.

van der Walt, J. P., and D. Yarrow. 1984. *South African J. Botany* **3**:340.

Yamada, Y., M. Arimoto, and K. Kondo. 1973. Coenzyme Q system in the classification of the Ascosporogenous yeast genus *Schizosaccharomyces* and yeast-like genus *Endomyces*. *J. Gen. and Appl. Microbiol.* **19**:353-358.

Yamada, Y., and K. Kondo. 1972. Taxonomic significance of the coenzyme Q system in yeasts and yeast-like fungi. In *Yeasts and Yeast-like Microorganisms in Medical Science*, K. Iwata (ed.). Proc. 2nd Int. Specialized Symp. Yeast. Tokyo, pp. 63-69.

Yamada, Y., M. Nojiri, M. Matsuyama, and K. Kondo. 1976. Coenzyme Q system in the classification of the ascosporogenous yeast genera *Debaryomyces*, *Saccharomyces*, *Kluyveromyces* and *Endomycopsis*. *J. Gen. and Appl. Microbiol* **22**:325-337.

Yarrow, D., and S. A. Meyer. 1978. Proposal for amendment of the diagnosis of the genus *Candida* Berkhout nom. cons. *Int. J. Systematic Bacteriol* **28**:611-615.

CHAPTER
2

# YEAST GENETICS

Although yeasts were used for baking, brewing, and wine making since the dawn of civilization, their existence was not known until the discovery of the tiny "animalcules" by van Leeuwenhoek in 1680. The role of yeast as a fermenting agent was later recognized by Louis Pasteur in 1866. These studies concluded that viable yeast cells cause fermentation under anaerobic conditions during which the sugar in the medium is converted to carbon dioxide and ethanol. With these findings, the role of yeast in industrial fermentations began to unfold.

Over the centuries, brewers relied on the yeast cultures from a previous fermentation for the inoculation of succeeding brews. The techniques used at the time for brewing were so primitive that these cultures were heavily contaminated with wild yeast and bacteria. Consequently, the beer made with such cultures was inferior and was not of uniform quality. In 1881, this problem was finally resolved by Emil Christian Hansen at Carlsberg Laboratory following his introduction of the pure culture technique to brewing. He used a serial dilution technique to separate the contaminating wild yeasts from the brewing strains of *Saccharomyces cerevisiae*. This technique permitted the isolation and propagation of pure cultures of either bottom or top fermenting strains capable of producing beers of uniform and acceptable quality. The technique was later extended to the baking, distilling, and wine industries. With this unique contribution, Hansen laid the foundation for strain selection and strain improvement for a variety of industrial fermentation processes.

Pioneering studies on *Saccharomyces ellipsoideus* conducted by Winge

(1935) of the Carlsberg Laboratory led to the discovery of the haploid and the diploid phases of the yeast life cycle. These studies made it possible to elucidate the sexual cycle of *S. cerevisiae*. Winge and Laustsen (1938) later demonstrated the technique of achieving interspecific hybrids in yeast that eventually gave birth to the use of *S. cerevisiae* for genetic studies and breeding programs. These studies together with the discovery by Lindegren and his co-workers that yeasts are heterozygous for the mating-type allele marked the beginning of systematic approaches to strain development. Deliberate selection of heterothallic haploid yeast cultures that not only mate and form diploids but also produce spores with high viability resulted in the development of most of the laboratory strains that are widely used in research today.

Genealogy of commonly used laboratory strains is provided by Mortimer and Johnston (1986). Industrial strains differ significantly from laboratory strains. They sporulate very poorly, and when they sporulate they rarely produce tetrads. The spores are generally not viable under conditions generally recognized as conducive to spore germination. However, a sudden burst of knowledge during the last few decades in yeast genetics and molecular biology has led to the application of these principles developed with laboratory strains to the genetic manipulation of industrial strains.

## LIFE CYCLE

Most members of the genus *Saccharomyces* exist either in haploid (single set of chromosomes) or diploid (two sets of chromosomes) phases. In the life cycle of yeast, these two phases alternate with each other, although under special circumstances this process may be interrupted by the formation of triploid or tetraploid strains.

Industrial strains are usually diploid, triploid, polyploid, or in most instances aneuploid where the sexual cycle is rarely present. Propagation of yeast cells takes place vegetatively by a process termed *budding*. In this process, the nucleus divides into two by a mitotic division, resulting in the same number of chromosomes in the daughter cells as in the parent cell. It is normal for vegetative cells of industrial strains to propagate by budding as long as the conditions are not conducive for sporulation.

Many laboratory strains that can be maintained in their stable haploid form for many generations are called heterothallic. These haploid cells exist as one of two mating types, designated "a" or "alpha" cells, which are able to fuse with one another to form a diploid cell type. Such conjugation or mating occurs only between haploids of opposite mating types. The diploids containing the a/alpha cell type are unable to mate. Unlike the haploid strains, they can, however, undergo meiosis or reduction division and sporulate when starved for nitrogen or easily assimilable carbon. Under these conditions, which are conducive for sporulation, each diploid cell produces an ascus containing four spores, two of the mating type "a" and the other two of the mating type "alpha" showing a typical mendelian genetic segregation. Following germination, each haploid spore gives rise to "a" or the "alpha" cell type that is the stable form of the heterothallic species (Fig. 2-1).

Among certain cell populations where the entire culture is derived from a single spore, cell fusion and diploid formation can still occur and such strains of yeast are termed homothallic. These strains have an HO gene on the left arm of chromosome IV capable of bringing about a switch in the mating type either from alpha to a or a to alpha cell types, resulting in a population of mixed mating types with high frequency of switching after the initial budding cycle. Heterothallic species, on the contrary, have a nonfunctional "ho" allele. The mating reaction between a and alpha cells either in the heterothallic or homothallic strains is initiated by a cell agglutination reaction involving complementary oligopeptide pheromones excreted into the environment by each haploid cell type. Haploids of alpha mating type produce the sex factor called alpha factor that can initiate changes only on the a cells. Likewise, haploids of a mating type produce a different pheromone called a factor that can act only on the alpha cells. Diploid (a/alpha) cells neither produce nor respond to pheromones (Sprague, Blair, and Thorner 1983).

The mating type of haploid strains of yeast is determined by the genetic expression of the MAT locus located on the right arm of the chromosome III in close proximity to the centromere. Two other loci known as HML and HMR are found to the left and right of the MAT locus controlling the conversion from $MAT_a$ to $MAT_\alpha$ or vice versa. The $HML_\alpha$ and the $HMR_a$ loci are storage copies of mating-type information for alpha and a factors that are responsible for the switch of the mating type at the MAT locus.

The $MAT_a$ and $MAT_\alpha$ DNA segments, although to a great extent identical, have shown limited sequence homology within a narrow region flanked by X segment on the left and Z segment on the right as shown in Figure 2-2. This region, unique to $MAT_a$, is termed Ya, made up of 642 base pairs (bps). The corresponding segment for $MAT_\alpha$ is 747 bps long and has been designated as the alpha region. The genetic segments $Y_\alpha$ and $Y_a$ found within HML and HMR are essentially the alpha and the a genes respectively, identical to those found in the MAT locus but generally silent. It is thus clear that the alpha gene is longer than a gene by about $10^2$ bps. Consequently, in homothallic strains, a switch in mating type from a to alpha results in a change from a shorter to a longer segment at the MAT locus. During the switch, genetic material associated with $HML_\alpha$ brings about the mating type interconversion from $MAT_a$ to $MAT_\alpha$ and the

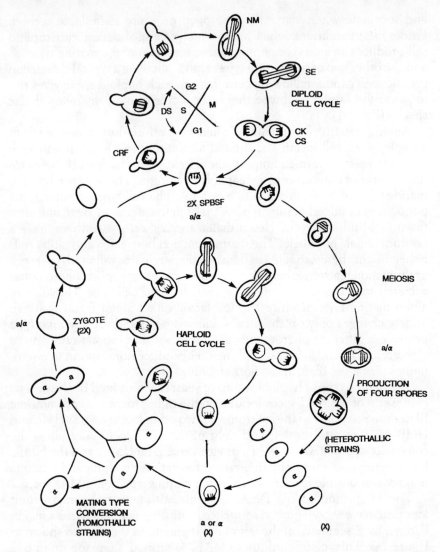

**Figure 2-1.** Budding cycle of yeast. SPBSF, spindle-pole-body satellite formation; CRF, chitin ring formation; NM, nuclear migration; SE, spindle elongation; CK, cytokinosis; CS, cell separation; DS, DNA synthesis; S, synthesis (DNA); G1, growth phase 1; G2, growth phase 2; M, mitosis. (*After Dawes 1983*)

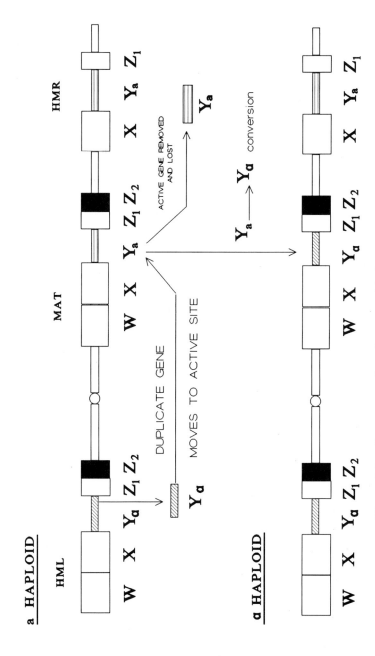

Figure 2-2. Casette-controlling element model of mating-type interconversion.

corresponding genetic material from HMR$_a$ allows mating-type interconversion from MAT$_\alpha$ to MAT$_a$. Expression of HMR and HML loci at other times is repressed by the products of four unlinked SIR (silent information regulator) genes designated SIR 1 through SIR 4.

Newly formed buds or germinating spores of homothallic strains rarely switch their mating types. Only those cells that have budded at least once are capable of switching the mating type and are termed experienced or competent cells. Most of these competent cells have shown a double-stranded break in chromosome III at the Y-Z junction of the MAT locus. This break is reported to be due to the activity of an endonuclease enzyme that may be a product of the HO gene. It has also been shown that HO-associated, double-stranded specific endonuclease is absent in new buds of homothallic cells. However, after one generation of growth, these yeasts acquire the endonuclease activity, providing them the ability to bring about the break necessary at the MAT locus for the mating-type interconversion.

The discovery of the sexual phase of the yeast life cycle stimulated yeast geneticists to initiate ambitious programs to develop improved strains for the yeast-based industries. However, the progress in such breeding studies is greatly hampered by differences found between industrial and laboratory strains. Most laboratory strains used in early experiments were haploid or diploid. On the contrary, many industrial strains are polyploids, particularly triploid, tetraploid, or aneuploid and have a DNA content between triploid and tetraploid values. These industrial strains propagate vegetatively by budding. Due to their multiple gene structure, these polyploids are genetically more stable and less susceptible to mutational forces than haploid or diploid laboratory strains. Hence, the use of genetically stable polyploid strains of yeast in brewing, baking, distilling, and wine making is no accident.

## TECHNIQUES APPLICABLE TO YEAST STRAIN DEVELOPMENT

In nature, wild yeasts are adapted to grow on sugary media such as those found in overripe fruits. Such yeasts are constantly in a genetic balance with the environment. These wild yeasts show limited use in commerical operations. However, improvement of yeast strains to perform definite functions can subsequently be achieved by critical study of a large number of different cultures followed by selection of the cultures that exhibit the desired characteristic. Such selected strains of commercial importance may find it difficult to compete with wild strains of yeast in a natural environment. These highly sensitive strains should therefore be handled with utmost care to ensure their genetic stability.

All industrial strains should satisfactorily perform the tasks for which they have been specially selected. For example, the baker's yeast should have the ability to cause the dough to rise quickly even under high osmotic pressure as in the sweet dough system. Bottom fermenting yeasts should ferment the wort to give a balanced flavor profile and settle to the bottom following the completion of the fermentation. Even though the yeast strains available at the turn of the century for use in baking, brewing, distilling, and wine making were previously accepted, recent developments have placed a greater demand on these traditional strains. These developments have provided sound reason for strain improvement. The major factors of commercial importance are under genetic control and are potentially subject to improvement by genetic manipulation. The following section summarizes some of the genetic approaches that can be used for yeast strain improvement.

## Selection

Following the introduction of the pure culture technique to the brewing industry by Christian Hansen, there was interest within the industrial community to extend the concept to other yeast-related industries like baking, distilling, and wine making. These efforts led to the identification and use of single-cell clones from naturally occurring yeast populations that would perform the required function and meet the set standards in the respective industrial processes.

Even though this method is the simplest and earliest one known for strain improvement, potential for a major improvement for any useful character is restricted by the lack of wide genetic variations in a given yeast population. Hence, this method of strain improvement consists of the recognition of desirable characteristics and direct selection from a given yeast population. Although this technique is dependent on the selection of desirable variants from a given population, it has often failed to provide the desired genetic homogeneity in the strain with constant use, perhaps due to spontaneous mutations and more likely as a result of mitotic recombination during vegetative propagation. Most industrial strains of *Saccharomyces cerevisiae* used today in the brewing and wine industries are natural isolates screened for their desirable properties. For example, a wide range of characteristics of potential importance in brewing have been studied: flavor, aroma, efficient sugar utilization, ethanol tolerance, fermentation rate, flocculation characteristics, good handling characteristics, and genetic stability. Genetic variation in a newly selected strain can eventually cause product variability, which is often the case with most industrial strains. Whenever the range of variation becomes significantly high, strain selection can again be employed to regain product quality and consistency.

The selection can be carried out by use of traditional methods or by use of continuous culture. The latter method has demonstrated significant practical value for selecting desirable industrial strains of improved genetic stability. Thorne (1970), who grew a single cell culture from a lager strain under continuous culture conditions, observed that 50% of the population was mutated at least for one character that is important for brewing after a period of nine months. He also observed that the strain did not mutate in batch cultures where selective pressure is minimal (Thorne 1968).

## Selection Following Mutagenesis

The urgency that existed during World War II to develop superior antibiotic-producing strains of penicillin led to perhaps one of the first concerted efforts to select for improved strains by the application of mutagenic techniques. Although the progress was slow and the improvements realized were in small increments, the technique proved to be of considerable value for strain improvement. Over the past several decades many yeast geneticists have taken advantage of these techniques to develop novel strains for brewing, wine making, and to a lesser degree for baking.

Mutagenesis has not been extensively employed for the development of strains for the baker's yeast industry. Clearly, the biomass yield for an industrial microorganism depends on the interaction of its genetic makeup with the specific environment dictated by the design of the fermentation process. In this regard, the number of genes involved to bring about growth in yeast is generally numerous, and the variety of environmental conditions that could influence these genes is equally large. In considering the complexity of the system, mutagenic programs designed to increase sugar yields in industrial strains are probably nonproductive, especially in the case of yeast propagations where the yields are already approaching the theoretical yields.

Selection protocols have provided a reliable means to develop novel strains of yeast for industrial fermentations where flavor of the product is of primary importance. Included in this category are the industries that produce wine, beer, distilled spirits, and other fermented beverages. Selection after mutagenesis has been attempted for the development of strains for wine making with improved flocculation characteristics, nonfoaming ability, ethanol tolerance, minimal $H_2S$ or $SO_2$ production, less fusel alcohol production, and a balanced flavor profile for esters in the final wine. Some of these strains are presently in commercial use. Mutagenesis has also been used for the development of brewing strains with the ability to produce low levels of diacetyl, 2, 3 pentadione, fusel alcohols, and acetoin.

The mutagenic techniques available for strain improvement are numerous. The common mutagens used with yeasts have been ultraviolet (UV) light, N-methyl-N'-nitro-N-nitrosoguanidine (NTG), ethyl methane sulfonate (EMS), N-nitrosourea, nitrous acid, diethylstilbestrol, and so on. The mutations caused by these mutagens are generally recessive and are only expressed after gene conversion or mitotic recombination (Spencer and Spencer 1983).

The manner in which different mutagens induce different types of mutations is known fully at the molecular level. When UV light is absorbed by nucleic acid, the absorbed energy can cause alterations in the bond characteristics of the purines and pyrimidines. The pyrimidines are found to be more sensitive to such changes than purines. Altered bond characteristics lead to the formation of covalent bonds between the adjacent pyrimidines of the same DNA strand to form dimers. The thymine dimer is one that forms most readily. Dimerization interferes with the proper base pairing of thymine with adenine and may result in thymine pairing with guanine. This pairing will provide a T-A to C-G transition. Other photo products such as cytosine dimers are formed to a lesser extent and cause mutations by becoming deaminated to uracil dimers. Since uracil acts like thymine, this deaminating will eventually produce a G-C to A-T transition. The mutagenic effect by UV light can, however, be stabilized by ensuring that the irradiated culture is not exposed to photo reactions by plating immediately after mutagenesis, which will prevent the repair process prior to DNA replication (Jacobson 1981; Stanier, Doudoroff, and Adelberg, 1970). Use of UV light has been recommended as the method of first choice for strain improvement in yeast (Bridges 1976).

A strain of *Saccharomyces cerevisiae* maintained in the stationary phase for 48 hours is thinly streaked on YPD (1% yeast extract, 2% peptone, and 2% dextrose) agar and immediately exposed to UV light of varying dosages (e.g., 15, 20, 25 seconds) at a fluence rate of approximately 0.8 $Jm^{-2}s^{-1}$ as measured by the Lastarject dosimeter. The UV source may be a mercury vapor lamp emitting radiation in the 253-260 mu range. All experiments have to be carried out under yellow light to avoid photo reactivation. The irradiated culture is incubated at 30°C in a dark chamber, and the clones can be picked and streaked onto agar plates for further screening. It is generally recommended that survival curves be established for each industrial strain in order to determine its sensitivity to UV light. Such data can further help in the determination of the radiation dose required to achieve the desired level of mutagenesis in a strain development program. Some authors recommend that the mutant search be carried out at a 10% survival level (von Borstel and Mehta 1976).

Alkylating agents like NTG have been used extensively by many research workers to develop commercially useful yeast strains. It is generally known that NTG alkylates base residues in DNA (Lawley and Thatcher 1970), probably the most important adducts being $O^6$ methyl guanine (Sklar and Strauss 1980) and $O^4$ thymine (Singer 1975). As a result of these reactions, NTG produces a preponderance of G-C to A-T transitions. An interesting property of the alkylating agent NTG is its specific interaction with DNA at the replication fork. These induced base-pair substitution mutations have been known to occur preferentially at the replication forks. NTG also produces multiple mutations in closely linked genes (Guerola, Ingraham, and Cerda-Olmedo 1971). This phenomenon has been referred to as *comutation*.

Early stationary phase cultures of yeast are centrifuged and resuspended at about $10^8$ cells/ml in filter-sterilized tris (hydroxy methyl) amino methane-maleate buffer (50 mM, pH 7.8). The NTG is added at a final concentration of 20 $\mu$g/ml. The suspension is then incubated at 37°C for 15 minutes with gentle and intermittent shaking. The yeast is then washed and plated on appropriate media for subsequent screening.

Industrial strains are often homothallic and are generally polyploid or aneuploid as stated earlier. Mutations carried out with such industrial strains are mostly recessive. They are therefore not expressed due to heterozygocity. This problem can, however, be resolved if the strain can be induced to sporulate, and viable homothallic spores can be isolated. Recessive mutations can also be made to express in homothallic strains if the mutation is induced in a haploid spore (Esposito and Esposito 1969). This solution is possible because the self-diploidization that occurs in homothallic strains causes the recessive gene to become homozygous. This technique coupled with EMS mutagenesis has been used by Snow (1979) to improve the genetic characteristics of wine yeast.

### Hybridization

Hybridization is a technique that has made an invaluable contribution in the past to strain improvement and is by no means obsolete. Pioneering efforts of Winge and Laustsen (1937) led to the production of interspecific yeast hybrids by mating haploids of opposite mating types. This method of strain development is referred to as hybridization. However, when attempts were made to apply this technique directly to manipulate the industrial strains that are generally polyploid or aneuploid researchers encountered numerous difficulties because of the homothallic nature of industrial strains, lack of mating ability, poor sporulation, and poor spore viability.

These workers were able to hybridize the few that they were able to generate by placing the ascospore of one mating type in close proximity to an ascospore of the opposite mating type by use of a micromanipulator. Under favorable conditions, the two spores mated to produce a heterozygous diploid cell.

This technique was later improved by Lindegren (1949) based on his observation that ascospores from a single mating type heterothallic strain of *Saccharomyces cerevisiae* consistently produced stable haploid cultures contrary to those from homothallic strains. This important feature enabled Lindegren to hybridize two different haploid cultures of opposite mating type by simply mixing the cells together in an appropriate growth medium.

Sporulation efficiency depends on how well the yeast cells were in a physiologically active phase prior to onset of sporulation. Hence, it is necessary that the yeast cells be grown in a nutritionally rich medium to ensure that the entire population is well nourished. The use of 10% w/w (weight per weight) malt extract as a presporulation medium has shown satisfactory results for baker's yeast (Fowell and Moorse 1960). The effectiveness of several other presporulation media has also been reviewed by Fowell (1970).

Because of the problems of poor sporulation, industrial breeders were constantly searching for methods that would provide improved sporulation and spore viability. A standard procedure used by yeast geneticists to achieve this purpose is as follows: 10 ml of a suitable presporulation medium are inoculated with yeast and incubated at 30°C for 24-48 hours with occasional shaking. Yeast cells are then isolated by centrifugation, washed at least twice with distilled water, and then resuspended in 0.5 ml distilled water. A 4-mm inoculating loop that holds about 0.01 ml of aqueous suspension is used to inoculate a slant containing 0.5% sodium acetate, 1% KCl, and 12.5% agar. After inoculation, the slant is incubated at 25°C for 3-5 days.

The sporulation efficiency of industrial strains is relatively low as has been pointed out previously. The separation of spores from vegetative cell suspensions can be accomplished by shaking the culture with paraffin oil. The spores enter into the oil phase. The ascus wall may now be ruptured mechanically, or by use of certain enzymes, and the individual spores may be isolated by micromanipulation. Selective heat treatment may also be used as a method of isolating haploid spores due to the fact that the spores are more heat-resistant than the vegetative cells. If the entire population is suspended in sterile water and held at 58°C for 2 to 4 minutes, the vegetative cells will be killed and only the spores will survive. These suspensions can then be streaked on agar plates and incubated at 30°C for 2 days. A number of small, well-spaced colonies can then be picked from these plates. The larger colonies are avoided as they may have arisen from vegetative cells that may have survived heating.

The mass mating technique, which was introduced by Lindegren, is commonly used as a technique to produce hybrids and involves the mixing together of a large number of haploid cells of opposite mating type on a nutrient medium. The resultant mixture is left overnight at 16°C and then examined for the presence of zygotes, which is determined by transferring the culture directly to gypsum in order to induce sporulation in diploid hybrids. According to Lindegren and Lindegren (1943), such asci are only produced by "legitimate" diploids.

Lindegren's technique of mass mating has been criticized by Winge and Roberts (1948) due to the possibility that some, or perhaps all, of the diploids may arise as a result of self-diploidization. Efforts were then made to modify Lindegren's mass mating technique with the intention of minimizing errors and reducing the time spent for the development of hybrid strains. This intention has subsequently been achieved by preparing a mixture of the two mating types and by subculturing it two or three times to encourage the hybrid cells to outgrow the haploid cells. Single large cells are then isolated by the use of micromanipulation and transferred to tubes containing malt extract wort for further propagation. These hybrids, when sporulated, have always been found to segregate as a and alpha spores so that it is almost certain that these have arisen as a result of mating and not self-diploidization. In a strain development program, these hybrid strains can then be screened for the desired character.

**Protoplast Fusion**

Over the past few years, several novel techniques have been applied to the genetic modification of yeast for use in commercial applications. Protoplast fusion represents one such technique to effect gene transfer. It is a useful addition to the techniques employed by yeast breeders for strain development.

An important application of protoplast fusion is the establishment of recombination through fusion in industrial strains that do not have the ability to mate naturally. Complementary biochemical mutants are used, and protoplasts are generated following the treatment of whole cells with a variety of lytic enzymes. Eddy and Williamson (1957) reported the first successful preparation of yeast protoplasts using enzymes isolated from the gut of snails, *Helix pomatia*. An osmotic stabilizer is essential to provide osmotic support for the protoplasts, and the fusion is enhanced by the addition of polyethylene glycol (PEG) and a calcium salt. With further improvements, this technique has emerged as the method of choice to improve industrial strains of *S. cerevisiae* during the past decade (Trivedi, Jacobson, and Tesch 1986; Stewart, Russell, and Panchal 1983; van Solingen and van der Plaat 1977).

The protoplast fusion technique has been used in the past to genetically improve both brewer's and baker's yeasts. The baker's yeast industry has been interested in improving a number of characteristics important for baking. These include osmotolerance, high bake activity, and high maltase activity. However, most of these characteristics are either polygenic or are not well understood at the molecular level. Therefore, many of these characteristics cannot be improved by cloning or by use of DNA transformation techniques because these approaches are commonly employed to transfer well-defined genes, probably very few at a time. Protoplast fusion disregards the necessity for gene identification and isolation and does not require the use of molecular probes. Furthermore, fusion between different genera of yeast is possible, which may further expand the genetic diversity available in strains over those that can be obtained by interspecies hybridization. However, protoplast fusion lacks the specificity required to modify industrial strains with high precision.

Procedures and basic techniques applied in protoplast fusion for strain improvement of baker's yeast were first described by van Solingen and van der Plaat (1977). The strains to be used in the fusion experiment were characterized by the heterozygocity at the mating-type locus where neither the parental strains nor the petites could be mated naturally. The two parental strains are propagated in a medium containing dextrose or sucrose as the sole source of carbon. Both parental strains are used to isolate spontaneous petites (respiratory-deficient) mutants. These petite colonies are distinguished from the *grande* colonies by size, shape, and color. One technique useful in identification is to use an agar overlay containing the dye 3-5-triphenyl tetrazolium chloride. This dye stains the normal *grande* colonies pink, while petites remain white. The petites are then isolated and subcultured in agar media containing 1% yeast extract, 2% peptone, and 3% glycerol (YP GLY). The petites will not show substantial growth on glycerol. Therefore, this step offers a final confirmation of the respiratory-deficient nature of the isolated mutants.

Two respiratory-deficient mutants selected from each of the parental strains are then grown in YPD broth for 48 hours, harvested by centrifugation, and washed twice with sterile water. Both washed cultures are then suspended in a hypertonic buffer and treated separately with the enzyme zymolyase ($\beta$-glucuronidase). This treatment removes the cell wall, leaving the protoplasts intact. The two types of protoplasts are then washed separately with hypertonic buffer, gently resuspended in fusion buffer, and then combined in a 1:1 ratio on the basis of protoplasts. Under such conditions, the protoplasts adhere and then their membranes, their cytoplasm, and finally their nuclei fuse together. The fusion buffer contains polyethylene glycol (m.w. 4000-6000), sorbitol, and a calcium salt. Following a 30-minute incubation at 30°C, the protoplast mixture is gently centrifuged to remove the fusion buffer. The resulting pellet is resuspended in a recovery broth and incubated overnight. The recovery broth contains 1% yeast extract, 0.8 M sorbitol, and 0.2% dextrose. Following incubation in the recovery broth, the regenerating protoplasts are plated on hypertonic glycerol agar containing 1% yeast extract, 2% peptone, 3% glycerol, 0.8 M sorbitol, and 3% agar. Only fusion products that have recombined to complement each of the parental type's respiratory deficiency will propagate on this medium. Colonies of individual fusion products are then picked and used in subsequent screening tests (Fig. 2-3).

Several other modified techniques are now available in the field of

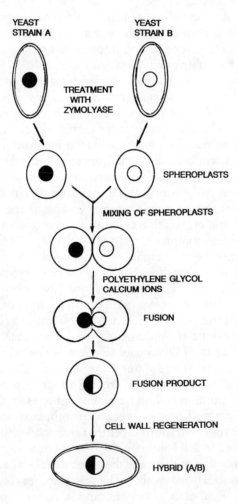

**Figure 2-3.** Spheroplast fusion of *Saccharomyces cerevisiae*.

fusion technology. One such approach that deserves mention is electrofusion. In polyethylene-glycol-mediated electrofusion, the fusion can be induced by an electric pulse between 20 and 1000 volts, depending on the distance of the electrodes for 2 to 100 milliseconds. The protoplasts first form chains in the pulsed electric field prior to the fusion. Electrofusion has been shown to significantly enhance the yield of intrageneric fusion products (Schnettler, Zimmerman, and Emeris 1984). In the past decade, this technique has been used extensively for strain improvement of industrial strains.

## Transformation and rDNA

Recombinant DNA (rDNA) technology, or DNA cloning, has gained unprecedented recognition in the past decade for providing indefinite scope and precision to the efforts to modify genetic constitution of microorganisms to increase their usefulness to humankind. It has also offered a means of overcoming the nonspecificity inherent in protoplast fusion.

The term *DNA cloning* refers to isolating a specific DNA fragment present in the genome and amplifying it or making as many copies of it as possible to obtain the amount of product needed for detailed biochemical analysis. DNA cloning is generally achieved by linking or recombining the isolated DNA fragment of interest with the specific plasmid DNA capable of functioning as a vector that can direct its own replication and that of the DNA segment tied to it within a host microorganism. With the introduction of the new genetic material to the host organism, there is likely to be a corresponding change in the protein amount or function, or in most instances, in the appearance of certain critical proteins or enzymes not originally found in the organism. In the majority of cases, the major outcome can probably be the hyperproduction of certain useful products that the organism did not produce prior to the genetic change, or an improvement of activity related to a biological reaction. Such an alteration of heredity of a microorganism by the use of rDNA technology is a major development in the field of genetic engineering.

The development of appropriate plasmid-cloning vectors has made it possible to clone virtually any yeast gene for which a mutation can be identified. A limitation that is often encountered is the availability of suitable yeast mutant strains as recipients. The general approach has been to introduce clone banks through the use of plasmid-cloning vectors into mutant strains of interest and subsequently to select and screen for complementation of the mutant phenotype. Transformation carried out in this manner has permitted the genetic improvement of industrial yeasts by the introduction of appropriate genes into the test organism from external sources.

Over the years, *Escherichia coli* has been considered by far the most powerful cloning system, yet it has disadvantages that may complicate the production of certain proteins for the food or pharmaceutical industries. *E. coli* is not an organism commonly used in the fermentation industry. Hence, if *E. coli* is to be used, there is a need for additional studies for scale-up and development work. It is also known that only a few eucaryotic genes can be expressed in *E. coli* because of the differences that exist in the posttranscriptional and posttranslational modifications between procaryotic and eucaryotic organisms. Additionally, certain cell components in *E. coli* can act as an endotoxin that could perhaps be harmful to humans. Unlike eucaryotic organisms, this organism is also unable to secrete proteins into the growing medium, which is a serious disadvantage for those who seek a continuous protein production process with minimal cost for downstream processing. As a result of these problems inherent in *E. coli*, yeast has begun to receive prominence during the last decade as an alternate host system for genetic manipulation studies. Yeast is an organism that is commonly used in fermentations and is free of endotoxins; it is a eucaryote that has the capability to excrete substances into the environment. It is the most studied eucaryotic microbial system and has attracted attention as a model for higher eucaryotic systems because of its easy amenability to genetic manipulations.

Hinnen, Hicks, and Fink (1978) were the first to demonstrate that yeast can be successfully transformed with DNA supplied from an exogenous source with appropriate alteration of the phenotype of the recipient cell. These investigators used a chimeric ColEl plasmid carrying the yeast LEU2 gene (pYeleu 2) to transform spheroplasts of yeast strains carrying the Leu2 double mutation. The pYeleu2 sequence became integrated at the LEU2 locus of chromosome III, forming a stable transformant. The spheroplasts were plated, and the selection was made on the basis that the $Leu2^-$ recipient will grow on leucine-deficient medium only if transformed with the vector that carried the LEU2 gene. To stably maintain a DNA fragment in a yeast cell, that fragment must be linked to vector DNA, which contains the region encoding DNA replication and inheritance. Most vectors are designed to confer upon host cells a unique property that would facilitate the selective enrichment of the yeast population containing the vector. These cells often carry a gene for amino acid metabolism that allows the selection for the factor in the host lacking that gene. Four such vector classes have been developed and elegantly engineered for cloning genes in yeast (Botstein and Davis 1982). These have been named for the way in which they are maintained in the yeast following transformation.

The first class of vectors directs integration of the hybrid vectors into the host DNA based on the presence of DNA sequences in the vector that are homologous to those present in the host, and such vectors are designated

YIP (yeast-integrating plasmid). The second class includes those that carry the elements of the 2u circle plasmid and is termed YEP (yeast-episomal plasmid). The third class of vectors is capable of existing in the host cell in high copy numbers due to the presence in its DNA of an autonomously replicating sequence (ars), and these plasmids are often referred to as YRP (yeast-replicating plasmid). The fourth class includes those plasmids that contain a functional centromere region that makes the region behave like a minichromosome so that it acts like a normal host chromosome during cell division. These plasmids are called YCP (yeast-centromeric plasmid). A complete description of the different systems available for transformation is beyond the scope of this review. The reader who is interested in the subject can find a more complete treatment of this subject in Stiles et al. (1983).

These cloning vectors are often engineered to contain a single recognition site specific to at least one restriction endonuclease enzyme. Hence, such an enzyme cleavage can result in the linearization of the circular DNA molecule without affecting the genetic elements of the plasmid. These cleavages occur in a staggered fashion so that resulting cuts carry complementing single-stranded termini. These staggered ends created by the specific restriction endonuclease enzyme are expected to be homologous with the ends of the DNA fragments of the "target" gene obtained by the treatment of the same restriction enzyme. These target genes are thus capable of ligating with the vector DNA by virtue of their "sticky ends" in the presence of the ligase enzyme, resulting in the regeneration of the circular DNA, and thereby yielding a novel hybrid molecule.

The discovery of restriction enzymes has been a major breakthrough in rDNA technology. It has provided the ability to dissect and clone DNA with a high degree of precision. These special endonucleases are generally produced by certain microbes as a type of mechanism for defense against invading organisms. The enzymes present in the microbe are able to cut invading DNA molecules like those associated with bacterial viruses, at specific sites, thus making them ineffective by reducing their DNA to fragments. The same sites that might occur within the host genome are generally protected by covalent modification of one of the bases within the active site. The number of restriction enzymes so far identified from microbial sources exceeds 200. These have greatly enhanced our ability to introduce new genetic information into yeast cells.

Transformation involves the introduction of genetic material from a donor cell into a recipient cell through the use of any of the transforming vectors described previously. Figure 2-4 outlines the principal steps of the gene-cloning operation.

Yeast cells are commonly rendered competent to take up exogenous DNA by the removal of the yeast cell wall by enzymatic digestion. As a first

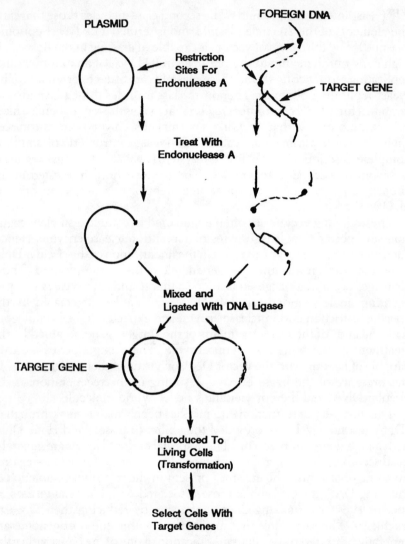

**Figure 2-4.** Transformation of yeast. The introduction of recombinant plasmids into an industrial yeast followed by selection.

step, the yeast cells suspended in 1.2 M sorbitol are treated with a cell wall digesting enzyme such as zymolyase at 2.5 mg/ml at 30°C. Protoplasting is monitored under a microscope by removing 20-$\mu$l samples and suspending in 10% sodium dodecyl sulfate as a diluent. Under phase contrast microscopy, protoplasts appear black whereas the whole cells appear much more refractile. Within 40-50 minutes, a 90% protoplasting rate can be achieved. When adequate protoplasting has occurred the protoplasts are concentrated by centrifugation at 1000 × g for 5 minutes and washed twice with 1.2 M sorbitol. These protoplasts are generally suspended and stored prior to use in YPD containing 1.2 M sorbitol, 10 mM $CaCl_2$ and 10 mM Tris hydrochloride at pH 7.5.

An alternative approach suggested by Ito et al. (1983) does not require the need to remove the cell walls. This novel approach makes the yeast cells competent by treatment with lithium or caesium salts. Liposomes have also been used to transport DNA in an encapsulated form into spheroplasted cells (Panchel et al. 1984). This technique has been particularly useful when the introduction of a large dose of DNA into a host cell is necessary.

The introduction of recombinant plasmids into industrial strains of yeast requires the use of either an auxotrophic host with a wild-type allele of the auxotrophic marker in the plasmid, or the plasmid used for the transformation must carry a marker that is selectable against a wild-type polyploid background. The amount of the genetic material required for the transformation will depend on the experiment, the yeast strain, and the transforming vector.

The competent yeast spheroplast and the recombinant plasmid are mixed under conditions similar to those used in protoplast fusion and incubated for one hour at ambient temperature. Following the transformation, it is necessary to identify the yeast cells that have received the target genes. The success of the transformation often depends on the efficiency of the screening or the selection method employed. A common approach is to transfer the yeast spheroplasts that have been treated with recombinant plasmids into tubes containing 30 ml of melted regeneration agar at 47°C, which are selective media that support growth only of those organisms that have taken up the target gene. For example, the transformants harboring the gene for $\beta$-galactosidase required for lactose metabolism may be selected by growing the transformants in media where the sole carbon source is lactose. The content is gently mixed and quickly poured into sterile plates where the transformants become visible after 3-5 days of incubation. Efficiencies of transformation depend on the effectiveness of the cloning vector and the conditions employed during the transformation.

Many specific targets for strain development that are amenable to a genetic approach can be identified. Often these traits are governed by small segments of the S. cerevisiae genome, which in most instances do not exceed one or two genes. Potentially, a wide range of useful characteristics could be introduced into industrial strains of yeast. However, most yeast characteristics important to industrial strains are known to be complex, and the precise understanding of these genes and their control mechanisms are a prerequisite for successful application of rDNA technology. Hence, an extensive data base must be developed before some of the more imaginative gene manipulations can be attempted, particularly with regard to the genetic makeup of their specific characters and regulations.

Some traits have already been introduced into baker's and brewer's yeasts to improve the usefulness of these strains in their respective applications. These developments are described briefly in the following sections.

## BAKER'S YEAST

Strain selection and strain improvement in the baker's yeast industry are aimed at achieving two important goals: achieving superior bake activities and producing a low-cost baker's yeast product without affecting product quality. Under the first category possible targets that are likely to receive consideration in a strain improvement program are rapid maltose fermenting ability (lean dough yeast), improved osmotolerance (sweet dough yeast), rapid fermentation kinetics, and better freeze/thaw tolerance. Baker's yeast production can also be made more cost-effective by using novel strains that have rapid melibiose-utilizing ability (when beet molasses is used as a substrate) or rapid lactose-utilizing ability (when whey is used as a substrate).

### Improvement of Fermentation Characteristics

*Rapid Maltose Adaptation*

Types of baker's yeast have changed in recent years to accommodate extensive marketing strategies, and the traditionally used strains have now become inadequate to satisfy the baker's complex demands. The baking industry characterizes dough systems by the level of sugar that is included in the respective formulations. Lean dough formulations often include flour, water, shortening, salt, and yeast with no added sugar. In this dough system, the main source of fermentable sugar is produced when the amylolytic or diastatic enzymes ($\alpha$ and $\beta$ amylases) that are naturally present in the flour catalyze the hydrolysis of damaged starch granules. The predominant sugar that is formed by such enzymatic reaction is maltose. Special strains of yeast with strong maltose-fermenting ability must, therefore, be selected to leaven such lean dough systems.

Generally, the maltose-utilizing enzymes within yeast are induced by the maltose present in the medium following the reduction of any glucose present to minimal concentrations (<0.1%). The maltose utilization by *Saccharomyces cerevisiae* strains require the induction of two proteins: $\alpha$-D-glucosidase (maltase) and maltose permease. In the absence of glucose, few molecules of maltose can enter the cell, perhaps by nonspecific transport mechanism, to induce the synthesis of maltase and maltose permease, which in turn allow the maltose to enter the cell rapidly. Each maltose molecule that enters the cell is cleaved to two glucose molecules by the maltase enzyme that are then metabolized by the cell. The lag period for the appearance of the two enzymes is entirely strain-dependent.

The development of genetically modified yeast strains that utilize maltose under repressing conditions would be of importance to a lean dough system. It could be achieved by developing strains that either exhibit constitutive maltase or maltose permease activity or by increasing the copy number of MAL genes to take advantage of their additive effect.

The presence of any one of the family of five unlinked MAL genes (MAL 1-4 and MAL 6) confer on yeast the ability to utilize maltose as a carbon source. Studies have suggested that each MAL locus is composed of two linked genes designated $MAL_p$ and $MAL_g$ (Naumov 1971, 1976). Naumov (1976) and ten Berge, Zoutenelle, and van de Poll (1973) also concluded that the $MAL_p$ gene is associated with the regulatory function. In addition, studies of Federoff et al. (1982) and Needleman and Michels (1983) have demonstrated that the $MAL_g$ gene encodes the maltase structural gene. More recent studies of Needleman et al. (1984) have suggested that the $MAL_g$ segment may include two maltose-inducible transcripts: one for maltase ($MAL_s$) and the other for maltose permease ($MAL_t$), both possibly being regulated by the $MAL_p$ segment of the MAL complex.

Mutations in the MAL system have been shown to confer constitutivity on the synthesis of the maltase enzyme in yeasts (Khan and Eaton 1971). Although many of these strains were unsuitable for use in commercial baking, when these strains were crossed with baking strains with poor lean dough characteristics the hybrid strains were found to ferment maltose without the usual lag period. These quick strains of yeast are described in British patents 868,621 (Burrows and Fowell 1961a), 868,633 (Burrows and Fowell 1961b), and 989,247 (Koninklijke Nederlandsche Gist en Spiritusfabriek 1965). Recent studies conducted in the United States, France, and the Netherlands have shown that these strains could be further improved by the incorporation of other useful baking characteristics (e.g., osmotolerance, improved baking activity, etc.) either by classical hybridization or protoplast fusion.

A recent development in the baker's yeast industry has been the introduction of instant active dry yeast (IADY) or quick-rising yeast with unique baking characteristics. In the construction of these novel strains, geneticists have often considered constitutive synthesis of maltase and maltose permease as an important trait for achieving improved stability and improved fermentation characteristics in the final yeast. These studies have demonstrated that rapid adaptability to maltose not only improves lean dough activity but also provides greater stability during drying. Accumulation of trehalose during growth enhanced by the presence of constitutive MAL genes (mal$^c$) may account for improved tolerance of the yeast to drying. Strains with such genetic makeup have accordingly shown good stability during drying together with acceptable leavening properties in lean dough systems (Giesenschlag and Nagodawithana 1982). However, strains with constitutive synthesis of the MAL system enzymes have, obviously, very little effect on dough systems where sucrose or glucose has been included in the formulation.

*Osmotolerant Yeast*

Most traditional baking strains of yeast show a decline in leavening activity when the sugar concentration in a dough system exceeds a certain critical concentration, as in the case of a sweet dough formulation. Similarly, salt concentrations in the 2-2.5% range can cause considerable inhibition in yeast activity, which is a common osmotic phenomenon observed with yeast in high sugar or high salt fermentations. Danish pastry, coffee cakes, and doughnuts, with their sugar concentrations as high as 25%, belong to this category.

The inhibition by high osmotic pressure, as in the sweet dough system cited above, is due to the physicochemical effect of high numbers of molecules and not to a specific inhibitory effect of the molecules as in the case of ethanol inhibition. The poor performance by traditional baking yeast in high sugar systems is commonly compensated for by the use of higher levels of yeast (7-10% on the flour basis) to achieve comparable leavening activity in the dough system. Since the baking industry can benefit by use of a more economical way of leavening of sweet dough systems, there exists a need to develop baking strains with improved osmotolerance.

The studies reported here represent an attempt at understanding the adaptation of osmotolerant yeast to higher sugar concentrations based on physical observations and, to some extent, on the biochemical processes that do take place within the cell in response to the osmotic effect. Since the underlying genetics associated with osmotolerance are not fully understood, it has become difficult to take full advantage of the newly developed genetic engineering techniques for strain improvement. This difficulty does not, however, preclude the use of certain random or "shotgun" approaches to bring together the essential genes clustered in a completely unpredictable manner.

Although the genetic systems associated with osmotolerance in yeast is not fully known, the following features have been found desirable for achieving a high degree of osmotolerance in yeast when subjected to high osmotic pressure. According to these observations, for a yeast to be osmotolerant, it must have: (1) the ability to accumulate high levels of intracellular glycerol rapidly to equilibrate with the high osmotic environment, (2) a high trehalose level as a source of carbon and probably to maintain membrane integrity, and (3) low invertase activity to maintain low osmotic pressure, especially in those systems where sucrose is used as a sugar source. A review of the literature on the progress made to date should provide leads to future research work on strain development.

*Effect of Glycerol.* Over the past few decades several investigators have demonstrated that different proteins in osmophillic yeast cells are basically the same as those in osmosensitive yeast strains (Onishi 1963; Anand and Brown 1968; Brown and Simpson 1972). Thus, there is no question that the environment within the cell must be altered to maintain the regular metabolic functions from the otherwise harmful effects of high solute concentration in the surrounding medium. It, therefore, follows that some intracellular property such as composition must be significantly different in the two types of yeast. In fact, it has long been noted that those yeast strains that are capable of producing and retaining high levels of polyhydric alcohols have a remarkable tolerance to high solute concentrations.

Much of the information available on osmoregulation of yeast is the result of work done on the osmotolerant yeast *S. rouxii*. These strains grow at water activity (Aw) levels in the range of approximately 0.6 to 1.0 Aw. More sensitive *S. cerevisiae* strains grow in sugars at Aw levels exceeding 0.9. *S. rouxii* responds to low water activity by the rapid production and subsequent retention of glycerol within the cell. At very low water activity, the plasma membrane of *S. rouxii* is known to become remarkably impermeable to glycerol. *S. cerevisiae*, on the contrary, produces a high level of glycerol under similar conditions, but does not have the ability to retain it within the cell due to the high permeability of the plasma membrane to glycerol (Brown and Edgley 1980). There is also sufficient experimental data to show that glycerol transport in *S. rouxii* is active, but it is not in *S. cerevisiae* (Brown 1974).

Considerable effort has already been made to determine the mode of synthesis and the factors that influence glycerol production in osmotolerant and osmosensitive yeasts. There is a significant difference between osmotolerant and osmosensitive yeast strains with respect to the regulatory control and mode of glycerol synthesis, although both types produce glycerol under high osmotic conditions (Brown and Edgley 1980). *S. cerevisiae* is well known for its glycerol production via the glycolytic pathway using NADH as a cofactor. In contrast, glycerol production by *S. rouxii* appears to depend primarily on NADPH generated by the pentose phosphate pathway. It is thus apparent that the two systems are regulated genetically by two different complex reactions. Furthermore, the number of genes associated with each system, besides being different, could perhaps be too complex for the application of genetic manipulation for strain improvement.

The use of hybridization techniques for the development of highly osmotolerant yeasts has proved difficult because of the poor mating and sporulating characteristics of osmophillic *Saccharomyces* spp. However, Kosikov and Miedviedieva (1976) have successfully isolated mutants with improved osmotolerance by mutagenesis followed by the use of enrichment techniques. However, these mutants were unsuitable for use in hybridization with commercial strains because of their inability to produce ascospores. Despite these problems, Gunge (1966) and Windisch, Kowalski, and Zander (1976) have been able to obtain osmotolerant yeast strains for commercial use by the application of the hybridization techniques.

The feasibility of successfully constructing a stable hybrid between highly fermentative *S. cerevisiae* and osmotolerant *S. mellis* strains has recently been demonstrated by Legmann and Margolith (1983) by use of the protoplast fusion technique. Although this hybrid strain was tested in the distilling industry, there are no reports of its applicability in other industrial fermentation processes. More recently, Spencer et al. (1985) has been able to construct a stable hybrid between *S. diastaticus* and *S. rouxii* by protoplast fusion. This hybrid has performed exceptionally well as a sweet dough strain. Similarly, Jacobson and Trivedi (1986) reported the successful development of osmotolerant strains with good sweet dough activity for use in IADY production by use of the protoplast fusion technique.

The general understanding one has about osmotolerance is entirely based on physical and, to a minor extent, biochemical observations. Hence, the existing theory on osmotolerance is that accumulation of glycerol and other polyhydric alcohols to a concentration commensurate with extracellular Aw is the principle mechanism by which these yeasts adjust to high osmotic pressure. While we continue to use the classical hybridization techniques to achieve highly osmotolerant industrial strains, efforts must be directed toward the development of genetic techniques to gain a complete understanding of the regulation of the gene complex associated with osmotolerance in order to take advantage of this newly developed technology for strain improvement.

*Trehalose Accumulation.* Trehalose is a nonreducing disaccharide commonly present in yeast as a storage carbohydrate. This sugar, made up of two glucose units, exists in the negligible to 20% range on a dry weight basis depending on the stage of growth. The stability of this sugar in yeast can be due to the storage of the disaccharide within a membrane-bound vesicle to protect it from the trehalase enzyme present in the cytoplasm. Oura, Suomalainen, and Parkkinen (1974) reported that aeration without feed at the end of a baker's yeast propagation increased the level of storage carbohydrate within the cell, primarily trehalose, with simultaneous improvement in the storage stability of the yeast. A relatively high degree of importance is now attributed to trehalose and its actual concentration is currently an important criterion for evaluation of osmotolerance and membrane stability of active dry baker's yeast. A high trehalose concentration, together with glycerol formed during adaptation, may serve as critical stabilizers for the regular functioning of the yeast under high osmotic conditions.

Trehalose is also known for its role in the stabilizing of cell membranes in anhydrobiotic organisms. This role is well demonstrated not only in dried baker's yeast but also in a variety of other dessication-tolerant systems such as the macrocysts of the slime mold *Dityostelium*, cysts of *Artemia salina*, and some nematode worms and larvae that are known to survive after complete dehydration (Crowe, Crowe, and Chapman 1984).

When biological membranes are subjected to dehydration, in the absence of stabilizers, certain changes could occur within the membrane during the phase transition that cause permanent damage to the membrane (Crowe, Crowe, and Chapman 1984). These investigations have suggested the possibility of hydrogen bonding between OH groups of the trehalose molecule and the phosphate groups of the phospholipids, thereby providing a protective effect to the membranes. These H- bonds may replace similar H bonds between lipids and water that are naturally occurring in a hydrated membrane. Trehalose molecules may thus be functioning as the water molecule in maintaining the integrity of the dried plasma membrane.

As presented earlier in this chapter, trehalose accumulation or TAC(+) phenotype is found to be closely associated with maltose utilization in yeast (Panek et al. 1979; de Oliveira et al. 1981). Thus, MAL genes must be in the constitutive mode (MAL$^c$) in order to achieve the expression of the TAC(+) phenotype. Any mutation that would alter the constitutive MAL allele to the maltose-inducible or maltose nonfermenting (mal) state results in a decline or complete loss of trehalose accumulation during growth. Thus, it is confirmed that the TAC(+) phenotype is directly linked to MAL$^c$ genes and not associated with different but closely linked genes.

Hence, trehalose and its influence on osmoregulation can be regarded as a complex character regulated by a large and indeterminate number of genes whose biochemical roles may remain unknown. As long as these genetic systems remain unidentified, classical genetic technique followed by screening will continue to be the only procedure that can be successfully applied for strain improvement.

*Effect of Invertase Enzyme.* Yeasts belonging to the species *S. cerevisiae* are generally rich in invertase, an enzyme capable of splitting sucrose into a mixture of glucose and fructose (invert sugar). Two types of invertases have been described in yeast: (1) an insoluble enzyme associated with cell wall, periplasmic space, or outer surface of the cytoplasmic membrane and

(2) a soluble enzyme located exclusively within the yeast cell. The initial reaction in sucrose utilization is the cleavage of the sucrose to the monosaccharides outside the cell by the action of the external invertase enzyme. The products formed, which are glucose and fructose, are transported into the cell by facilitated diffusion.

Six unlinked polymeric genes (SUC 1-5 and SUC 7) have been identified. Any one of the six genes is capable of conferring sucrose utilization ability to the yeast (Carlson, Osmond, and Botstein 1981). Each SUC gene encodes an internal nonglycosylated invertase enzyme and an external invertase enzyme that becomes glycosylated during its passage to the yeast cell surface. The number of SUC genes in industrial strains can vary as a function of the number of genes present in each set of chromosomes and as a function of the ploidy of the organism. Most industrial strains do not carry all six SUC genes. The external invertase enzyme is largely under glucose regulation and at low glucose concentration its activity is known to increase by a factor of 1000 (Gascon and Lampen 1968).

Gene dosage effect on invertase activity has been demonstrated by Grossmann and Zimmermann (1979). Hence, excessive levels of invertase enzyme production, perhaps due to the presence of multiple copies of the SUC gene, or due to its constitutivity, adversely affect yeast activity and consequently, its leavening power in a high-sugar system. This phenomenon is primarily due to the rapid doubling of the osmotic pressure as a result of the hydrolysis of the sucrose molecule to two hexose molecules. It can become important in high-gravity molasses fermentations or in sweet dough systems in the baking industry.

In order to minimize the effect of invertase on sweet dough systems, some baker's yeast manufacturers treat their yeast cream with acids such as sulfuric or phosphoric acids to inactivate excessive invertase enzymes prior to pressing. Clement and Hennette (1982) have described in their patented process for obtaining improved baking strains a technique for obtaining haploid strains with a minimal level of invertase activity for use in hybridization studies. They obtained low-invertase-producing mutants by treating maltose-adapted haploid and diploid strains with mutagenic agents such as EMS or NTG. The mutants screened using O-dianisidine followed by a more elaborate colorimetric assay were hybridized with a less osmotolerant baking strain to achieve osmotolerant strains of yeast for use in the baking industry.

*Rapid Fermentation Kinetics*

In the selection of commercial strains for use in yeast-related industries an attribute that is considered vitally important for improved productivity is the ability of the yeast to ferment fast with efficient utilization of the fermentable sugar present in the medium. In baker's yeast propagations, an important objective is to achieve high growth rates and greater productivity in biomass. Such selected strains with high productivity must also satisfy the baker with respect to their leavening ability. The precise mechanism by which a large number of genes control strain growth and fermentation characteristics and how different yeast strains acquire different fermentation characteristics are still unclear.

For application of classical hybridization or protoplast fusion techniques for strain improvement, an absolute understanding of the genetic control of a given character is not important. Under these conditions genetic recombination followed by an efficient screening procedure can serve as an effective method to successfully generate desired genetic variations. The use of hybridization for the development of industrial strains has extensively been used by early geneticists. One such breeding scheme that included the crossing of a commercial strain of baker's yeast and a brewer's strain for the development of baking strains with improved baking activity has been described by Burrows (1979). In this study, the hybrids with improved fermentation activity were again crossed with different baking strains having high fermentation activity but poor adaptability to maltose. By use of this technique, it was possible to improve the fermentation activity further at both early and late stages of the dough fermentations.

A gene dosage effect of MAL genes in the brewing strain *S. uvarum* has been described by Stewart, Goring, and Russell (1977). Hybridization studies involving the construction of isogenic triploids using the MAL-2 system have illustrated that multiple copies of the gene would be useful in shortening the wort fermentation time. The technique can also be applied to develop baking strains with improved lean activity. A similar but more complex system has been identified in yeasts that contained a different combination of SUC genes. It would thus be important to consider introducing the optimal gene dosage into the genotype when developing strains for the distilling or baking industries.

The protoplast fusion technique has shown the greatest potential for recombination in yeast when there is inadequate understanding of the genetics associated with a complex character such as the fermentation activity in yeast. In one such study, Jacobson and Trivedi (1986 and 1987) constructed a fusion product exhibiting superior leavening characteristics in lean, sweet, and regular dough baking systems. These strains are currently used in the baking industry.

A substantial level of research effort is currently in progress to determine the genetics associated with fermentation activity in yeast. The studies on the rate-limiting steps of the glycolytic pathway in different yeast strains could perhaps provide the information necessary for genetic manipulation in order to achieve faster fermentation activity. On the contrary, the regulation could well be more complex than what has been already described. For example, elimination of one limiting step in a pathway may lead to the appearance of a second limiting step at a different location in the same or different but related pathway, perhaps nullifying the effect of the previous improvement. Hence, an approach that could lead to identifying, cloning, or amplifying genes associated with the rate-limiting steps in the glycolytic pathway may not necessarily improve the carbon flow and, correspondingly, improve the fermentation characteristics of the yeast.

In yeast metabolism, the regulation of carbon flow has come under close scrutiny in the past few decades. Although most of the reaction steps in the glycolytic pathway are reversible and catalyzed by the same enzyme in both glycolytic and gluconeogenic directions, three of the steps are catalyzed by metabolically antagonistic, obligate glycolytic and obligate gluconeogenic enzymes (Lehninger 1975). Of these three reactions, conversion of fructose-6-phosphate to fructose-1,6-bisphosphate by the enzyme phosphofructokinase is unique because this reaction in the glycolytic direction leads to the hydrolysis of adenosine triphosphate (ATP) whereas the reverse (gluconeogenic) reaction catalyzed by fructose-1,6-bisphosphatase results in the generation of ATP. The simultaneous action of these two antagonistic enzymes could lead to an energy-wasting "futile cycle." (Holzer 1976). A net reduction of the energy level of the cell can be expected by the continued operation of such a cycle. Most biochemists speculate that such an energy-wasting mechanism in place could lead to increased fermentation rates. Many possibilities are now available for introducing characteristics into yeasts, and this new technology should offer the possibility of manipulating yeast to test the validity of this exciting theory. However, such a system may not exist in nature because, for survival of the organism, there are complex regulatory systems built in that preclude wasteful "futile cycling."

The efficiency of fermentation can be improved by increasing the tolerance of the yeast to a variety of environmental factors. Yeast strains specially selected for high-gravity brewing should not only be tolerant to high sugar concentrations in the wort but also must have the ability to maintain high fermentation activity even at high alcohol concentrations (Nagodawithana 1986). Such characteristics are also useful for the distilling industry. Similarly, by using strains with increased tolerance to temperature or by appropriately controlling the flocculation characteristics of the yeast it is possible to influence the overall fermentation rate. Although such genetic changes in yeast have hitherto been difficult to achieve by use of conventional techniques, the new technology in genetic engineering should offer valuable support in the construction of novel strains with better fermentation performance.

*Frozen Dough Yeast*

The frozen dough industry has grown steadily over the years as a result of the quality improvements that have been made to meet the diverse needs of the baker. They have provided several advantages to the consumer and the frozen dough manufacturer.

The effect of subzero temperature and freeze/thaw conditions on yeast have been extensively reviewed (Mazur and Schmidt 1968; Mazur 1970; Lorenz 1974). According to these studies, the cooling velocity has been found to affect the physical and chemical events that take place within the cell during freezing. When the cooling rate is slower than optimal, the major cell damage is due to the prolonged exposure to concentrating solutes in the surrounding medium. Under conditions where the cooling rate is optimal, the period of exposure to the solutes in the surrounding medium is correspondingly shorter and the survival rate is maximal. At higher cooling rates, the intracellular ice begins to form with rapid decline in viability. Hence, prevention of ice crystal formation within the cell is a prerequisite for a higher survival rate during freezing (Nagodawithana and Trivedi 1990).

There is sufficient evidence to indicate that the composition and integrity of the plasma membrane are important factors for the enhancement of freeze tolerance in yeast (Watson, Arthur, and Shipton 1976). According to these findings, these yeasts are capable of adjusting the unsaturated fatty acid index with temperature to alter their membrane fluidity and function in order to perform normally. Our current understanding of the genetics and physiology of freeze tolerance is inadequate, and any new knowledge that could be made available should increase the probability of successful strain improvement.

The genetic elements that confer osmotolerance to yeast may also impart, at least partially, the ability to withstand freeze/thaw conditions. This belief is based on the observation that high osmotolerant yeast such as *S. rosei* and *S. rouxii* have high viability following exposure to repeated freeze/thaw conditions. These osmotolerant yeasts show poor performance as a leavener in dough systems. However, such strains have been used in hybridization studies to improve osmotolerance in regular baking strains of yeast. Due to the apparent similarity that exists between osmotolerance and freeze tolerance, it seems likely that any genetic improvement made to enhance osmotolerance in baking strains should also improve the strain's ability to tolerate freeze/thaw conditions. Interspecific and intraspecific hybrids have been constructed by crossing baking strains of S. cerevisiae with such yeasts as S. logos, S. uvarum, S. rosei, and S. rouxii. This strategy could perhaps be of value in the development of strains for the frozen dough industry.

Hybridization has been used for the development of special strains with freeze tolerance and some hybrid strains are already patented (Nakatomi et al. 1985). The haploid strain that Nakatomi and colleagues designated as FD 612 was developed by selective hybridization between a baking strain and a hybrid between S. uvarum and an unspecified Saccharomyces species (IFO 1416) presumably having the genetic makeup that is responsible for freeze tolerance. The stability of the resulting hybrid has been improved by carrying out a series of backcrosses with the parental baking strain.

Likewise, there is some evidence to suggest that alcohol tolerance is also closely associated with freeze tolerance in yeast. In this regard, an ethanol-tolerant strain applicable to the frozen dough industry has also become the subject of a patent claim (Kawai and Kazuo 1983).

Genetic information associated with freeze tolerance in yeast is not available in the literature. Hence, as long as the specific genetic elements pertinent to freeze tolerance remain unidentified, it will be almost impossible to directly apply the recently developed genetic engineering techniques for strain improvement. There is thus a definite need to identify and characterize the important structural and regulatory genes related to freeze tolerance to take full advantage of the newly developed recombinant DNA technology.

## Improvement of Process Economics

### Melibiose-Utilizing Baker's Yeast

In a typical baker's yeast propagation, the fermentable sugar for yeast growth is derived from a medium that contains a mixture of cane and beet molasses. The ratio of the two molasses types used is governed by the economics and general availability of the two types. In addition to sucrose, which is predominant in both molasses, beet molasses contains the sugar raffinose in the 0.5-2.0% range. Only one-third of the molecule is utilized by baker's yeast. The $\beta$-fructosidase (invertase) enzyme present in the baker's yeast yields fructose and melibiose from raffinose. The fructose formed is readily assimilated by the yeast. Baker's yeasts are unable to utilize the disaccharide melibiose due to the lack of the enzyme $\alpha$-galactosidase (melibiase). Clearly, the productivity and process economics could be markedly improved by use of baker's yeast strains that have the added capability of utilizing melibiose.

At least six unlinked polymeric genes (SUC 1-5 and SUC 7) can be identified in yeast as described previously. Only one of the six SUC genes is required for the fermentation of sucrose or one-third of the raffinose molecule. In addition to the SUC genes, the brewing strain *S. uvarum* has the MEL genes (Gilliland 1956) that provide the ability to completely utilize both melibiose and raffinose sugars. Hence, the utilization of melibiose by baker's yeast should not only improve the yeast yield based on unit weight of molasses used but it should also result in a saving in sewer charges through the reduction of the BOD load in the effluent stream.

Hybrid baker's yeast strains capable of fermenting melibiose have already been developed by hybridizing baker's yeast strains and lager (*S. uvarum*) strains of yeast that readily ferment melibiose (Lodder, Khoudokormoff, and Langejan 1969; Kew and Douglas 1976; Winge and Roberts 1956). The two parental strains were sporulated and the $MEL^+$ *S. uvarum* spores of one mating type were crossed with the *S. cerevisiae* baker's strain of the opposite mating type according to the classical hybridization technique developed by Lindegren and Lindegren (1943). All hybrids were then selected for their ability to ferment melibiose rapidly. In order to conserve the baking characteristics of the parent genotype, several backcrossings with the parental baking strain were often necessary.

Poor sporulation and very low spore viability of industrial baking strains have caused serious difficulties in the use of hybridization techniques for strain development. Nevertheless, Universal Foods Corporation has circumvented this problem by use of protoplast fusion (Jacobson and Jolly 1989). Their experiment utilized protoplast fusion technique to develop $MEL^+$ industrial hybrids capable of improving the molasses yield in yeast propagations. In this study, a genetically marked laboratory strain, 1453-3A, was fused with an industrial baker's strain, FE1-4. The fusant identified as MBL 7 has consistently shown higher molasses yields than the baker's yeast parent. However, this hybrid performed poorly in baking applications.

Unlike most other complex biochemical characters in yeast, specific genetic systems pertinent to melibiose utilization appear to be well defined. The enzyme $\alpha$-galactosidase or melibiase is encoded by a family of MEL genes (Kew and Douglas 1976). Sumner-Smith et al. (1985) have successfully sequenced one MEL gene and its flanking regions. These authors have identified an 18-amino-acid N-terminal signal sequence for the melibiase enzyme to be transferred across the membrane.

The melibiose utilization pathway appears to be under the control of the GAL regulatory system (Post-Beittenmiller, Hamilton and Hopper 1984). Baker's yeast, like most other galactose-utilizing organisms, require three important enzymes to convert galactose to glucose-1-phosphate.

These are referred to as Leloir pathway enzymes. These three enzymes are galactokinase, galactose-1-phosphate uridyltransferase, and uridine diphosphogalactose-4-epimerase, which are encoded by GAL 1, GAL 7, and GAL 10 genes, respectively. The enzyme phosphoglucomutase produced by the yeast constitutively converts glucose-1-phosphate to glucose-6-phosphate, thus permitting its entry into the glycolytic pathway. The induction of the three galactose-metabolizing enzymes in the Leloir pathway is regulated by the highly coordinated interaction of galactose with the products of two major regulatory genes, GAL 4 and GAL 80 (Kew and Douglas 1976). The GAL 4 gene product is responsible for induction of the Leloir pathway genes and is inhibited by the GAL 80 protein. The GAL 80 protein is thought to bind to the GAL 4 protein, or at its site of action in the absence of galactose. Kew and Douglas (1976) have also shown that GAL 4 and GAL 80 proteins regulate transcriptional expression of the $\alpha$-galactosidase gene (Fig. 2-5). Although the induced and basal modes of MEL 1 expression are GAL 4-dependent, the unique behavior of the MEL 1 gene reflects a different mode of regulation in the presence of melibiose.

The use of rDNA technology for the construction of MEL$^+$ baker's yeast strains has been reported in the literature (Liljestrom-Suominen, Joutsjoki, and Korhola 1988; Ruohola et al. 1986). A hybrid plasmid has been developed by cloning the MEL 1 gene from *S. uvarum* into a shuttle vector like YEp 13. However, when the hybrid plasmid was transformed into commercial baker's strains, transformation efficiency was significantly low, and few that transformed were highly unstable under nonselective conditions.

The problem of instability was greatly minimized by the development of a new vector containing the LEU 2 gene that has the ability to complement LEU 2 mutation in the yeast. Such vectors integrate at the site of homology to result in a direct duplication of the homologous sequence, thereby stabilizing the transformed gene. The site-directed integration and selection for LEU 2 was used to target MEL 1 to the LEU 2 locus of chromosome III. Certain transformants have shown only a minor loss in MEL+ phenotype even after 80 generations on nonselective media. This phenotype is surprisingly stable in view of the fact that these integrants contained the hybrid plasmid tandemly integrated into a single copy of the chromosome III, providing an improved $\alpha$-galactosidase activity to the transformant. Although the presence of bacterial DNA contained in the hybrid plasmid is essential for the functioning of the shuttle vector, marketing considerations have resulted in a reluctance to introduce such bacterial genes into industrial strains. While taking advantage of rDNA techniques, it may be preferred to avoid the presence of bacterial DNA to ensure the safety of the people who consume the products made by the strains constructed by this technique.

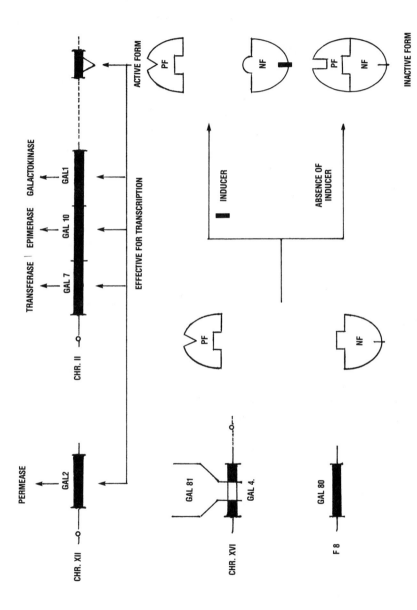

**Figure 2-5.** Regulation of the galactose-metabolizing enzymes in the Leloir pathway.

## Lactose-Fermenting Baker's Yeast

According to the traditional method of baker's yeast production, the principal raw material that supplies the fermentable sugar for yeast growth is beet and/or cane molasses. Although this method has been the practice for 60 years, the quality of molasses has been declining during the last few decades due to the introduction of novel technologies to the sugar-processing industry to achieve improved process efficiency. The progressive decline in molasses quality with increased pricing for molasses during the recent past has prompted baker's yeast manufacturers to search for cheaper substrates that would make the price of yeast less dependent on the price and quality fluctuations of molasses.

Currently, whey is a cheaper substrate than molasses on the basis of sugar. Lactose is the principal sugar present in whey. However, baker's yeast, *S. cerevisiae*, does not have the required enzyme system to assimilate lactose sugar as a carbon source. Yet, there are a few yeast species belonging to *Kluyveromyces marxianus* that utilize lactose, but these strains are unsuitable for use in baking because of poor gassing characteristics (Bruinsma and Nagodawithana 1987). There is, thus, a definite need to develop new strains of baker's yeast with the added capability to utilize lactose in order to take full advantage of the cheaper whey that is available from the cheese industry.

Lactose is a disaccharide like maltose whose uptake and utilization is under the control of the polymeric gene system. However, unlike sucrose and melibiose, which are hydrolyzed outside the cell membrane, the lactose sugar like maltose is first transported into the cell before it is hydrolyzed within the cell matrix. Hence, for lactose fermentation, a minimum of two gene products is necessary: a permease to transport the lactose molecule across the cell membrane and the $\beta$-galactosidase enzyme to hydrolyze the lactose to the two assimilable sugars, glucose and galactose.

The structural gene for $\beta$-galactosidase has been identified and designated LAC 4 (Sheetz and Dickson 1981), and the one for lactose permease is LAC 12 (Sreekrishna and Dickson 1985). The induction of $\beta$-galactosidase enzyme in *K. lactis* involves a complex regulatory system whose mechanism has already been partially elucidated. It is, however, apparent that the induction of the LAC 4 gene by lactose is regulated at the transcriptional level (Lacy and Dickson 1981). Based on their findings, Dickson et al. (1981) hypothesized that regulation is partly due to the action of a LAC 10 gene product that has a negative control over the transcription of the LAC 4 gene. Hence, the function of LAC 10 is analogous to the negative regulator GAL 80 gene of *S. cerevisiae* as described previously. Although LAC 10 is not closely linked to the LAC 4 gene, both loci have been

## BAKER'S YEAST 71

mapped to chromosome II. Subsequent studies with deletion mapping of the 5' regulatory region of the LAC 4 gene have revealed the possible presence of a region necessary for the binding of a positive regulator.

Following an extensive search, the positive regulator gene in *Kluyveromyces lactis* was identified as the LAC 9 gene, which is analogous to the GAL 4 gene in *Saccharomyces* spp. for the galactose-melibiose regulator. The LAC 9 gene has subsequently been cloned and sequenced (Wray et al. 1987). These investigators reported that when LAC 9 was introduced into a GAL 4 defective mutant of *S. cerevisiae* it complemented the mutation and activated the galactose-melibiose regulator. Unlike the GAL 4, the presence of LAC 9 provided constitutive expression of the galactose-melibiose gene system. Although the strain exhibited catabolite repression in the presence of glucose, the repression exhibited by LAC 9 was less severe than GAL 4. The structural differences that these authors have observed between LAC 9 and GAL 4 could perhaps account for the functional differences observed in the two regulatory genes (Fig. 2-6).

A technique that has been used extensively in the recent past for genetic manipulation of industrial strains is protoplast fusion. In spite of successful protoplast fusion among intrageneric partners, intergeneric fusions have met with little success to date. In all those attempts at fusing the protoplasts of the lactose-utilizing yeast *K. lactis* with those of *S. cerevisiae* there was the inherent problem of genetic instability of the fusion product (Stewart 1981). This problem may presumably be due to the fact that genetic recombination is still dependent on the degree of DNA homology between the strains involved in the fusion.

With inadequate understanding of the regulatory controls associated with the lactose-galactose regulator, the chances of constructing hybrid strains using genetic engineering techniques are rather slim. Nevertheless, at least two research institutions have applied for patent coverage for their lactose-fermenting yeast strains of baker's yeast (Dickson and Sreekrishna 1986; Yocum and Hanley 1987).

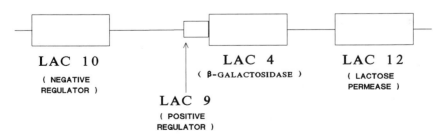

Figure 2-6. LAC 4 and LAC 12 gene control in *K. lactis*.

## BREWER'S YEAST

### Improvement of Carbohydrate Utilization

Both ale (*S. cerevisiae*) and lager (*S. uvarum*) strains have the ability to ferment the sugars glucose, fructose, sucrose, maltose, and maltotriose present in a regular wort. Brewer's yeasts are unable to utilize maltotetrose, higher dextrins, and starches. The degree to which the two yeast types ferment raffinose is used as a criterion to distinguish the two commercial strains. Top-fermenting *S. cerevisiae* ferment only the fructose moiety of the trisaccharide due to the presence of the invertase enzyme, which results in one-third utilization of the raffinose molecule. The lager strains, in contrast, are capable of utilizing the entire raffinose molecule due to their ability to split both bonds in the trisaccharide molecule because of the presence of the two enzymes $\beta$-fructosidase (invertase) and $\alpha$-galactosidase (melibiase).

An area of research that is of significant interest to the brewing industry involves carbohydrate utilization by the brewing yeast. It often includes the improvement of the fermentation rate of a brewery fermentation that can be brought about by increasing the rate at which maltose and maltotriose in the malt wort is metabolized. In addition, construction of brewing strains with the ability to ferment dextrin and starch would significantly increase the efficiency of the brewing process as well as provide the ability to produce a low-calorie (light) beer. The genetics associated with these traits will be discussed in the following section.

### *Maltose Utilization*

Maltose is the predominant fermentable sugar in maltwort. The fermentation of maltose sugar by *Saccharomyces* strains requires at least two enzymes: $\alpha$-D-glucosidase (maltase) and maltose permease. When cells are grown in the presence of glucose, or under glucose-repressing conditions, only a very low level of maltose-utilizing enzymes is present in the cells. The addition of maltose at very low concentrations of glucose leads to a coordinate increase in both maltase and maltose permease levels. Generally, a lag period is present for the two enzymes in wild strains after the initiation of the induction, which is entirely strain-dependent.

Like certain other disaccharides, maltose is under the control of polymeric gene systems (Winge and Roberts 1950). It is now known that *Saccharomyces* species contain at least five unlinked related genes: MAL 1-4 and MAL 6. The presence of any one of these genes gives the ability to ferment maltose as a carbon source. Most haploid strains that have the ability to utilize maltose contain only a single functional copy of one of the MAL genes.

Hybridization experiments involving the construction of isogenic triploids have demonstrated that a gene dosage effect can be achieved using the MAL 2 system. In these studies, yeasts with the genotype MAL 2/MAL 2/MAL 2 fermented the wort faster than the strains with genotypes MAL 2/MAL 2/malo or MAL 2/malo/malo. Hence, a brewing strain with multiple copies of the MAL gene would be useful to achieve a faster fermentation rate (Panchel et al. 1984).

The MAL genes have been found to be complex loci consisting of at least two factors required for maltose utilization. Naumov (1971, 1976) reported that each MAL locus was made up of two linked genes termed $MAL_p$ and $MAL_g$. Naumov (1976) and ten Berge et al. (1973) proposed that $MAL_p$ gene is associated with a regulatory function. Subsequent studies by Federoff et al. (1982) and Needleman and Michels (1983) showed that $MAL_g$ gene encodes the maltase structural gene. Needleman et al. (1984) subsequently reported that the $MAL_g$ segment, in fact, contained two maltose-inducible transcripts: one for maltase ($MAL_s$) and the other for maltose permease ($MAL_t$), both regulated by the transcriptional level by the $MAL_p$ gene product.

*Starch Utilization*

The wort used in a regular brewery fermentation contains about 60-70% fermentable sugar on a dry solids basis. These fermentable sugars include glucose, fructose, sucrose, maltose, and maltotriose. The maltotetrose and dextrins that constitute 30-40% form the unfermentable fraction of the wort. The latter fraction passes through the regular brewery fermentation, unaffected by the yeast to eventually end up as the calorie-contributing component of the final beer.

The public has become increasingly calorie conscious. In response to this public need, the brewing industry introduced a low-calorie beer made by eliminating the fraction of the unfermentable carbohydrates from the regular-type beer. The modern brewer has been able to develop this product by use of the enzyme amyloglucosidase in the brewing process.

The combination of characteristics possessed by the enzyme amyloglucosidase makes it the enzyme of choice in the production of low-calorie beer. This enzyme hydrolyzes starch at the $\alpha$-1-4, $\alpha$-1-6 linkages and, to some degree, at $\alpha$-1-3 linkages to yield glucose. Currently, this enzyme is used in the commercial production of both ethanol and high-fructose corn syrup. Commercial preparations of amyloglucosidase enzyme are derived from strains of *Aspergillus* and *Rhizopus*. Although these starch-degrading enzymes are not present in brewing strains of yeast, an exocellular enzyme has been reported in strains of *S. diastaticus*. Over the years, several attempts have been made to use *S. diastaticus* strains to improve the efficiency of the brewing process, but in all such instances the fermentations were incompletely attenuated, and the beer made was unpalatable.

During the last two decades, there has been considerable interest in transforming brewing strains of yeast with amyloglucosidase genes (DEX genes) from a variety of sources to improve the efficiency of the wort carbohydrate utilization. However, the successful transformants have so far failed to produce palatable beer.

The amyloglucosidases are extracellular enzymes that remove glucose units from the nonreducing end of the starch molecule. Tamaki (1978) demonstrated that dextrin utilization capability was due to the presence of any one of the STA genes (STA 1, STA 2, and STA 3). These genes were later found to be allelic to the DEX genes (DEX 1, DEX 2, and DEX 3) as described by Erratt and Nasim (1986). The amyloglucosidase gene is described as having a putative signal sequence of 24 amino acids and 4 introns of about 60 base pairs each.

Several attempts have been made in the past to construct brewing strains capable of reducing the dextrin portion of the regular beer with the objective of producing a low-calorie beer. Although genetic manipulation of brewing strains by the protoplast fusion technique has been successful, fermentations carried out with such fusion products were incompletely attenuated, and the beer produced was unpalatable. This problem was due to the heavy medicinal phenolic aroma from 4-vinyl guaiacol produced as a result of the enzymatic decarboxylation of ferulic acid derived from malt or hops. More recent studies (Goodey and Tubb 1982) have shown that the dextrin-utilizing property was associated with the phenolic off-flavor due to the presence of a nuclear gene designated POF (phenolic off-flavor) coding for the ferulic acid decarboxylating enzyme. Appropriate mating studies followed by tetrad analysis have shown that the DEX and the POF genes could segregate independently, making it possible to select haploid strains having the ability to utilize dextrin without imparting the phenolic off-flavor (Fig. 2-7).

In the past few years, the transformation technique has presented an opportunity to transfer specific genes to brewing strains. This approach has made it possible to overcome the nonspecificity inherent in protoplast fusion. The transformation techniques have already been described elsewhere in this chapter. The reader desiring more extensive information is encouraged to refer to Botstein and Davis (1982) and Dahl, Flavell, and Grosveld (1981). Following the introduction of the vector containing the desired foreign gene into the recipient, the transformants are screened to select for those having received the target gene.

Most plasmids commonly used in yeast cloning contain yeast chromosomal DNA, a fragment of the 2 $\mu$m plasmid, and a certain level of

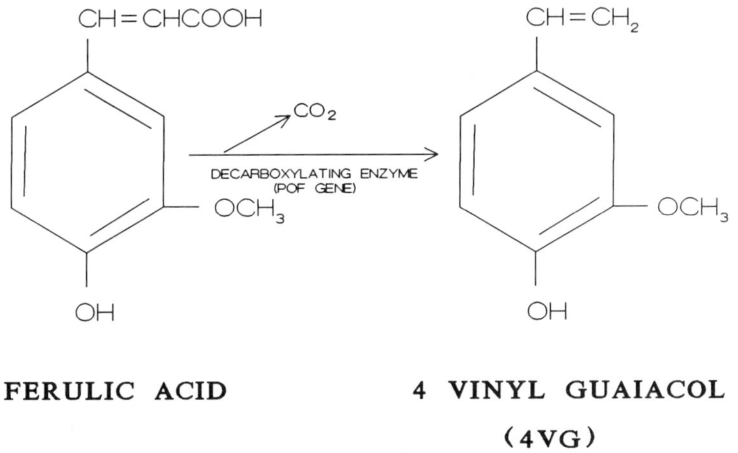

FERULIC ACID      4 VINYL GUAIACOL
(4VG)

Figure 2-7. Production of 4-vinyl guaiacol from ferulic acid.

bacterial DNA to provide the ability for the plasmid to replicate in *E. coli* in order that the plasmid reconstruction and preparation of adequate DNA can be made possible for transformation experiments. However, a serious problem that is often encountered is that some of these bacterial genes may be capable of producing polysaccharide endotoxins that can be harmful to humans. Hence, it may well be preferred to avoid the presence of bacterial DNA in brewing strains either as plasmids or when integrated in the yeast genome.

To overcome the problems that can be caused by bacterial genes, Kielland-Brandt et al. (1979) developed a shuttle system that carried only the yeast DNA. These researchers constructed a chimeric plasmid by ligating a mixture of fragments of yeast DNA following the action of restriction endonucleases to linear fragments of 2 $\mu$m yeast DNA plasmid. Such plasmids provided the geneticists and molecular biologists with the ability to transform histidine auxotrophic yeast mutants to prototrophy. As an alternative method, Yocum (1986) designed jettison vectors that remove sequences flanked by the direct duplication of the primary integrant structure. In such studies, the commercial gene of interest is cloned at the edge of the yeast integration sequence (HO locus). This system provides the added advantage of achieving repeated rounds of integration and jettison to finally introduce the commercial gene to additional HO loci. This system has been applied to transfer an *Aspergillus niger* amyloglucosidase gene to a brewing strain to be tested in low-calorie beer production. Despite these extensive studies no evidence has so far been found of any suitable dextrin-utilizing brewing strain for the production of light beer of acceptable quality. However, these positive advances should eventually lead to the construction of suitable brewing strains for low-calorie beer production.

## Killer Yeast

Bevan and Makower (1963) first reported the existence of certain strains of yeast in nature capable of producing toxins that could kill other sensitive strains. These are termed "killer" strains of yeast. These killer strains that have the killer property to which they themselves are immune, are not, however, effective against other fungi and bacteria. Maule and Thomas (1973) isolated strains of killer yeast contaminants in stirred continuous fermentations that were lethal to both ale and lager strains. Such contaminations have resulted in loss of viability and consequent displacement of the brewery strain, resulting in the production of an unpalatable beer.

The killer factor has been reported to contain a glycoprotein (Bussey 1972; Palfrey and Bussey 1979) that exerts its lethal effect by changing the integrity of the plasma membrane of sensitive strains (Middelbeck et al. 1980). These researchers have also observed an inhibition of the active transport system for amino acids into the yeast.

These killer strains of *Saccharomyces* contain a cytoplasmic double-stranded RNA (dsRNA) that codes for an extracellular toxin as well as for an immunity factor that gives the ability for the infected cell to protect itself from the toxin. Two main types of dsRNA molecules have been identified in yeasts. MdsRNA has been found to encode for the extracellular toxin as well as for the immunity factor that protects from destruction. It is found only in the killer strains of *Saccharomyces*. LdsRNA has been detected in most of the yeast strains, including brewing strain and killer strains, and its principal function is to code for the capsid protein that surrounds the MdsRNA to form a viruslike particle. The killer strains generally contain about 12 copies of MdsRNA per cell, and these intracellular viruslike particles are extrachromosomally transmitted during cell division. The two dsRNA molecules with regulation from the chromosomal gene are responsible for the expression of the killer and the resistant phenotypes as well as for the maintenance or the replication of the viruslike particle (Wickner 1976).

The mechanism by which the killer phenomenon is expressed is yet unclear. A model proposed by Bostian et al. (1983) suggests a mechanism whereby the membrane structure of the toxin-producing strain is changed to acquire immunity to its own extracellular toxin while making it lethal only to the sensitive strains of yeast present in the medium. According to this model, a toxin polypeptide sequence (m.w. 32,000) is produced by the in vitro translation of the MdsRNA. During protein synthesis, the leader peptide region remains attached to the rough endoplasmic reticulum. The gene product is then cotranslationally transported across the endoplasmic reticulum membrane into the lumen. The leader peptide portion is cleaved by enzymes exposing the N-terminal of the mature toxin. Glycosylation is assumed to take place within the C-terminal resistance-determinant region. Subsequent translation through the Golgi apparatus and the accumulation of glycosylated toxin precursors presumably occur in secretory vesicles. Such vesicles fuse with the cytoplasmic membrane, resulting in the cleavage and release of the toxin to the surrounding medium. The glycosylated resistance determinant remains exposed on the exterior of the cytoplasmic membrane and confers immunity to the cell against its own toxins. The model of toxin precursor structure, synthesis, and maturation is shown in Figure 2-8.

Generally, modified brewing strains containing the killer character should provide two important advantages to the brewer. First, these strains would be immune to their own toxin and related killer yeasts, and second, the production of the toxin would prevent the growth of many wild yeasts. Such strains could, thus, be useful in older plants where wild yeast contamination is frequent, or in newly designed plants to minimize the cost for sterilization and pasteurization for process economics.

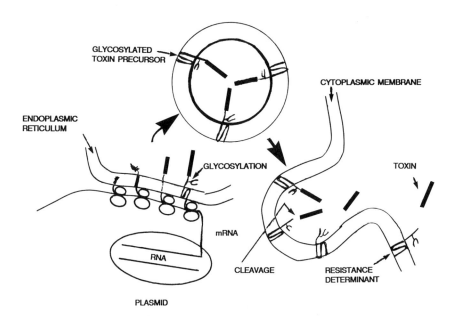

**Figure 2-8.** Model for the maturation of killer toxin and the mode of development for the resistance determinant. (*After Bostian et al. 1983*)

Many possibilities are now available to bring about the transfer of the cytoplasmically inherited killer character from laboratory strains of *S. cerevisiae* brewing strains. Young (1981) first outlined a procedure that was capable of successfully transferring the cytoplasmic killer factor to a brewing strain without altering its nuclear genome (Fig. 2-9). According to this procedure, sometimes referred to as *rare mating,* spontaneously arising respiratory-deficient (petite) prototrophic nonkiller mutants of commercial brewing yeasts were crossed with respiratory-sufficient auxotrophic haploid laboratory killer strains bearing the Karl mutation. The Karl mutation bears the alleles for defective nuclear fusion, and only one of the parental strains should carry the Karl allele to exhibit defective karyogamy. The heterokaryons formed are unstable, and these segregate to yield cytoductants (or heteroplasmons) containing the nucleus of one of the parents with cytoplasmic contribution from both parents. From the hybrids produced by rare mating, cytoductants containing the killer factor and the nucleus of the brewing parent can be identified by screening for prototrophic respiratory-sufficient strains that contain the killer immunity phenotype.

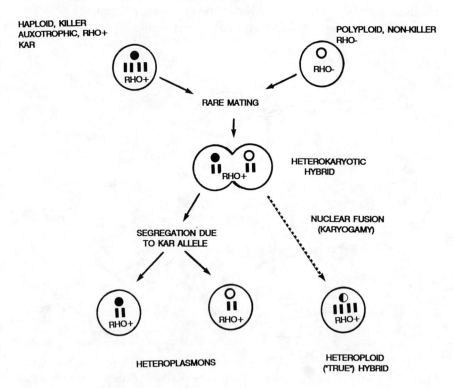

**Figure 2-9.** A method of producing yeast with killer character using haploid strains carrying the Karl mutation. (*After Young 1981*)

It is possible that the newly constructed industrial strains would show differences from the parent strain for certain traits due to the presence of the mitochondria introduced from the laboratory strain. Nevertheless, the triangle taste tests have shown no difference in the quality of beers produced by such genetically manipulated strains and the controls. Furthermore, as yet, no region of mitochondrial DNA has been identified as important for any given characteristic. However, if the type of mitochondria is found to be important for beer quality, the desired mitochondria for the industrial strain can be inserted into laboratory strains prior to use in rare mating. Although the beer produced by the test strains is not affected by the killer factor organoleptically, the head space analysis has shown a significant reduction in the fusel alcohol content in the test beers when compared with the control.

Transformation of the killer factor has also been suggested as a probable technique for constructing industrial strains capable of purging killer and nonkiller-contaminant yeasts (Thomas et al. 1987). In this regard, Bussey, Vernet, and Sdicu (1987) transformed a $K_2$ killer strain with a YEP vector containing a cDNA copy of the $K_1$ killer/immunity gene. The transformant that contained $K_1/K_2$ killer genes exhibited stable inheritance and was lethal not only to sensitive strains but also to $K_1$- or $K_2$-containing strains in the medium. However, these authors have not made any reference to the flavor characteristics of the beer made with these strains.

**Flocculation**

*Flocculation*, as commonly used in the brewing industry, is defined as the reversible aggregation of the dispersed yeast cells into clumps, and subsequent separation from the suspending medium. This process generally takes place toward the latter half of a brewery fermentation. In the case of lager fermentations, the floc sediments rapidly toward the latter part of the brewery fermentation or the floc may rise and collect as a yeast head at the surface of the fermented medium if the yeast happens to be a top-fermenting yeast. The latter phenomenon characteristic of ale yeast is due to the buoyant nature of the yeast clumps following the entrapment of carbon dioxide bubbles in such cell aggregates.

The degree of flocculence is one of the important characteristics considered in selecting a good brewing yeast. It is highly important that the selected strain remains in suspension during the fermentation and spontaneously settles out of the beer late in the fermentation. A strain character of this nature gives the ability to the brewer to minimize or completely avoid certain expensive process operations like centrifugation and clarification. Premature separation of yeast cells results in an incompletely fermented sweet beer with less intense flavor profile. A more powdery (less flocculent) type of yeast produces a drier beer with yeasty flavor, often creating difficulties in downstream processing. Hence, brewers prefer to work with yeast strains having moderate flocculence.

Although flocculation is under genetic control, it is to some extent influenced by physical and chemical factors. Studies on the genetic aspect of flocculation began in the early 1950s with Gilliland's (1956) observation from mating studies with two nonbrewing strains of *S. cerevisiae* that differed only in their flocculation properties. These studies showed that a single gene in the yeast genome is responsible for the flocculence in yeast. Thorne (1951) performed similar mating studies with brewing strains and demonstrated that flocculence was an inherited character that was dominant over nonflocculence. These studies also showed that at least three pairs of polymeric genes (FLO) are responsible for the flocculence, with the possible existence of a suppressor gene that could have a negative influence on the expression of the FLO genes. These pioneering studies were finally confirmed nearly a quarter century later by Lewis, Johnston, and Martin (1976) and Anderson and Martin (1975). These researchers concluded that flocculation is under the genetic control of multiple gene systems and that it is modulated by the influence of additional modifier genes.

Variation in ploidy, which is a characteristic of brewing strains, has already been mentioned elsewhere in this chapter. These strains sporulate poorly, and the few that sporulate rarely form tetrads, and the viability of these spores is unusually poor. Hence, the use of the hybridization technique for the analysis of flocculation characteristics has proven difficult. Nevertheless, the first dominant flocculent genes, FLO 1 and FLO 2, were identified from commercial brewing strains (Lewis, Johnston, and Martin 1976). Flocculating gene FLO 4 was subsequently identified in laboratory strains (Stewart and Russell 1977). The genes originally described as FLO 1, FLO 2, and FLO 4 are, in fact, allelic and constitute one single locus designated as FLO 1 in chromosome I (Russell et al. 1980). A recessive gene, referred to as flo 3, has been identified on a mutant of a nonflocculant laboratory strain (Johnston and Lewis 1974). Although modifier genes are present in certain crosses, these do not, by themselves, determine flocculence; instead, they influence the degree of flocculation in association with the flo 3 gene.

An additional gene, which was dominant for flocculation and distinct from FLO 1, was discovered in a brewing strain. The gene was later designated FLO 5 (Russell and Stewart 1979). Unlike FLO 1, which was centomere-linked, sufficient evidence was not available to indicate whether FLO 5 or flo 3 were centomere-linked. Two additional recessive flocculating genes, flo 6 and flo 7, have also been identified in laboratory strains (Johnston and Reeder 1983). More recently, an additional flocculating gene, FLO 8, has been identified in a *S. diastaticus* strain on chromosome VIII (Yamashita and Fukui 1983).

The precise mechanism of the flocculation is still not well understood, which may well be due to the effect of the environment on different strains not necessarily sharing a single mechanism for flocculation. Flocculence is genetically controlled and expressed as changes on the surface of the yeast cell wall structure. A theory to which most researchers agree is that flocculence is associated with a change in the proportion of protein in relation to the level of mannan making up the cell wall. Under flocculating conditions, cells may link to each other by the formation of certain bridges through the interaction of ions with the exposed carboxyl groups of the cell wall peptides of adjacent cells. Another view is that $Ca^{++}$ ions bridge between two $PO_4$ groups of cell wall mannan of two adjacent cells (Stewart and Russell 1981). However, the preferred model to explain flocculation at the molecular level would seem to be that FLO genes code for polypeptide chains that are associated with the structural integrity of the cell wall, and that lytic enzymes that are coded by various modifier genes modify the intensity of the flocculation by varying the degree of exposure of carboxyl groups or any other anionic group present on the cell surface. Since this model does not apply fully to all yeast strains, it is at this stage appropriate to state in general terms that flocculation is dictated by a balance between biosynthesis and biodegradation of cell wall components such as proteins and carbohydrates. Such cell surface modifications can take place in the presence of certain enzymes whose production is under genetic regulation by the cell.

## REFERENCES

Anand, J. C., and A. D. Brown. 1968. Growth rate patterns of the so-called osmophillic yeasts in solution of polyethylene glycol. *J. Gen. Microbiol.* **52**:205-212.

Anderson, E., and P. A. Margin. 1975. Sporulation and mating of brewer's yeast. *J. Inst. Brew.* **81**:242-247.

Bevan, E. A., and M. Markower. 1963. The physiological basis of the killer character in yeast. *Int. Congr. Genet. 11th Proc.* **1**:202-203.

Bostian K. A., C. Jayachandran, S. and D. J. Tipper. 1983. A glycosylated protoxin in killer yeast: Models for its structure and and maturation. *Cell.* **32**:169-180.

Botstein, D., and R. W. Davis. 1982. Principles and practice of recombinant DNA research with yeast. In *The Molecular Biology of the Yeast Saccharomyces: Metabolism and Gene Expression,* J. N. Strathern, E. W. Jones, and J. R. Broach (eds.). Cold Spring Harbor Laboratories, New York, pp. 607-636.

Bridges, B. A. 1976. Mutation induction. In *Second International Symposium and Genetics of Industrial Microbiology,* K. D. MacDonald (ed.). Academic Press, New York, pp. 7-14.

Brown, A. D. 1974. Microbial water relations: Features of the intracellular composition of sugar-tolerant yeasts. *J. Bacteriol.* **118**:769-777.

Brown, A. D., and M. Edgley. 1980. Osmoregulation in yeast. In *Genetic Engineering of Osmoregulation*, D. W. Rains, R. C. Valentine, and A. Hollander (eds.). Plenum Press, New York, pp. 75-90.

Brown, A. D., and J. R. Simpson. 1972. Water relations of sugar-tolerant yeasts: The role of intracellular polyols. *J. Gen. Microbiol.* **72:**589-591.

Bruinsma, B., and T. W. Nagodawithana. 1987. Comparison of *S. cerevisiae* and *K. fragilis* in gas production, dough rheology and bread making. Unpublished data.

Burrows, S. 1979. Baker's yeast. Microbial mass. In *Economic Microbiology*, vol. 4, A. H. Rose (ed.). Academic Press, New York, pp. 31-64.

Burrows, S., and R. R. Fowell. 1961a. Improvement in yeast. British Patent 868,621.

Burrows, S., and R. R. Fowell. 1961b. Improvement in yeast. British Patent 868,633.

Bussey, H. 1972. Effect of yeast killer factor on sensitive cells. *Nature* **235:**73-75.

Bussey, H., T. Vernet, and A. M. Sdicu. 1987. Mutual antagonism among killer yeasts: Competition between $K_1$ and $K_2$ killers and a novel cDNA based $K_1$-$K_2$ killer strains of *S. cerevisiae*. *Can. J. Microbiol.* **34:**38-44.

Carlson, M., B. C. Osmond, and D. Botstein. 1981. SUC genes of yeast: A dispersed gene family. In *Cold Spring Harbor Symp. Quant. Biol.* **45:**799-812.

Clement, P., and A. L. Hennette. 1982. Strains of yeast for bread making and novel strains of yeast thus prepared. U.S. Patent 4,318,930.

Crowe, J. H., L. M. Crowe, and D. Chapman. 1984. Preservation of membranes in anhydrobiotic organisms: The role of trehalose. *Science* **223:**701-703.

Dahl, H. H., R. A. Flavell, and F. G. Grosveld. 1981. The use of genomic libraries for the isolation and study of eucaryotic genes. In *Genetic Engineering 2*, R. Williamson (ed.). Academic Press, London, pp. 49-63.

Dawes, I. W. 1983. Genetic control and gene expression during meiosis and sporulation in *Saccharomyces cerevisiae*. In *Yeast Genetics*, J. F. T. Spencer, D. M. Spencer, and A. R. W. Smith (eds.). Springer-Verlag, New York, pp. 29-64.

de Oliveira, D. E., E. C. C. Rodrigues, J. R. Mattoon, and A. D. Panek. 1981. Relationship between trehalose metabolism and maltose utilization in *S. cerevisiae*. II. Effect of constitutive MAL genes. *Genetics* **3:**235-242.

Dickson, R. C., R. M. Scheetz, and L. R. Lacy. 1981. Genetic regulation of yeast mutants constitutive for $\beta$-galactosidase mRNA. *Mol. Cell. Biol.* **1:**1048-1056.

Dickson, R. C., and K. Sreekrishna. 1986. $LAC^+$ *Saccharomyces cerevisiae* plasmids, production and use. Eur. Patent Appl. EP 0206571.

Eddy, A. A., and D. H. Williamson. 1957. A method of isolating protoplasts for yeast. *Nature* **179:**1252-1253.

Erratt, J. A., and A. Nasim. 1986. Allelism within the DEX and STA gene families of *Saccharomyces diastaticus*. *Mol. Gen. Genet.* **202:**255-256.

Esposito, M. S., and R. E. Esposito. 1969. Genetic control of sporulation in *Saccharomyces*. I. The isolation of temperature sensitive sporulation deficient mutants. *Genetics* **61:**79-89.

Federoff, H. J., J. D. Cohen, T. R. Eccleshall, R. B. Needleman, B. A. Buchferer, J. Glacalone, and J. Marmur. 1982. Isolation of a maltase structural gene from *S. carlsbergensis*. *J. Bacteriol.* **149:**1064-1070.

Fowell, R. R. 1970. Sporulation and hybridization of yeast. In *The Yeasts*, vol. 1, A. H. Rose and J. S. Harrison (eds.). Academic Press, New York, pp. 303-383.

Fowell, R. R., and M. E. Moorse. 1960. Factors controlling the sporulation of yeasts. I. The presporulation phase. *J. Appl. Bacteriol.* **23:**53-68.

## REFERENCES 83

Gascon, S., and J. O. Lampen. 1968. Purification of the internal invertase of yeast. *J. Biol. Chem.* **243**:1567-1572.

Giesenschlag, J., and T. W. Nagodawithana. 1982. Effect of reserve carbohydrate on the stability of yeast. Unpublished data.

Gilliland, R. B. 1956. Maltotriose fermentation in the species differentiation of *Saccharomyces*. *Compt. Rend. Trav. Lab. Carlsberg. Ser. Physiol.* **26**:139-148.

Goodey, A. R., and R. S. Tubb. 1982. Genetic and biochemical analysis of the ability of *S. cerevisiae* to decarboxylate cinnamic acid. *J. Gen. Microbiol.* **128**:2615-2620.

Grossmann, M. K., and F. K. Zimmermann. 1979. The structural genes of internal invertases in *S. cerevisiae*. *Mol. Gen. Genet.* **175**:223-229.

Guerola N., J. L. Ingraham, and E. Cerda-Olmedo. 1971. Introduction of mutations by nitrosoguanidine. *Nature* **230**:122-125.

Gunge, N. 1966. Breeding of baker's yeast: Determination of the ploidy and an attempt to improve practical properties (Japan). *J. Genet.* **41**:203-214.

Hinnen, A., J. B. Hicks, and G. R. Fink. 1978. Transformation of yeast. *Proc. Natl. Acad. Sci. USA* **75**:1929-1933.

Holzer, H. 1976. Catabolite inactivation in yeast. *Trends Biochem. Sci.* **1**:178-180.

Ito, H., Y. Fukuda, K. Murata, and A. Kimura. 1983. Transformation of intact yeast cells treated with alkali cations. *J. Bacteriol.* **153**:163-168.

Jacobson, G. K. 1981. Mutations. In *Biotechnology*, vol. 1, H.-J. Rehm and G. Reed (eds.). Verlag Chemie, Weinheim, pp. 280-304.

Jacobson, G. K., S. O. Jolly. 1989. Yeasts, molds and algae. In *Biotechnology*, vol. 7b. H.-J. Rehm and G. Reed (eds.). Verlag Chemie, Weinheim, pp. 279-313.

Jacobson, G. K., and N. B. Trivedi. 1986. Improved yeast strains, method of production and use in baking. U.S. Patent Appl. 818,852.

Jacobson, G. K., and N. B. Trivedi. 1987. Yeast strains, method of production and use in baking. U.S. Patent 4,643,901.

Johnston, J. R., and C. W. Lewis. 1974. Genetic analysis of flocculation and tetrad analysis of commercial brewing and baking strains. In *Second Proceedings of the International Symposium on Genetics of Industrial Microorganisms*, K. D. MacDonald (ed.). Academic Press, London, pp. 339-355.

Johnston, J. R., and H. P. Reeder. 1983. Genetic control of flocculation. In *Yeast Genetics*, J. F. T. Spencer, D. M. Spencer, and A. R. W. Smith (eds.), Springer-Verlag, New York. pp. 205-224.

Kawai, M., and U. Kazuo. 1983. Doughs comprising alcohol resistant yeast. European Patent 78182.

Kew, O. M., and H. C. Douglas. 1976. Genetic co-regulation of galactose and melibiose utilization *Saccharomyces*. *J. Bacteriol.* **125**:33-41.

Khan, N. A., and N. R. Eaton. 1971. Genetic control of maltose fermentation in yeast. I. Strains producing high and low basal levels of enzymes. *Mol. Gen. Genet.* **112**:317-322.

Kielland-Brandt, M. C., T. Nilsson-Tillgrem, S. Holmberg, J. G. Litske Peterson, and B. A. Svenningsen. 1979. Transformation of yeast without the use of foreign DNA. *Carlsberg Res. Commun.* **44**:77-87.

Koninklijke Nederlandsche, Gist en Spiritusfabriek, N. V. 1965. British Patent 989,247.

Kosikov, K. V., and A. A. Miedviedieva. 1976. Experimental increase in the osmophillic properties of yeast. *Mikrobiologiya* **45**:327-328.

Lacy, L. R., and R. C. Dickson. 1981. Transcriptional regulation of *K. lactis* $\beta$-galactosidase gene. *Mol. Cell. Biol.* **1**:629-629-634.

Lawley, P. D., and C. J. Thatcher. 1970. Methylation of deoxyribonucleic acid in cultured mammalian cells by N-methyl-N-nitro-N-nitroso-guanidine. *Biochem. J.* **116**:693-707.

Legmann, R., and P. Margolith. 1983. Interspecific protoplast fusion of *S. cerevisiae* and *S. mellis*. *Eur. J. Appl. Microbiol. Biotechnol.* **18**:320-322.

Lehninger, A. L. 1975. *Biochemistry*, 2nd ed. Worth Publishers, New York.

Lewis, C. W., J. R. Johnston, and P. A. Martin. 1976. The genetics of yeast flocculation. *J. Inst. Brew.* **82**:158-160.

Liljestrom-Suominen, P. L., V. Joutsjoki, and M. Korhola. 1988. Construction of stable α-galactosidase-producing baker's yeast strain. *Appl. Environ. Microbiol.* **54**:245-249.

Lindegren, C. C. 1949. *The Yeast Cell. Its Genetics and Cytology.* Education Publishing, St. Louis, pp. 2-8.

Lindegren, C. C., and G. Lindegren. 1943. Selecting, inbreeding, recombining and hybridizing commercial yeasts. *J. Bacteriol.* **46**:405-419.

Lodder, J., B. Khoudokormoff, and A. Langejan. 1969. Melibiose fermenting baker's yeast hybrids. *Antonio van Leeuwenhoek Yeast Symposium* **35**:F9.

Lorenz, K. 1974. Frozen dough. *Bakers Digest* **4**:14-22.

Maule, A. P., and P. D. Thomas. 1973. Strains of yeast lethal to brewery yeast. *J. Inst. Brew.* **79**:137-141.

Mazur, P. 1970. Cryobiology: The freezing of biological systems. *Science* **168**:939-949.

Mazur, P., and J. J. Schmidt. 1968. Interaction of cooling velocity, temperature, and warming velocity on survival of frozen and thawed yeast. *Cryobiology* **5**:1-17.

Middelbeck, E. J., J. M. H. Hermans, C. Stumm, and H. L. Muytjens. 1980. High incidence of sensitivity in yeast killer toxins among *Candida* and *Tropicales* isolates of human origin. *Antimicrob. Agents Chemother.* **17**:350-354.

Mortimer, R. K., and J. R. Johnston. 1986. Genealogy of principal strains of the yeast genetic stock center. *Genetics* **113**:35-43.

Nagodawithana, T. W. 1986. Yeasts: Their role in modified cereal fermentations. In *Advances in Cereal Science and Technology*, vol. 8, Y. Pomeranz (ed.). American Association of Cereal Chemists, St. Paul, Minn., pp. 15-104.

Nagodawithana, T. W., and N. Trivedi. 1990. Yeast selection for baking. In *Yeast Strain Selection.* C. J. Panchel (ed.). Marcel Dekker, New York.

Nakatomi, Y., H. Saito, A. Nagashima, and F. Umeda. 1985. *Saccharomyces* sp. FD 612 and the utilization thereof in bread production. U.S. Patent 4,547,374.

Naumov, G. I. 1971. Comparative genetics in yeast. V. Complementation in the MAL gene in *S. cerevisiae* which do not utilize maltose. *Genetika* **7**:141-148.

Naumov, G. I. 1976. Comparative genetics of yeast. XVI. Genes for maltose fermentation in *S. carlsbergensis*. *Genetika* **12**:87-100.

Needleman, R. B., and C. A. Michels. 1983. A repeated family of genes controlling maltose fermentation is *S. carlsbergensis*. *Mol. Cell. Biol.* **3**:796-802.

Needleman, R. B., D. B. Kaback, R. A. Dubin, E. L. Perkin, N. A. Rosenberg, K. A. Southerland, D. B. Forrest, and C. A. Michels. 1984. MAL 6 of *Saccharomyces*: A complex genetic locus containing three genes required for maltose fermentation. *Proc. Natl. Acad. Sci. USA* **81**:2811-2815.

Onishi, H. 1963. Osmophilic yeasts. *Adv. Ed. Res.* **12**:53-94.

Oura, E., H. Suomalainen, and E. Parkkinen. 1974. Changes in commercial baker's yeast during its ripening period. *Fourth Intl. Symp. Proc. on Yeast* **B25**:125-126.

Palfrey, R. G. E., and H. Bussey. 1979. Yeast killer toxin: Purification and characterization of the protein toxin from *Saccharomyces cerevisiae*. *Eur. J. Biochem.* **93**:487-493.
Panchel, C. J., I. Russell, A. M. Stills, and G. G. Stewart. 1984. Genetic manipulation of brewing and related yeast strains. *Food Technol.* **38**:99-106.
Panek, A. D., A. L. Sampaio, G. C. Braz, S. V. Baker, and J. R. Mottoon. 1979. Genetic and metabolic control of trehalose and glycogen synthesis. New relationship between energy reserves, catabolite repression and maltose utilization. *Cell. Mol. Biol.* **25**:345-354.
Post-Beittenmiller, M. A., R. W. Hamilton, and J. E. Hopper. 1984. Regulation of basal and induced levels of the MEL 1 transcript in *S. cerevisiae*. *Mol. Cell. Biol.* **4**:1238-1245.
Ruohola, H., P. Liljestrom, T. Torkkeli, H. Kopu, P. Leitinen, N. Kalkkuben, and M. Korhola. 1986. Expression and regulation of the yeast MEL 1 gene. *FEMS Microbiol. Lett.* **34**:179-185.
Russell, I., and G. G. Stewart. 1979. Spheroplast from brewer's yeast strains. *J. Inst. Brew.* **85**:95-98.
Russell I., G. G. Stewart, H. P. Reader, J. R. Johnston, and P. A. Martin. 1980. Revised nomenclature of genes that control yeast flocculation. *J. Inst. Brew.* **86**:120-121.
Schnettler, R., U. Zimmermann, and C. C. Emeris. 1984. Large scale production of yeast hybrids by electrofusion. *FEMS Microbiol. Lett.* **24**:81-85.
Sheetz, R. M., and R. C. Dickson. 1981. LAC 4 is the structural gene for $\beta$-galactosidase in *Kluyveromyces lactis*. *Genetics* **98**:729-745.
Singer, B. 1975. The chemical effects of nucleic acid alkylation and their relation to mutagenesis and carcinogenesis. *Prog. Nuc. Acid. Res, Mol. Biol.* **15**:219-284.
Sklar, R., and B. Strauss. 1980. Role of UVrE gene product and of inducible $O^6$-methylguanine removal in the induction of mutation by N-methyl-N'-nitro-N-nitrosoguanidine in *E. coli*. *J. Mol. Biol.* **143**:343-362.
Snow, R. 1979. Towards genetic improvement of wine yeast. *Am. J. Enol. Vitc.* **30**:33-37.
Spencer, J. F. T., and D. M. Spencer. 1983. Genetic improvement of industrial yeasts. *Ann. Rev. Microbiol.* **37**:121-142.
Spencer, J. F. T., D. M. Spencer, C. Bizeau, A. V. Martini, and A. Martini. 1985. The use of mitochondrial mutants in hybridization of industrial yeast strains. *Curr. Genet.* **9**:623-625.
Sprague, G. F., L. C. Blair, and J. Thorner. 1983. Cell interactions and regulation of cell type in the yeast *S. cerevisiae*. *Ann. Rev. Microbiol.* **37**:623-660.
Sreekrishna, K., and R. C. Dickson. 1985. Construction of strains of *S. cerevisiae* that grow on lactose. *Proc. Natl. Acad. Sci. USA* **82**:7909-7913.
Stanier, R. Y., M. Doudoroff, and E. A. Adelberg. 1970. *The Microbial World*, 3rd ed. Prentice-Hall, Englewood Cliffs, N.J.
Stewart, G. G. 1981. Genetic manipulation of industrial yeast strains. *Can. J. Microbiol.* **27**:973-990.
Stewart, G. G., J. E. Goring, and J. Russell. 1977. Can a genetically manipulated yeast strain produce palatable beer? *J. Am. Soc. Brew. Chem.* **35**:168-178.
Stewart, G. G., and I. Russell. 1977. The identification, characterization and mapping of a gene for flocculation in *Saccharomyces* sp. *Can. J. Microbiol.* **23**:441-447.
Stewart, G. G., and I. Russell. 1981. *Yeast Flocculation in Brewing Science*, vol. 2, J. R. A. Pollock (ed.). Academic Press, London, pp. 61-92.
Stewart, G. G., I. Russell, and C. Panchal. 1983. Current developments on the genetic manipulation in brewing yeast strains. A Review. *J. Inst. Brew.* **89**:170-188.
Stiles, J. J., L. Clark, C. Hsiao, J. Carbon, and J. R. Broach. 1983. Cloning of genes into yeast

cell. In *Methods in Enzymology—Recombinant DNA*, vol. 101, R. Wu, L. Grossman, and K. Moldave (eds.). Academic Press, New York, pp. 290-325.

Sumner-Smith, M., R. P. Bozzato, N. Skipper, R. W. Davies, and J. E. Hopper. 1985. Analysis of inducible MEL 1 gene of *S. carlsbergensis* and its secreted products, alpha-galactosidase (melibiase). *Gene* **36**:333-340.

Tamaki, H. 1978. Genetic studies of ability to ferment starch in *Saccharomyces*: Gene polymorphism. *Mol. Gen. Genet.* **164**:205-209.

ten Berge, A. M. A., G. Zoutenelle, and K. W. van de Poll. 1973. Regulation of maltose fermentation in *S. carlsbergensis*. I. The function of the gene MAL 6 is recognized by MAL 6 mutants. *Mol. Gen. Genet.* **123**:233-246.

Thomas, D. Y., N. A. Skipper, P. C.K. Lou, S. Lolle, and H. Bussey. 1987. Production and secretion of proteins and polypeptides in yeast. Can. Patent 479062.

Thorne, R. S. W. 1951. Some aspects of yeast flocculence. *3rd Eur. Brew. Conv. Proc. Congr.* Brighton, Elsevier, Amsterdam, pp. 21-34.

Thorne, R. S. W. 1968. Some observations on yeast mutation during continuous fermentation. *J. Inst. Brew.* **74**:516-524.

Thorne, R. S. W. 1790. Pure yeast cultures in brewing. *Process Biochem.* **4**:15-16.

Trivedi, N. B., G. K. Jacobson, and W. Tesch. 1986. Baker's yeast. In *CRC Critical Reviews in Biotechnology*, vol. 24, issue 1, G. G. Stewart and I. Russell (eds.). CRC Press, Boca Raton, Fla., pp. 75-109.

van Solingen, P., and J. B. van der Plaat. 1977. Fusion of yeast spheroplasts. *J. Bacteriol.* **130**:946-947.

von Borstel, R. C., and R. D. Mehta. 1976. Mutation and selection systems for yeast. In *Microbiology*. D. Schlessinger (ed.). American Society of Microbiology, Washington, D.C., pp. 507-509.

Watson, K., H. Arthur, and W. A. Shipton. 1976. Leucosporidium yeasts: Obligate psychrophiles which alter membrane-lipid and cytochrome composition with temperature. *J. Gen. Microbiol.* **97**:11-18.

Wickner, R. B. 1976. Killer of *S. cerevisiae*: A double stranded ribonucleic acid plasmid. *Bacteriol. Rev.* **40**:757-773.

Windisch, S., S. Kowalski, and I. Zander. 1976. Dough raising tests with hybrid yeast. *Eur. J. Appl. Microbiol.* **3**:213-221.

Winge, O. 1935. On haplophase and diplophase in some *Saccharomyces*. *C. R. Trav. Lab. Carlsberg Ser. Physiol.* **21**:77-112.

Winge, O., and O. Laustsen. 1937. On two types of spore germination and on genetic segregation of *Saccharomyces* demonstrated through single spore cultures. *C. R. Trav. Lab. Carlsberg Ser. Physiol.* **22**:99-117.

Winge, O., and O. Laustsen. 1938. Artificial species hybridization in yeast. *C. R. Trav. Lab. Carlsberg Ser. Physiol.* **22**:235-247.

Winge, O., and C. Roberts. 1948. Inheritance of enzymatic character in yeast and the phenomenon of long term adaptation. *C. R. Trav. Lab. Carlsberg Ser. Physiol.* **24**:263-315.

Winge, O., and C. Roberts. 1950. The polymeric genes for maltose fermentation in yeasts and their mutability. *C. R. Trav. Lab. Carlsberg Ser. Physiol.* **25**:36.

Winge, O., and C. Roberts. 1956. Complementary action of melibiase and galactozymase on raffinose fermentation. *Nature* **177**:383-384.

Wray, L. A., M. M. Witte, R. C. Dickson, and M. I. Riley. 1987. Characterization of a positive regulatory gene LAC 9 that controls induction of the lactose-galactose regulon of *K. lactis*: Structural and functional relationships to GAL 4 of *S. cerevisiae*. *Mol. Cell. Biol.* **7**:1111-1121.

Yamashita, I., and S. Fukui. 1983. Mating signals control expression of both starch fermentation genes and novel flocculation gene FLO 8 in the yeast *Saccharomyces*. *Agric. Biol. Chem.* **12**:2889-2896.

Yocum, R. R. 1986. Genetic engineering of industrial yeasts. In *Bio. Expo. 86 Proceedings*. Butterworth Publishers, Stoneham, Mass., pp. 171-180.

Yocum, R. R., and S. Hanley. 1987. Genetically engineered lactose utilizing yeast strains. British Patent Appl. GB 2178431A.

Young, T. W. 1981. The genetic manipulation of killer character into brewer's yeast. *J. Inst. Brew.* **87**:292-295.

# BREWER'S

The history of beer brewing reaches into the oldest records of humankind. From documents and inscriptions found in Egyptian tombs, it appears that beer, once known as "barley wine," was produced by the ancient civilizations as much as five thousand years ago. Malted barley, which had been preserved for thousands of years and was used in brewing processes, has been recovered from such Egyptian tombs. The art of brewing then became familiar to the ancient civilizations of Asia. In ancient China beer was made from rice, whereas the early civilizations of India brewed certain beerlike beverages from malt. All these beers were produced without the use of hops except perhaps in China where hops were used at an early date to flavor the beer.

Originally, brewing of beer was part of the work of the household and, as such, was carried out primarily by women. In the ninth century, monasteries began to brew beer. The brewing of beer then turned into an industry that moved into the cities and began to develop primarily in the tenth and eleventh centuries. After these early beginnings, production of beer developed as an art and craft throughout the world in complete absence of scientific know-how.

Beer was a popular beverage in England in the eighth century. Numerous ale houses, which were already common in England during this period, brewed their own beer to sell to the public. Addition of hops to flavor the drink was first believed to have started in the 1400s, and at that time hops came to England by way of Holland. In the fourteenth century,

# BREWER'S YEAST

Germany was the principal seat of the brewing industry, and from that time until the sixteenth century the beer industry continued to flourish throughout Germany. By the fifteenth century, brewing had become well established both on the continent and in England.

In 1620, beer was first introduced to America as one of the provisions on board the Mayflower. English ale brewing traditions accompanied the early colonists. Until the middle of the nineteenth century, all the beer brewed in America was the type known today as ale. Lager beer, which had been brewed extensively in Germany since the thirteenth century, was introduced to America in the 1840s by the German immigrants who made their home in this country and who preferred the lager to that of ale brewed according to the English method. Although in the first decade after its introduction brewing of lager beer made slow progress, with the gradual development of the industry the production and consumption of the new beverage grew, finally replacing most of the heavy English beers. Thus, the predominantly popular American lager beer that exists today, with its emphasis on lightness, sparkle, and dryness, is a variation of the original German brew.

The beer that was known to ancient civilizations was very different from the beverage known by that name by all civilized people today. Our modern beer is indeed the product of a long development and, as we know it today, is a beverage made predominantly from malt by fermenting an extract of malted barley and hops using a special strain of yeast. More recently, the brewers have adopted starchy adjuncts as a partial replacement for the more expensive barley malt, mainly for economic reasons.

Final quality of the beer is determined by the precise control of those factors that generally contribute to it in a brewing process. The first important criterion is the quality of the brewing materials such as malt, hops, water, and other starch adjuncts. The proportion of these different components used in the brewing process have a direct impact on the flavor profile of the final beer. The second area includes the enzymatic hydrolysis, solubilization, extraction, and isomerization of the respective brewing ingredients and the separation of the soluble fraction from the particulate matter. In the brewing process, these procedures include milling, mashing, lautering, kettle boiling, and water treatment. The third factor is the fermentation process involving the conversion of the sugar substrate derived from malt and adjuncts to an acceptable beer through the use of special strains of yeast. The fourth factor involves the postfermentation processing including maturation, clarification, and packaging in order that the final beer is of acceptable flavor and physical stability. The fermentation process, which is highly critical for the final quality of the beer, will be discussed in detail in this chapter.

## GENERAL CHARACTERISTICS OF YEAST

Yeast plays an important role in the brewing process not only by converting the fermentable sugar in the wort to ethanol and carbon dioxide but also by producing a variety of volatile and nonvolatile constituents mostly in trace amounts that contribute to the overall flavor of the beer. According to the previous classification, two species of the genus *Saccharomyces* are often considered important for brewing depending on the type of brewing process used by the brewer.

Most beers are produced by the bottom fermentation process in which strains of the yeast *S. uvarum* are used. These yeasts have a tendency to remain suspended for a limited time in the fermentation medium while the fermentation is active, after which most of the yeast settles to the bottom of the fermenting vessel. This strain is predominantly used in Germany and North America for the production of lager beer. The other type of beer, referred to as ale, is produced by a top fermenting yeast like *Saccharomyces cerevisiae*. These strains tend to rise to the top of the fermenting medium and are generally skimmed off the surface of the fermenting liquid after the fermentation is completed. Ales are a popular kind of beer in Great Britain and Ireland. Despite these differences in fermentation characteristics, the two species *S. cerevisiae* and *S. uvarum* are now grouped under *S. cerevisiae* according to the most recent classification (Kreger-van Rij 1984). Yet, the brewing industry must continue to differentiate these two strains because of the differences that can be identified with respect to their brewing characteristics. Thus, the most recent nomenclature will not be adhered to when presenting the subject matter in this chapter.

The strains of *S. cerevisiae* used in most laboratory studies are either haploid or diploid. Results of studies conducted to determine the ploidy of brewing strains by comparing the DNA content per cell with that of a typical haploid strain suggest that such industrial brewing strains are polyploid, particularly triploid, tetraploid, or aneuploid, having a DNA content intermediate between triploid and tetraploid (Johnston and Oberman 1979). Although the total amount of chromosomal DNA detected by analytical methods in a haploid strain of yeast is equivalent to 17,000 kilobase pairs (kbp), roughly amounting to 15,000 genes, linkage relationships of only 586 genes have so far been mapped to the 17 chromosomes (Mortimer and Schild 1985).

The life cycle of a typical laboratory strain is comprised of alternating haploid and diploid phases rarely interrupted by the appearance of triploid or tetraploid strains. These strains of yeast belong to two classes based on their sexual behavior. Those yeast strains that have a long and stable haplophase are termed heterothallic. These strains have two mating types, known as "a" and "α," and mating can occur only when cells of the opposite mating type are brought into close proximity. The mating reaction between a and $\alpha$ cells is caused by a cell agglutination reaction following the appearance of complimentary oligopeptides on the cell walls of the two mating types. These oligopeptides are 12-13 amino acid residues in length and have the ability to arrest growth upon adhesion to their opposite mating type. Nuclear fusion takes place immediately following the fusion of the cytoplasmic contents of the two cells types. The zygote thus formed gives rise to the diploid phase. In heterothallic species, the diploid phase can be sustained indefinitely by mitotic division. However, in the presence of nonfermentable carbon sources, like acetate, yeasts that are starved for nitrogen soon give rise to four-spored asci following meiosis or reduction division. Certain crosses rarely form spores in tetrads, and instead the asci often give rise to one, two, or three spores. All the spores germinate under favorable conditions to give rise to the haploid cells completing the entire cycle.

The other class of yeasts that are termed homothallic are characterized by their ability to undergo fusion even though they are derived from a single spore (a or $\alpha$). These strains of yeast have the ability to switch their mating types during their vegetative growth, thus allowing cell fusion to occur as in the case of heterothallic strains. A brief outline of how homothallic strains switch their mating type is presented in Chapter 2.

Cells of homothallic yeast do not have the ability to switch mating types at the first budding. Probably after the first budding, a small population of the cells becomes competent to switch mating type, and thereafter the frequency of switching increases with every generation that follows until it is stabilized after several buddings.

Brewing strains of yeast are generally polyploids or often aneuploids having chromosome numbers that are not an exact multiple of the chromosome number found in the haploid strain. Under these conditions, such industrial strains do not have a mating type, they sporulate very poorly, and spores are often nonviable even under optimal germination conditions. Due to their high ploidy and low level of recombination, these brewing strains are genetically more stable and less susceptible to mutational forces. Selection of their genetically stable polyploid or aneuploid strains for brewing is obviously intended to maintain constant product quality and product consistency in brewery fermentations.

## YEAST CHARACTERISTICS IMPORTANT FOR BREWING

### Fermentation

The flavor profile that distinguishes a particular quality beer from another is largely governed by a number of flavor compounds present in the finished beverage at substantially low concentrations, most of which are produced by the action of the yeast on the different substrate components present in the fermenting wort. Over the years, brewers have found quite empirically that some strains of yeast are capable of producing beer of higher quality than that from other yeast strains despite the use of almost the same brewing materials. It is thus clear that genetics of yeast can play an important role in determining the flavor characteristics of the final beer.

Most beers manufactured throughout the world are broadly classified into two main types, namely ale and lager. This differentiation is based primarily on the strain of yeast used and the fermentation conditions applied during the fermentation. The lager beer is produced by the bottom fermentation process in which the strain of yeast, *Saccharomyces uvarum* (per previous classification), is used. This yeast has a tendency to remain suspended until the major part of the fermentable sugar in the wort is depleted after which it tends to settle to the bottom of the fermenting vessel. Lager fermentations are generally conducted at the 10°-15°C temperature range, even though the temperature optimum for *Saccharomyces uvarum* is somewhere in the neighborhood of 25°C. In general, a decoction or double mash system is used in the preparation of wort for the production of lager beer. The primary fermentation carried out for five to seven days is followed by the lagering or maturation period of several weeks at a temperature nearing 0°C. Such a long period of cold maturation is known to improve palatability and physical stability of the beer.

The other type of fermentation that is highly popular in Great Britain and Ireland is the ale type that produces ale beer. These are sometimes referred to as top fermentations because the strain of *Saccharomyces cerevisiae* used in these fermentations tends to rise to the top, forming a thick layer of yeast on the surface of the liquid. Top fermentations are generally conducted at higher temperature, about 15°-20°C, although the optimal temperature for the strain is approximately 30°C. In general, infusion-mashed worts are often used for this type of fermentation. The yeast layer is removed from the surface after the completion of the fermentation. The fermented beer is then run into casks and a limited amount of sugar and finings is added. The small amount of yeast that remains suspended in the beer provides a secondary fermentation in the presence of added sugar, thereby producing carbon dioxide that dissolves in the beer medium. The fish collagen often used as the fining agent forms a coagulant with the suspended yeast cells and ultimately settles to the bottom, leaving behind a clear ale. Further downstream processing of ale is similar to that of lager beer.

## Flocculation

*Flocculation* is defined in the brewing industry as the aggregation of the yeast cells in the later stages of the batch fermentation and their separation from the suspending liquid either by sedimentation, as in the case of bottom fermenting lager yeast, or by flotation, as exhibited by the top fermenting ale yeast. In a brewery fermentation, it is important that the fermenting yeasts remain suspended in the medium in order to achieve maximum conversion of the fermentables. However, to avoid expensive centrifugation of the fermented beer, it is an advantage to the brewer that the yeasts spontaneously aggregate and settle out of the beer as soon as the fermentation is approaching limit attenuation. Hence, the ability to flocculate and the stage at which the flocculation begins to manifest itself are important criteria considered in selecting yeast strains for brewing. Premature separation of yeast results in an incompletely fermented sweet beer with an abnormal flavor profile. Likewise, nonflocculant strains produce drier beer with unpleasant yeasty flavor. Often such fermentations show problems in downstream processing, especially in centrifugation and filtration.

Changes of flocculation that are mainly dictated by the genetic constitution of the cell are expressed as changes in cell wall structure. Although certain yeast strains have several FLO genes, there is no apparent additive effect with respect to flocculation when compared with certain other strains that contain only a single FLO gene. Accordingly, the occurrence of certain strains of above average flocculence was explained as due to the influence of additional modifier genes or cytoplasmic genetic factors (Stewart and Russell 1977). It is highly unlikely that there is a simple explanation for the phenomenon of flocculation. A preferred model is that FLO genes code for the polypeptide chains that are associated with structural wall molecules, and certain lytic enzymes encoded by various modifier genes may be controlling the degree of flocculation by varying the number of exposed carboxyl groups on the surface of the cell wall.

Flocculence in brewer's yeast, although genetically controlled, can be considered a surface phenomenon with every change in flocculence accompanied by changes in the cell wall structure. This subject has been reviewed by Rose (1980), Stewart and Russell (1981), and Rainbow (1970). The interested reader is encouraged to consult these reviews for further details.

Many theories have been advanced to account for the phenomenon of flocculation. The large number of theories presented attests to the difficulty of the problem. Early investigators first observed that certain anionic groups on the surface of the yeast cells already present or produced at the stationary phase of cell cycle form calcium bridges with similar ionic groups on adjacent cells with the ultimate formation of the yeast floc. The studies that followed were on the nature of the cell surface during flocculation, principles involved during flocculence, and the study of the physiochemical forces that exist within the floc structure as a result of flocculation.

Research carried out in recent years has shown that the anionic groups associated with carboxyl groups in acidic wall proteins and phosphodiester linkages on cell surface phosphomannans are critical for the formation of calcium bridges. Nevertheless, new evidence indicates an actual increase in flocculence with the removal of phosphodiester linkages from the yeast cell surface. These findings were subsequently confirmed by Jayatissa and Rose (1976), Beavan et al. (1979), and Stewart, Russell, and Garrison (1975). These researchers showed a linear relationship between the floc-forming ability and increase in the density of surface carboxyl groups.

Beavan et al. (1979) have also discussed the biochemical processes that could lead to a higher degree of flocculence with an increase in the surface charge density of the yeast cells. It is known that there is an increase in activity of lytic enzymes as the yeast cells in a fermentation reach the stationary phase. In these studies, the flocculating ability in yeast became apparent immediately following the induction of amidase, mannosidase, and protease C activities. Thus, it is highly probable that the change in the composition of the wall surface during the late stages of the fermentation could be due to the action of the lytic enzymes. However, this study failed to explain how the density of carboxyl groups on the surface of the yeast cell increased at the time of initiation of floc formation. It is probable that when cells enter the stationary phase of growth there is synthesis of new acidic wall proteins followed by an increase in carboxyl groups on the cell surface, thereby increasing the cell surface's ability to flocculate from the medium.

According to these observations, the preferred model to explain the molecular mechanism for flocculation would thus seem to be that FLO genes code the polypeptide chains that are associated with the structural wall molecules and that lytic enzymes modifying the intensity of flocculation by exposing carboxyl groups are specified by various modifier genes (Johnston and Reader 1983). There could also be an increase in acidic wall proteins in the stationary phase, increasing the possibilities for flocculation.

**Nutritional Requirements**

Malt extract or beer wort is one the oldest classical media used for growing yeast cultures. It contains nitrogen, mineral salts, and vitamins that are adequate for the sustenance of yeast growth and for fermentable carbohydrates to supply energy and carbon skeletons for biosynthesis. There is also an absolute requirement for molecular oxygen in commerical propagations of yeast. These yeasts can, however, exhibit some growth under completely anaerobic conditions.

*Nitrogen Metabolism*

All malt worts contain about 65-100 mg nitrogen per 100 ml of the medium of which about 30% are $\alpha$-amino nitrogen, about 20% are high molecular weight proteins, and about 40% are polypeptides. The remaining 10% represent the nitrogen in purines and other nitrogenous compounds. The extracellular proteolytic activity of yeast is negligible, and proteolysis does not occur in a wort medium due to the presence of the yeast unless autolysis of yeast occurs. Therefore, major attention has been focused on the amino acids in wort.

Although the nitrogen requirement for protein synthesis can be met by ammonium ions, when present, amino acids are a preferred source. Thus, when a mixture of amino acids is present in brewer's wort the yeast grows faster than when ammonium ions are the sole source of nitrogen (Thorne 1949).

Absorption of amino acids by brewer's yeast and the rate at which different amino acids are assimilated varies somewhat with the strain of yeast or the amino acid composition of the wort. The different amino acids can be divided into three or four groups, depending on their rate of assimilation. Table 3-1 shows such groupings for *S. cerevisiae* and *S. uvarum*.

Table 3-1. Classification of Amino Acids Based on Absorption Rates from Wort

|  | Jones and Pierce (1964) | Enari et al. (1970) |
|---|---|---|
| Group A (Fast absorption) | Asparagine<br>Serine<br>Threonine<br>Lysine<br>Arginine<br>Glutamic acid<br>Aspartic acid<br>Glutamine | Asparagine<br>Serine<br>Threonine<br>Lysine |
| Group B (Intermediate absorption) | Valine<br>Methionine<br>Leucine<br>Isoleucine<br>Histidine | Arginine<br>Aspartic acid<br>Glutamic acid<br>Valine<br>Methionine<br>Leucine<br>Boleacine |
| Group C (Slow absorption) | Glycine<br>Phenylalanine<br>Tyrosine<br>Tryptophan<br>Ammonia<br>Alanine | Histidine<br>Glycine<br>Phenylalanine<br>Tyrosine<br>Tryptophan<br>Ammonia |
| Group D (Little or no absorption) | Proline | Alanine<br>Proline |

There is good agreement between these results. Although the patterns of assimilation are similar for the above two brewing strains, it does not seem to be the case for other species of *Saccharomyces*. The use of amino acid absorption rates to differentiate between various species of this genus has been suggested by Brady (1965).

Only the L-forms of the amino acids can be utilized by yeast, with the exception of glutamic acid, aspartic acid, and asparagin in which both D and L forms are effective sources of nitrogen. Early investigators hypothesized that amino acids are assimilated intact by yeast and incorporated into its cell proteins. Dissemination only occurs when the cell is obliged to synthesize particular amino acids as they become deficient in the medium. On the contrary, results obtained by Jones, Pragrell, and Pierce (1969) demonstrated that amino acids are not directly incorporated into the proteins. With radiolabeled $^{15}N$ and $^{14}C$, Jones and co-workers observed the conversion of the amino acids into the corresponding keto acids. Other keto acids required by the cell for amino acid synthesis could also be derived from carbohydrate metabolism. These keto acids can then yield the corresponding amino acids by transamination when these amino acids are required for protein synthesis.

A complete nitrogen balance considering the nitrogenous materials of the wort, the beer, and the yeast crop does not seem to be available. However, certain conclusions can be drawn from the available data. At the start of the fermentation, the concentration of yeast is about 0.5 to 1.0 pound cream yeast per barrel, corresponding approximately to a level of 0.5 to 1 gram yeast solids per liter. At the lower pitching level, and for a fourfold to eightfold multiplication, 1.5 to 3.5 grams yeast solids per liter would be formed during a fermentation. At the nitrogen concentration of 7% in the yeast (based on solids) the supply of 104 mg to 245 mg nitrogen per liter of wort would be required. It has already been said that wort contains 650 to 1000 mg nitrogen per liter and that about 30% of this amount is in the form of amino nitrogen. It appears, therefore, that for the majority of fermentations there is adequate amino nitrogen available for yeast growth. Table 3-2 shows the concentrations of some amino acids in a wort and the corresponding levels that remain in the beer after the fermentation (Hardwick 1983).

Hudson (1967) has reported that in regular brewery fermentations up to 45% of the total nitrogen was taken up by the yeasts. Since this amount is also the upper limit of amino nitrogen (as percentage of total nitrogen), the question arises whether di-, tri-, or polypeptides are assimilated by yeast. Damle and Thorne (1949) showed that brewing yeast could assimilate peptides of low molecular weight, but this assimilation is again complicated by the excretion of polypeptides and protein into the young beer as demonstrated by MacWilliam and Chapperton (1969).

The question of the adequacy of the nitrogen supply becomes more

Table 3-2. Uptake of Amino Acids During Brewery Fermentation

| Amino Acid | In Wort | In Beer | Percent Assimilation |
|---|---|---|---|
| Aspartic acid | 50 | 10 | 80 |
| Asparagine, glutamine | 140 | – | 100 |
| Glutamic acid | 45 | 8 | 82 |
| Alanine | 88 | 24 | 73 |
| Isoleucine | 50 | 7 | 86 |
| Phenylalanine | 90 | 40 | 56 |
| Cysteine + cystine | Trace | Trace | 0 |
| Arginine | 100 | 9 | 81 |
| Threonine | 39 | 7 | 82 |
| Proline | 210 | 220 | 0 |
| Valine | 88 | 18 | 80 |
| Leucine | 111 | 13 | 88 |
| Lysine | 40 | 3 | 92 |
| Serine | 50 | 10 | 80 |
| Glycine | 28 | 13 | 54 |
| Methionine | 30 | 3 | 90 |
| Tyrosine | 80 | 29 | 64 |
| Histidine | 23 | 3 | 87 |
| Tryptophan | 30 | 10 | 67 |

*Source:* Hardwick 1983.

apparent with higher amounts of adjunct in the mash. Sucrose supplies no nitrogenous matter, and rice- and corn-derived products provide much lower levels of $\alpha$-amino nitrogen than malt. Also, the adequacy of the nitrogen supply is more critical with top fermenting yeast where higher rates of yeast growth and a greater yeast crop are achieved. Nitrogen uptake from worts is also strongly influenced by the degree of aerobicity during fermentation. A conspicuous feature commonly observed in high-gravity fermentations is the presence of a high level of $\alpha$-amino nitrogen in the final beer. Here again, an increase of wort concentration above the normal gravity level does not result in a proportional increase in the yeast crop. The assimilable nitrogen level in the high gravity wort tends to be significantly greater than that needed for yeast growth. The net result is an elevated level of $\alpha$-amino nitrogen in beer made by high-gravity brewing.

## Vitamins

Yeasts vary widely in their need for vitamins for their metabolism. For a given strain, these may differ greatly between active respiration and growth

YEAST CHARACTERISTICS IMPORTANT FOR BREWING 99

Table 3-3. Function of Certain Essential Vitamins in Yeast Metabolism

| Vitamin | Metabolic Function | Analogs |
|---|---|---|
| Biotin | Carboxylation reactions Protein, nucleic acid, carbohydrate, fatty acid | Desthiobiotin, biocytin, and biotin-D-sulfoxide |
| Thiamine ($B_1$) (TPP) | Decarboxylation of pyruvate rearrangements in pentose cycle, transketolase reactions, isoleucine and valine biosynthesis | Thiazole, pyrimidine |
| D-pantothenic acid | Coenzyme A, acetylation reactions | $\beta$-alanine |
| Nicotinic acid | (Under anaerobiosis) as coenzymes in oxidation/reduction reactions | |
| Riboflavin | (Under anaerobiosis) as coenzymes in oxidation/reduction reactions | |
| Inositol | Membrane phospholipids (structural) | Mesoinositol |

on the one hand and alcoholic fermentation on the other. Some genera such as *Hansenula anomala* can grow normally in a completely vitamin-free medium, because unlike other strains these microbes have the necessary biochemical pathways to synthesize all the required vitamins for their sustenance.

Almost all vitamins (except mesoinositol) required by yeast function as a part of a coenzyme serving a catalytic function in yeast metabolism. The function of the various essential vitamins for yeast metabolism are outlined in Table 3-3. Atkin et al. (1962) examined the vitamin requirements of actively growing yeasts with media in which the yeast was allowed to multiply for an extended period of time. The defined medium contained 25 mg inositol, 25 mg calcium pantothenate, 0.025 mg biotin, 0.5 mg thiamine hydrochloride, and 0.5 mg pyridoxine hydrochloride per liter, in addition to the customary nitrogen, carbon, and mineral sources. All of the 58 strains of brewer's yeast investigated required biotin. Of the 53 lager strains, 23 required the addition of pantothenic acid; 4 required only biotin. Five top fermenting ale strains required pantothenic acid, inositol, and either vitamin $B_1$ or pyridoxine in addition to biotin. There was no evidence to indicate that yeast acclimatized to media in which certain vitamins were missing, since the dependence on a vitamin did not change after several transfers.

Biotin, which is the most commonly required vitamin, participates in all carboxylation reactions. Among the biotin analogs, D-desthiobiotin, biocytin, and biotin-D sulfoxide can replace the need for biotin in yeast. Biotin takes part in all major biochemical reactions that involve protein synthesis, in nucleic acid synthesis, in carbohydrate metabolism, and in the synthesis of fatty acids. Its deficiency is shown clearly by poor growth and damaged plasma membranes.

Brewer's yeast can synthesize thiamine (vitamin $B_1$) since the individual moieties of the thiamine — thiazole and pyrimidine — are supplied in the wort medium. Thiamine, as thiamine pyrophosphate, participates in the decarboxylation of pyruvate and in the rearrangement reactions of the pentose cycle. It also takes part in isoleucine and valine biosynthesis and in all transketolase reactions. Its presence is most essential for achieving optimal rates in brewery fermentations. Most brewer's yeasts require D-pantothenic acid for initiation of growth for which, sometimes, $\beta$-alanine could be substituted. Pantothenic acid is a component of coenzyme A and is involved in all acetylation reactions. In media deficient in pantothenic acid, some brewer's yeasts produce $H_2S$ in significant amounts.

Nicotinic acid is not considered a growth requirement when yeast is propagated under highly aerobic conditions, but it cannot be synthesized under anaerobic conditions. A similar phenomenon has been observed with riboflavin. Niacin and riboflavin are components of coenzymes participating in most oxidation/reduction reactions. There are few strains of brewer's yeasts incapable of synthesizing their own requirement for inositol. Growth characteristics of such strains are improved by supplementation with mesoinositol. This vitamin, which is the only vitamin without catalytic function, serves a structural function as a component of phospholipids involved in the synthesis of membranes that provide stability to the cell.

Brewer's wort is a rich source of vitamins and other growth factors. The fermenting yeast depletes wort of these substances so that the final beer often contains substantially less of them. The important vitamins critical for yeast growth that are found in wort are summarized in Table 3-4. However, the actual levels of these vitamins that occur in worts may vary with the level of adjuncts used in the preparation of the wort. Some of these vitamins are found in the malt in bound form and are generally liberated by the enzyme action during mashing.

Table 3-4. Vitamins in Sweet Wort

| Vitamin | Nutrient Level in Wort |
| --- | --- |
| Biotin | 0.56 $\mu$g/100 ml |
| Thiamine ($B_1$) | 60 $\mu$g/100 ml |
| Calcium pantothenate | 45-65 $\mu$g/100 ml |
| Nicotinic acid | 1000-1200 $\mu$g/100 ml |
| Riboflavin | 33-46 $\mu$g/100 ml |
| Inositol | 9.3 mg (free) and 18.9 mg (total)/100 ml |
| Pyridoxine, pyridoxal, and pyridoxamine | 85 $\mu$g/100 ml |

## Minerals

Worts generally used for brewing seem to contain sufficient quantities of mineral nutrients to allow yeast growth and fermentation. In light of recent studies with culture media of known chemical composition, six major mineral elements are essential for yeast growth. In order of importance these are phosphorus, magnesium, potassium, sodium, manganese, and sulfur, which are generally present in growth media as phosphates or sulfate salts. Only the trace elements—copper, iron, and zinc—are essential for yeast metabolism. Such minerals become integral parts of the proteins as activators of enzymes and function as cofactors in biochemical reactions within the cell. They also function as stabilizers for proteins.

Concentration of magnesium as supplied with the brew water coupled with that available in malt appears to be sufficient for yeast growth. Magnesium is essential for the growth of yeast because it acts as an enzyme activator in a number of biochemical reactions that involve phosphate transferases and a number of decarboxylases.

According to MacWilliam's review (1968), the phosphorus content of wort is in the 450-900 mg $P_2O_5$ per liter range, one-half to two-thirds of which is in the form of organic phosphates. Phosphorus in the form of phosphates is essential for growth and energy metabolism of yeast. The requirement of phosphorus for yeast is related to the availability of nitrogen in the medium. As a rule of thumb, one part of $P_2O_5$ is required for every three parts of nitrogen.

The uptake of potassium ($K^+$) is by an active transport mechanism and is associated with the transfer of $H^+$ ions from within the cell. The presence of glucose is essential to this active process. The cation exchange occurs more rapidly under aerobic conditions than under anaerobic conditions. Potassium concentrations of 0.018 M gave optimal growth for yeast. Calcium is apparently not essential for the growth of yeast cells, but it plays a role in the improvement of growth and fermentation. Calcium sulfate (gypsum) is often added during mashing to obtain a drop in wort pH, which follows the precipitation of some organic phosphate compounds. Excess calcium leads to excessive reactions with phytin, other phosphate compounds, and nucleic acid. Yeast cells take up only minimal amounts of sodium.

The requirement for some of the trace metals has not been accurately determined. Most of the studies with copper, iron, and other metals have been associated with determining the toxicity of the respective metals. Metals like Co, Cr, Cu, Fe, Mn, Ni, and Zn are not toxic at the 5 ppm level. Zinc concentrations may sometimes be limiting yeast growth in brewer's wort. Densky, Gray, and Buday (1966) showed that worts contain from 0.05 ppm to 0.10 ppm zinc and there was noted growth stimulation if these worts were supplemented with zinc to a level of 0.5 ppm. Frey, DeWitt, and Bellomy (1967) found from 0.04 ppm to 0.07 ppm zinc in worts and obtained stimulation of both fermentation and growth at 0.5 ppm. Levels of 500 ppm zinc were found toxic to yeast.

### Fermentation of Wort Sugars

A standard brewing wort in North America contains 11-12% soluble extract of which 90% is soluble carbohydrate. In general, 70-72% of the total solubles in such worts is fermentable sugar, and the remaining soluble solids include predominantly dextrins of a molecular size greater than maltotriose. The unfermented carbohydrates pass through the fermentation unaffected and contribute to the caloric content of the final beer. The principal fermentable carbohydrates in the wort include glucose ($\sim 14\%$), fructose ($\sim 1\%$), sucrose ($\sim 6\%$), maltose ($\sim 59\%$), and maltotriose ($\sim 20\%$). Pentose sugars, which are significantly minor (0.03%) in brewer's wort, are not metabolized by brewer's yeast.

The fermentation pattern prevailing during a regular fermentation of brewer's wort is shown in Figure 3-1 (Montreuil, Mullet, and Scriban 1961). Similar patterns have been established by many authors who investigated the sequence in which sugars in wort were fermented. According to their data, sucrose disappeared first from the wort, which indicated the disappearance of the disaccharide through hydrolysis by the enzyme invertase. Glucose and fructose are utilized rapidly, followed by the assimilation of maltose. When all such fermentable sugars are utilized, the yeast begins to assimilate the maltotriose slowly. Maltotetrose, isomaltose, higher polysaccharides, and pentoses are not fermented by brewer's yeast. A typical beer contains no monosaccharides, a trace of maltose, some maltotriose, and a relatively large proportion of dextrins.

Uptake of glucose and fructose into the yeast cell occurs by a process termed *facilitated diffusion*. It is a passive uptake without any mediation of energy. Both sugars are directly incorporated into the glycolytic pathway immediately after their entry into the cell. Sucrose is rapidly hydrolyzed on the surface of the yeast cell due to the presence of the invertase enzyme trapped between the membrane and the cell wall, or on the outer layers of the cell wall, releasing the products of hydrolysis—glucose and fructose—into the medium. The hydrolysis of sucrose by invertase is two hundred times faster than the rate of fermentation of the substrate by the yeast cell. Because of the rapid rate of hydrolysis of the sucrose molecule, it behaves like a mixture of glucose and fructose in the presence of yeast.

Maltose fermentation becomes especially important for brewer's yeast for its proper performance in a brewer's wort. Unlike sucrose, maltose and

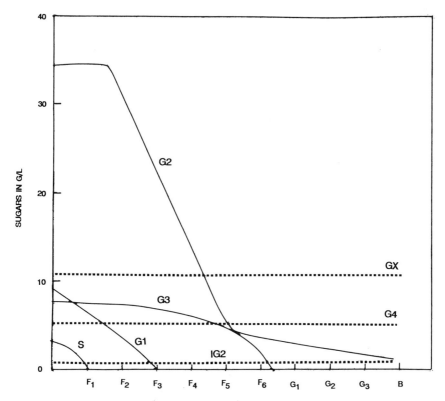

**Figure 3-1.** Changes in the concentration of various wort sugars during the primary fermentation, during the secondary fermentation, and their concentration in the beer. $F_1$, $F_2$, etc. = days of primary fermentation; $G_1$, $G_2$, $G_3$, = weeks of secondary fermentation; B = beer; S = sucrose; G = glucose; G2 = maltose; G3 = maltotriose; G4 = maltotetrose; IG2 = isomaltose; GX = glucosans. (*After Montreuil, Mullet, and Scriban 1961*)

maltotriose present in the wort are taken up intact by the cell by an active process, and the sugars are later hydrolyzed within the cell. Maltose utilization by an unadapted cell requires a certain induction period after the glucose concentration in the medium falls below a critical level (0.1%). The fermentation of maltose by *Saccharomyces* strains requires at least two enzymes: $\alpha$-1, 4-D-glucosidase (maltase) and maltose permease. The entry of maltose and maltotriose into the cell of *Saccharomyces* is brought about by specific permeases. It is possible that in certain strains malto permeases are constitutive, and in other strains they are inducible (Harris and Millin 1963). Transport of maltose across the membrane is unidirectional and the intracellular hydrolysis is followed by the immediate phosphorylation of the liberated glucose molecules as they enter the glycolytic pathway.

The presence of any one of the family of five unlinked MAL genes (MAL 1, MAL 2, MAL 3, MAL 4, and MAL 6) confers on yeast the ability to utilize maltose as a carbon source. However, most haploid strains that have the ability to utilize maltose contain only a single functional copy of one of the MAL genes. It is often referred to as the dominant gene. Genetics associated with maltose utilization by yeast are described in Chapter 2.

The development of yeast strains that utilize maltose under repressing conditions would be highly beneficial in brewery fermentations. The genetic approach to achieve this goal has been to develop strains that exhibit constitutive maltase and maltose permease synthesis. There have also been efforts to increase the copy number of MAL genes to take advantage of their additive effect.

Maltotriose fermentation in brewer's wort begins later than that of maltose. Some strains of brewer's yeast do not ferment maltotriose at all. Obviously, rapid and complete fermentation of maltotriose results in beer of low extracts. In a typical brewery fermentation, the brewer expects at least 70% of the maltotriose present in the wort to be completely utilized by the yeast. The efficiency of yeasts in fermenting maltotriose depends greatly on their nitrogen content. More flocculent yeasts show a lower rate of maltotriose fermentation than less flocculent yeasts relative to their rate of fermentation of glucose.

Different types of brewer's yeasts are indistinguishable from each other by their size, shape, colony, appearance, or cell structure. The top fermenting *S. cerevisiae* are able to hydrolyze raffinose with the use of $\beta$-fructofuranosidase (invertase) present on the surface of the cell as shown in Figure 3-2.

This process enables the yeast to utilize only the fructose portion of the raffinose molecule. It results in the utilization of one-third of the raffinose

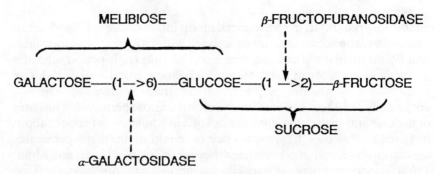

**Figure 3-2.** Action of $\beta$-fructofuranosidase (invertase) and $\alpha$-galactosidase (melibiase) on the sugar raffinose.

molecule, leaving behind the melibiose residue. It is known that at least six unlinked polymeric genes (SUC genes) provide the ability to produce the $\beta$-fructosidase enzyme. The presence of at least one of these SUC genes is adequate to confer sucrose-utilizing ability to a yeast. In addition to these SUC genes, bottom fermenting S. uvarum is known to carry a MEL gene (Gilliland 1956) that confers the ability for the yeast to utilize melibiose. Hence, in the presence of both $\beta$-fructosidase and $\alpha$-galactosidase (melibiase) enzymes, brewer's yeast (S. uvarum) is capable of utilizing the entire raffinose molecule. This procedure is used as a diagnostic test to distinguish bottom-fermenting S. uvarum from top-fermenting S. cerevisiae according to the previous classification (Lodder 1970).

As previously reported, maltotetrose and higher glucose polymers are not fermented by brewer's yeast and can be recovered quantitatively in the final beer. The percent of various fractions from G5-G13 is the same in wort and in beer made from the same wort. A high percentage of these dextrins are branched-chained oligosaccharides containing one or more $\alpha$ (1-6) linkages.

Most brewers customarily use unmalted cereal adjuncts in up to 40% of the total grist for mashing. Adjuncts are carbohydrates with the desired composition and properties that complement the principal brewing material, barley malt. The use of adjuncts not only provides a cost advantage to the brewer but also gives the brewer the ability to produce more acceptable light-colored beer of higher stability. These attributes are of special importance in packaged beer, which may be subjected to extremes of temperature, rough handling, and extended storage at the marketplace.

Some of the adjunct materials used extensively today are those derived from rice or corn, although unmalted barley, wheat, or sorghum are also used in rare instances. Generally most of the adjunct grains used are preprocessed to some degree. Corn may be used as ground corn or grits, as flakes or pregelatinized starch granules, or as corn syrup. Rice, rice grits, and other large cereal grits reduced to the smallest particle size by milling are also commonly used in brewing.

## BREWING PROCESS

### Malting

Since antiquity, the principal raw material used by the brewer in the production of beer has been barley. However, in its native form, the barley grain is unacceptable for direct use in the brewing process. The modification that is generally done to make it suitable for brewing is known as malting. Although other cereal malts are usually designated as wheat malt, rye malt, and so on, the term *malt* generally refers to barley malt. This product is generally preferred for the production of beer, because in contrast to wheat and other cereal malts the liquified and saccharified extract can be readily separated from the husks and other insoluble materials.

The traditional malting process involves the initiation of the germination of the barley grain so that the carbohydrate reserves available for the development of the embryo have undergone some minimal modification by the enzyme developed within the seed. As a result of the complexity of malting, only a brief outline on the subject will be presented in this chapter; the reader desiring more extensive information can find a more comprehensive treatment in Briggs et al. (1981) and Broderick (1977).

Three distinct phases are apparent in a conventional malting process. These include steeping of the barley; germination of the grain under controlled conditions of temperature, humidity, and aeration; and finally drying of the germinated grain, which is referred to as *kilning*. The grain is first conveyed into steep tanks previously filled with water at approximately 10°-15°C. The water is replaced several times during the steeping cycle in order to wash the barley grains free of dirt and other extraneous matter. These steeping tanks are equipped with aeration devices to aerate the wetted grains. As the steeping operation progresses, the grain absorbs moisture and the dominant embryo inside the kernel becomes biologically active. The steeping is continued until the moisture content of the grain reaches 42-45%, generally in two-and-one-half days. The germinating barley grains are then transferred to germination chambers where malt is spread evenly on the floor of drums or compartment systems. In general, too high or too low a moisture content results in irregular germination of the barley seed.

The grain at the outset begins to release hormones like gibberellins capable of stimulating the aleurone layer within the seed to synthesize the necessary enzymes to hydrolyze carbohydrates, proteins, and the enzymes associated with energy metabolism. At the outset, the amylases already present in the barley begin to degrade the starches to assimilable sugars, and the proteases act on the proteins, converting them to peptides and amino acids. Under aerobic conditions, the respiratory enzymes in the aleurone layer oxidize the sugars and provide energy to produce more enzymes to support the germination process. If adequate aeration is not provided during this germination period, the embryos may be killed due to the toxic effect of certain metabolites such as ethanol and acetaldehyde accumulated within the tissues of the growing embryo.

The progress of germination is monitored based primarily on the texture of the endosperm and the length of the acrospire. The acrospire grows to a length of 75-100% of the full kernel length within four to five days. The development of the acrospire and rootlets take place at the expense of the endosperm carbohydrates and amino nitrogen. This process should proceed to achieve sufficient biochemical alteration in the grain without allowing excessive growth and respiration to achieve optimal extract yield. When the modification within the barley seed is judged optimal, the product is called *green malt*. The next stage in the processing is kilning.

## Kilning

The germinated barley, or green malt, has to be dried to a low moisture content under carefully controlled conditions to preserve the enzymatic activity developed during the germination. Kilning chambers are similar to the germinating chambers except that heated dry air instead of cool humid air is passed through the grain beds. Temperature within the kiln is controlled rigidly to minimize the denaturation of malt enzymes during the drying process. The rootlets formed during germination become dry and brittle during kilning and are removed on a scalper. The malt is finally cleaned and conveyed to storage bins to await shipment.

The final malt thus produced contains 4-5% moisture. The reduction of moisture content under controlled conditions arrests the respiration and biological activity of the germinating seed. It also reduces the chances of bacterial spoilage significantly and preserves the partially modified sugar substrate and the hydrolytic enzymes that would later become critical in the mashing process. At the end of the kilning process, a temperature as high as 80°C is used, and this treatment is known to provide the desired color together with flavor and aroma components to the final beer.

## Milling

Prior to mashing, the malted barley has to be ground in a mill to reduce the particle size to a coarse flour with minimal damage to the husks. This process is termed *dry milling*, and under controlled conditions the brewer is able to maintain an optimal extract yield while producing an efficient filter bed in the lauter tub to achieve the most optimal filtration characteristics.

In dry milling, the dry malt is crushed in a roller drum between three or more rollers set to a desired clearance, tension, and speed. Optimal extract yield and lauter time are achieved by suitably adjusting the operating conditions of the mill. The main objectives of milling are (1) to split open the grain, mostly lengthwise to expose the starchy endosperm, (2) to provide access for enzymes to reach all the constituents in the crushed grain, and (3) to achieve optimal particle distribution with minimum fines. Brewers who operate a lauter tub prefer a relatively higher level of coarse grit than flour, whereas those who operate mash filters desire a low level of coarse grit and a high percentage of flour.

Although dry milling has been the general practice in the past, wet milling has gained importance in the brewing industry in recent years following the recognition of its effectiveness in the corn wet milling industry. According to this procedure, the whole malt grains are soaked in hot water, or steamed until the moisture content is raised to 30-35%, at which point the kernels have lost their brittleness. During the grinding process, the endosperm material in the wet malt is squeezed out more or less intact without seriously damaging the husks. Oversteeping should be avoided to prevent material from clinging to the surface of the rollers. Wet milling provides the ability to attain uniform crushing to a slurry of fine grits, providing high extract yield and desired filtration characteristics.

## Mashing

Since brewing strains of yeast cannot ferment starch and other polysaccharides in the malt, various methods have been developed for converting these polysaccharides into fermentable sugar. In brewing, the process of converting starch or dextrins to fermentable sugar is termed *mashing*. During this operation the soluble materials are extracted from the ground malt and from adjuncts that had already been partially solubilized in a previous step, and the remaining insoluble components are converted to soluble assimilable products under controlled conditions with the help of the enzymes already present in the malt.

Mashing in the brewing process basically involves two important enzymatic reactions. The first is the conversion of starch to fermentable sugar in the presence of $\alpha$- and $\beta$-amylases already present in the malt. The other reaction is the conversion of proteins and polypeptides to amino acids and smaller peptides. The fermentable sugars formed during the mashing process are used by yeast to produce ethanol and carbon dioxide. Some of the amino acids formed are important for yeast growth at the early stages of the fermentation. The amino acids that are unused or excreted into the media by yeast during fermentation together with small peptides formed during mashing contribute to the characteristic flavor and the foaming ability in the beer. Thus, it is clear that the mashing procedure is of considerable importance for achieving optimal fermentation characteristics.

The two most popular mashing procedures used in Europe are infusion and decoction. These systems are somewhat different from the "double mashing," the most commonly employed procedure by the North American brewing industry. The interested reader will find further discussion of the three systems in Briggs et al. (1982).

The extraction procedure is carried out at a single temperature, usually about 65°C in the case of infusion mashing. This system is generally suitable for well-modified malts. The heated brewing water is made to directly heat up the malt as it enters the mash tun. The mash is loaded into the tun, and after standing for a period ranging from 15 minutes to 2 hours, the wort is withdrawn through the discharge pipes located in the true base of the tun. Wort in the trough is recycled at the beginning, and once the wort is bright it is no longer recycled but discharged into a holding vessel.

In the decoction mashing process, a portion of the malt mash is boiled to gelatinize the starch completely. The boiled portion of the mash is then used to raise the temperature of the main mash to the conversion temperature of approximately 65°C. High yields of fermentable sugars can be obtained because of higher susceptibility of gelatinized starch to the amylases present in the malt. However, the malt used in the main mash should supply the necessary enzyme requirement because the enzymes in the boiled portion have been inactivated. The decoction system requires a separate lauter tun or mash filter press for further processing to obtain clear wort.

The double mash system used extensively in North America requires the preparation of two separate mashes: one from an all-malt grist and the other from a mash containing predominantly adjuncts with a small proportion of malt to provide the necessary enzymes to initiate the conversion. The two mashes are then combined to attain the composition that is required by the brewer in the final wort (Figure 3-3). This procedure requires the use of well-modified, highly diastatic malt due to the use of a high proportion of adjuncts.

Most brewers, particularly in North America, customarily use a high proportion of starchy adjuncts as up to 40-50% of the total grist used in mashing. When the double-mashing system is used, the adjuncts, such as rice or corn grits, are mixed with 10% of the total malt in a vessel referred to as the *cooker*. They are mixed with water at about 50°C, and the temperature of the medium is then raised to boiling temperature during a period of 1 to 1.5 hours, and the boiling is continued for an additional 10-15 minutes with agitation. This heat treatment initially gelatinizes the starchy adjuncts and forms a viscous mass that is highly susceptible to action by amylase enzymes from the malt fraction. The final objective in this operation is liquification, which occurs due to the partial hydrolysis of the starch content in the mash. The incompletely hydrolyzed starches require a second conversion for complete hydrolysis.

While the preparation of the adjunct malt mash is in progress the mashing cycle for the main mash has to be initiated in the main mashing vessel. The bulk of the ground malt, which amounts to 90%, is mixed with

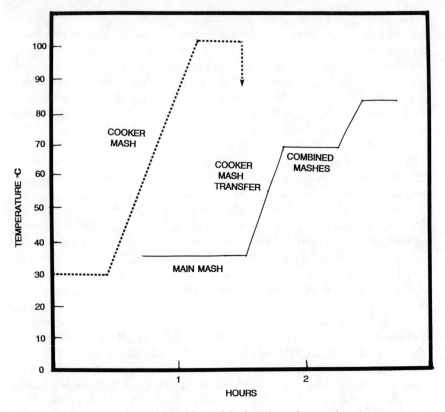

**Figure 3-3.** Typical "double-mash" schedule. (*After Hough et al. 1971*)

water at about 45°C and maintained at this temperature for approximately an hour to achieve optimal proteolysis. This process is called *protein rest*. Under these conditions, the proteolytic enzymes present in the malt degrade the proteins and polypeptides to predominantly amino acids. At the end of the protein rest the main mash is mixed with the boiling cooker mash containing the high proportion of adjunct, and the heat content in the cooker mash raises the temperature of the main mash to about 67°-68°C. There is a critical rate for the increase of temperature in the resulting mash, and it is generally controlled by the rate at which cooker mash is pumped into the main tank. A slower rate gives rise to a higher level of fermentable sugar, whereas a faster heating may account for a higher level of dextrins in the final wort. The saccharification is conducted for 20-30 minutes at 67°-68°C, and then the mash temperature is increased to 75°-80°C as the final "mashing-off" step. The particulate matter in the mash is then separated from the clear liquid by use of mash filter presses or the lauter tub.

The clarified liquid is then transferred to the brewing kettle where it is hopped and boiled sometimes with liquid adjuncts. Wort boiling is essential for brewing because it brings about certain desirable changes in the wort that eventually improve the quality and stability of the final beer. For example, wort boiling (1) sterilizes the wort medium, eliminating spoilage organisms; (2) inactivates any residual amylases or other hydrolytic enzymes, providing stability to the final beer; (3) helps in the extraction and isomerization of hop components, thereby providing the important bitter flavor components to the beer; (4) eliminates protein compounds from the wort (hot break), preventing haze formation; (5) eliminates certain undesirable volatile flavor and aroma compounds by steam distillation; (6) provides ability to achieve a higher concentrated wort; (7) helps to achieve the desired color in the wort; and (8) converts the organic sulfur compounds to $H_2S$, which can be boiled off in the process.

The precipitate that is formed subsequent to kettle boil is termed *trub*. It generally settles and can be separated from the hot clear liquid. Failure to remove these complex components would result in a beer of very poor colloidal stability. The clear wort is then cooled to 10°C, which then leads to the *cold break*. Those that are proteinaceous in nature aggregate and form a floc that is termed the *cold trub*. After the hot and cold breaks, the wort has a golden yellow color and is now ready for fermentation.

## Fermentation

In brewery fermentations, it is the objective of every brewer to produce an acceptable beer by converting the fermentable sugar present in the wort primarily into ethanol and carbon dioxide, using a specially selected strain of brewing yeast. Even though this process seems simple, many other reactions that take place during a fermentation contribute to the overall quality of the final beer. These biochemical changes that occur during a fermentation generate certain flavor components often in trace amounts to produce a distinct flavor profile that eventually characterizes the beer. It is, however, vitally important to conduct the fermentation under highly controlled conditions using a special strain of brewer's yeast with good fermentation characteristics to avoid the formation of certain flavor components that may be objectionable to a beer drinker. The biochemical pathway leading to the production of ethanol in an alcoholic fermentation at molecular level will not be discussed in this chapter. The interested reader should find the required information in any standard biochemistry text (Lehninger 1975). However, a section on the biochemistry of brewing will be presented later in this chapter, describing the relevant metabolic pathways that lead to the production of certain flavor and aroma compounds characteristic of beer.

The composition of worts has been reviewed by MacWilliam (1968). A specific gravity of 1.047 to 1.050 is desired in the United States for premium beer and 1.044 to 1.047 for less expensive beers (Lewis 1968a). This specific gravity corresponds to about 11° to 12.5° Balling or Plato according to the common terminology used in the brewing trade. (°Balling and °Plato, although not identical, are close enough to be used interchangeably for practical purposes.) The soluble substances of the wort are called *extract*.

The following data give an approximation of the gross composition of 100 ml of wort made with an all-malt mash: total solids, 10-12.5 g; total carbohydrates, 9-11.3 g; nitrogenous compounds (protein, ammino acids, amines, etc.), 0.4-0.5 g; lipids, 5-7 mg; tannin 200-300 mg; ash, 0.15-0.2 g. During the fermentation the extract value of the wort diminishes due to the utilization of the sugar by yeast. However, because of the ethanol formed during this process, determination of specific gravity and expression as °Plato indicates an extract value lower than the true value. The results of this measurement in the presence of ethanol is sometimes called *apparent Plato*. In order to determine the true extract value, one must distill the ethanol and bring the resultant liquid to the original volume of the beer by adding water.

The wort medium that has been heated before fermentation should contain no viable yeast. However, if this medium is allowed to remain unpitched for an extended period of time, it becomes rapidly contaminated with bacteria and wild yeast. It is possible that the low level of contamination of the heated wort comes from the vessels and equipment used in the process. A pure culture of yeast is introduced immediately as the first step in the fermentation process in order to suppress the growth of contaminants. The condition and nature of the pitching yeast is critical for the production of an acceptable beer. Hence, a brief outline of the techniques used by most brewers to propagate and maintain the pitching yeast will be presented next.

*Pitching Yeast*

During the development of the beer-brewing industry in the Middle Ages, the brewer relied on the yeast cultures from a previous fermentation to pitch his brews. These cultures generally contained bacteria and wild yeast in addition to the brewer's yeast. As a result of the presence of these contaminants the beer made from such brews was inferior and was not of uniform quality. This situation was finally resolved around 1881 by the adoption of the pure culture technique developed by Emil Christian Hansen of the Carlsberg Laboratory in Denmark. Most of the bigger breweries isolated and maintained their pure culture strains, and a few who did not have adequate facilities for microbiological work relied on independent laboratories for their services. Different brewing cultures are now available from several culture collections throughout the world.

Top-fermenting yeasts belong to the species *Saccharomyces cerevisiae* and bottom-fermenting yeasts to *Saccharomyces uvarum*. On occasion, other species have been used, such as *Brettanomyces* species for the production of top-fermenting export beer. The flavor of the beer produced depends on the metabolism of these respective yeasts. Of particular importance are the biochemical reactions that produce strong flavoring agents such as alcohol esters, aldehydes, ketones, and sulfur-containing compounds during an alcoholic fermentation. These may include both desirable and undesirable components basically dependent on the type of yeast strain used in the fermentation. Hence, yeast strains are selected in the brewing industry based on (1) their ability to provide a balanced flavor and aroma in the final beer, (2) rapid fermentation kinetics, (3) growth characteristics that do not exceed certain standards, and (4) their moderate flocculation characteristics during fermentation.

Several standard techniques are employed by different brewers in the selection of yeast cultures derived from a single cell. Yet there are others who may prefer several cells of a known culture population, a single colony on a plate, or several (7-10) colonies on a plate of uniform size. Some of these selections may be from the pitching yeast used previously. Selection from several colonies of uniform size should minimize the chance of introducing yeast strains that have undergone genetic variation within a population. Establishment of such stock cultures should finally be based on the results obtained in bench-top studies carried out under brewery conditions. The true-to-type nature of these brewing strains can be confirmed after examining such characters as degree of attenuation, flavor profile and aroma of the beer, and growth and flocculation characteristics of the yeast.

The early stages of the propagation of brewer's yeast are carried out under strict pure culture conditions. In this case, the vessels are sterilized under pressure, the media are sterilized, and cell transfers are made under strictly aseptic conditions. The propagation of the selected yeast is initiated by inoculating a flask of sterile aerated wort with yeast from the stock culture. With sufficient agitation, the culture will grow to be ready for a transfer to a larger flask in three to four days. It is usual practice in brewery propagations to increase the volume by about tenfold at each succeeding step in the propagation. The temperature during the propagation is maintained at 10°C for bottom-fermenting yeast whereas that for top-fermenting yeast is around 20°C.

A modern yeast propagation system in a brewery, using two yeast

## 114  BREWER'S YEAST

propagation vessels in parallel, has been described by Curtis and Clark (1960) as shown in Figure 3-4. The wort sterilizing and holding tank has a capacity of 40 barrels, and the yeast propagators have a working volume of 20 barrels each. Both vessels can be charged from the wort tank and enough head space can be left in the propagators to allow for a foam head. The wort is heated at 102°C for 10 minutes. The starter culture for inoculation of each 20 barrels of yeast propagator consists of 10 liters of pure culture prepared in the previous step in the seed train. Intermittent aeration is carried out 18 hours after the inoculation of the propagators. The practice of incremental feeding has not found favor in the production of brewer's yeast as it has with baker's yeast. In general, 90 g of pressed yeast inoculated into the yeast propagator would yield 60 kg of wet yeast within a 36-hour propagation. These propagations are completed in such a manner as to pitch the seed yeast, thus made immediately to a plant fermentation. Unlike the baker's yeast propagations, brewer's yeast propagations are small and are usually operated intermittently, only at times of need.

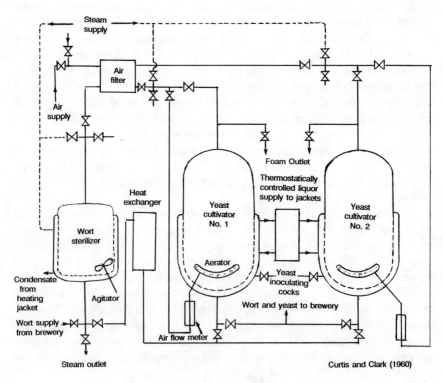

**Figure 3-4.**  Yeast culture plant. (*After Curtis and Clark 1960*)

It is customary to pitch yeast at rates that vary between 0.6 to 1.2 L per hectoliter. This pitching yeast may have the consistency of a thin slurry or a heavy paste, and its actual content of yeast solids or of yeast cells is highly variable. Thus, there are variations in the number of yeast cells actually pitched into brewery fermentations and variable fermentation rates may result. Therefore, a determination of the yeast content and the condition of pitching yeast becomes important.

Yeast content in a slurry may be determined by centrifuging, filtration, turbidiometry, direct microscopic count, or determination of yeast solids. The first two methods are extensively used, but are also the least reliable. Rainbow (1968) found the turbidiometric method useful as a rapid routine check and suggested the determination of yeast solids as a reference method. By direct microscopic count, it has been found that a pitching rate of 387 g of yeast per 100 L of beer is roughly equivalent to $15 \times 10^6$ cells per ml.

Generally, there is a sixfold to eightfold yeast growth during a commercial brewery fermentation. The yeast crop that is formed is commonly used again four to ten times for pitching succeeding brews. Despite this common practice in breweries, only a small portion of the available yeast will actually be used for pitching succeeding batches. In the case of bottom-fermenting yeast, it is customary to discard the yeast from the lowest sedimented layer since it contains *trub*, which are the insoluble particles separated from the wort during the fermentation. The uppermost layers are also discarded because of the probable presence of less flocculent yeast. A similar approach is often adopted in selecting top-fermenting yeast for succeeding brews. The surplus yeast is either discarded, dried into an inactive dried brewer's yeast for use in food or feed industries, or used for the production of extracts and autolysates as flavoring agents in the food industry.

The yeast that has been set aside for pitching succeeding batches can be stored in the form of a slurry, or it may be separated by pressing at 40-50 psi and kept as a press cake (25-30% solids). It is customary to store wet pitching yeast in any form at about 2°C to retain optimal viability. Brewers often use the methylene blue staining test for monitoring viability during storage. This test, which is generally applicable to yeast populations that have a high viability, assumes that living yeast cells will resist the entry of the dye into the cell whereas the dead cells will not. This test is, however, less sensitive when applied to a weaker population of yeasts. Even at these specified low temperatures an increase in the number of cells that can be stained with methylene blue in a yeast population can be observed. Furthermore, the fermenting power of the yeast decreases during storage, but autolysis does not take place until the yeast has been stored for 10 days or more despite its storage at low temperature.

It is often advisable to wash yeast to reduce the level of contaminating bacteria and to remove protein particles and dead yeast cells. Such an operation is usually carried out first by lowering the pH of the slurry to 2.5 to 3.0, generally with phosphoric acid or sulfuric acid, and then washing it with water. Ammonium persulfate has also been used as a disinfectant. It is very rare that brewers use antibiotics to suppress the bacterial contaminants.

The presence of wild yeast in the pitch is obviously a potential cause of faulty fermentations. Thus, it is standard practice to monitor for the presence of wild yeast at each stage of the brewing process. Samples of pitching yeast are examined not only to ensure satisfactory viability but also to check for the presence of wild yeast. Unfortunately, there is no one medium that does not suppress the brewery strain while suppressing the growth of the contaminating yeasts. Some of the tests used for detecting wild yeast will be described elsewhere in this chapter.

*Primary Fermentation*

The primary fermentation can be thought of as consisting of two distinct metabolic phases in sequence with some degree of overlap at the region where the two phases meet. During the initial phase, there is aerobic growth of yeast until most of the oxygen in the wort is exhausted. The metabolism of yeast in the second phase is overwhelmingly fermentative due to the depletion of oxygen in the preceding phase so that much of the fermentable sugar in the wort that is absorbed is degraded to ethanol and carbon dioxide. Of particular importance in the entire fermentation process are the biochemical reactions that proceed simultaneously, yielding a variety of flavor and aroma compounds that are important to enhance the specific character of the beer.

Historically, bottom-fermentation systems were used to produce lager beers in Europe and the top-fermentation system was typical of the United Kingdom and was traditionally used for the production of the beer. Presently, lager beer made by the bottom-fermentation process is the most extensively consumed beer throughout the world. The basic ingredients in both types of fermentations are sterile aerated wort and brewer's yeast.

In bottom fermentations, the brewing wort is cooled to a temperature of 8°-10°C and aerated with sterile air to provide a dissolved oxygen concentration of 8-10 ppm. If left unpitched, there exists a possibility for the wort to be contaminated by wild yeast and bacteria. Thus, it is essential to pitch these worts immediately at the rate equivalent of 10-15 million yeast cells per ml. The standard practice is to inject the pitching yeast slurry under controlled conditions directly into the aerated wort stream as it enters into the fermentor. Under the standard level of pitching, fourfold to fivefold multiplication of yeast can be observed within the initial two to three days of fermentation provided the original wort is sufficiently aerated.

The level of pitch added to the fermentation greatly influences the rate of fermentation and consequently the time at which the limit attenuation is reached. A higher pitching rate can be employed if available nitrogen in the wort is limited since less yeast growth takes place and less nitrogen is needed to sustain the growth. Although standard pitch rate is approximately 200 to 400 g yeast per 100 L which amounts to approximately 5 to $10 \times 10^6$ cells per ml, at significantly higher pitching rates (3 to 4 lb/barrel) the flavor tends to be bitter and more yeasty.

In general, American brewer's begin their fermentation at slightly higher temperature than their European counterparts (10°C vs. 6°C) during bottom fermentations even though the final temperature reached in both is roughly the same (5°C). The first visible sign of fermentation appears 12-24 hours after the wort is pitched. Small bubbles begin to appear on the surface of the wort, and a white creamy head known as *krausen* is formed. As the fermentation proceeds, the temperature continues to rise, and at about 13°-14°C cooling will be applied to maintain a steady temperature. The fermentable sugars, mainly glucose, maltose, and maltotriose, are transformed biochemically by the yeast cell primarily to ethanol and carbon dioxide.

When the fermentation is highly active, at 60-72 hours, the foam has a slightly yellowish color due to precipitation of colloidal substances derived from malt and hops. The brown spots that are conspicuous on the surface of the foam head are due to the oxidized hop resins and complexes made up predominantly of tannin. This type of heavy floc appearing at high fermentation activity is termed *high krausen*. There is a tendency at this stage for the temperature of the fermentation to rise to a maximum point of 13°-15°C due to the intense metabolic activity of the yeast. Further increase in temperature is prevented by cooling to a final temperature of about 10°C. Fermentation rates gradually decline with the gradual depletion of fermentation sugar, and the primary fermentation is judged complete when the decline in specific gravity becomes minimal (Fig. 3-5). Yeast becomes less active and tends to settle rapidly. Krausen begins to collapse and recede due to the reduction in the release of carbon dioxide. The primary fermentation is generally considered complete in seven to nine days in a lager-type fermentation. At this stage, the fermentation medium has a relatively thin cover of fine foam on the surface.

It is a general practice to partially separate the yeast from the beer at the end of the primary fermentation by decantation for the bottom-fermentation yeast or by skimming the top layer floating on the surface in the case of top fermentations. A portion of the recovered yeast is used to pitch subse-

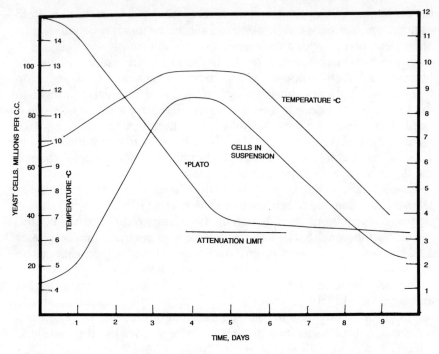

**Figure 3-5.** Variation of °Plato, cells in suspension, and temperature during a typical brewery bath fermentation.

quent brews. In actual practice yeasts are rarely used for more than 5 to 10 succeeding fermentations. Increasing contamination, loss of viability, loss of fermentation activity, and change in the degree of flocculence are some of the reasons advanced for the loss of yeast stability. When the yeast is judged unacceptable for repitching it becomes surplus yeast. Yeast is used for the production of food flavors or in feed formulations as described in Chapter 8.

*Secondary Fermentation*

During the primary fermentation, most of the principal flavor and aroma compounds critical for final beer flavor are formed. These include higher alcohols, esters, sulfur compounds, and organic acids that eventually characterize a specific beer. At this stage, the beer is highly immature and is commonly referred to as *green beer*. The composition of this immature beer has to be modified appropriately to bring about the acceptable balanced flavor profile natural to mature beer. This maturation process is carried out by a secondary fermentation termed *lagering*.

The young beer is moved to lager cellars where the temperature of the beer is gradually lowered to 3°-6°C. The green beer is matured in closed tanks in the presence of 6-10 million cells/ml yeast population that passes to the medium from the primary fermentation.

The maturation process that involves a secondary fermentation is designed to bring about several modifications to the green beer to improve its palatability. Of the several changes that take place during the aging process, five principal process functions can be considered critical for improved product quality (Coors 1977): (1) flavor maturation, (2) carbonation, (3) standardization, (4) chill proofing and stabilization, and (5) clarification. It should be emphasized, however, that flavor maturation is often considered by many brewers as the most critical of the five functions brought about by the aging process.

There are many complex changes that take place during flavor maturation of which three have been identified as highly significant because of their impact on beer flavor and aroma. These compounds include diacetyl, $H_2S$, and acetaldehyde. All three compounds are sometimes known to impart undesirable flavor or aroma due to their presence in the beer exceeding the threshold values. Hence, one of the primary goals of the secondary fermentation is to lower the diacetyl and $H_2S$ produced during the primary fermentation to values lower than the threshold values in order for the beer to be acceptable. The diacetyl and $H_2S$ concentrations are reduced by providing sufficient time for the residual yeast to act during the secondary fermentation. In a regular beer, the diacetyl level must be reduced to less than 0.1 ppm in order to reach the acceptable range. If it exceeds 0.15 ppm, the beer produces a strong buttery character. Yeast reduces the excess diacetyl in the beer during the secondary fermentation by assimilation.

Green beer contains detectable levels of hydrogen sulfide, and the threshold value for this compound in beer is in the 5- to 10-ppb range. $H_2S$ is present in the wort, but it is rapidly driven off during the kettle boil. Sulfur-containing volatile compounds, including $H_2S$, are formed by certain strains of yeast during the primary fermentation. The level of $H_2S$ is considerably reduced during beer maturation, which is presumed to take place as a result of the scrubbing action of carbon dioxide bubbles when produced in a beer fermentation. The pressure in the lagering vessels is generally released from time to time to allow some $CO_2$ gas to escape. Entrained with the released gas are volatile compounds of which $H_2S$ is a major component. Research to date has failed to show any transformation of $H_2S$ to other sulfur compounds with less objectionable flavor characteristics primarily due to the inaccuracy of the test procedures for testing these sulfur compounds at very low concentrations.

Acetaldehyde is an intermediate in the Embden-Meyerhof pathway linked to the alcohol dehydrogenase (ADH) system. It functions as the precursor for the ethanol in a normal brewery fermentation. The acetaldehyde formed by the decarboxylation of pyruvic acid is able to oxidize NADH and, in doing so, is itself reduced to ethanol in a reaction catalyzed by ADH enzymes. Despite the high affinity ADH has for acetaldehyde, some acetaldehyde molecules escape from the ADH system and leak into the environment. During lagering, part of the acetaldehyde in the media disappears, perhaps due to yeast metabolism (Coors 1977). Accordingly, the concentration of acetaldehyde in a beer that has undergone maturation seldom exceeds the threshold level of approximately 50 ppm.

Conditioning of beer refers to the saturation of beer with carbon dioxide gas. Part of the carbon dioxide saturation occurs during the primary fermentation, which may amount to approximately 1.5 volumes of $CO_2$ per volume of beer. In order to provide the required $CO_2$ level of approximately 2.8 volumes, additional carbonation can be achieved by krausening. It is a natural process that entails the production and saturation of $CO_2$ by the secondary fermentation process. Traditionally, after introducing wood chips into the lagering tanks, these are filled with fermented green beer to about 80% of their capacity, and krausen (a fermenting wort) containing 12-14% dissolved solids is pumped into the tank. It is a calculated amount of wort that will provide the desired carbonation when all the sugars are completely fermented. The tank is then sealed and connected to pressure regulators to maintain the desired pressure. The temperature in combination with pressure within the tank determines the $CO_2$ saturation in the beer. The krausening process generally lasts three to four weeks during which time the product matures, making the beer more palatable.

Conditioning of beer by a secondary fermentation is an expensive process that requires long periods to achieve satisfactory results. Thus, only a few brewers in North America use this procedure. In order to make the process more cost-effective, krausening has been replaced by many brewers by artificial carbonation using $CO_2$ gas. In lager breweries, this gas is derived from the $CO_2$ collected from a previous primary fermentation. This process is carried out under high pressure and low (0°C) temperature in conditioning tanks by injecting highly purified $CO_2$ gas into the filtered beer.

Bottom fermentations have, by tradition, a secondary fermentation that is commonly carried out at about 8°-10°C. After the secondary fermentation has been completed, the medium is chilled to about 0°C and held for sufficient time to allow the yeast and other protein/tannin complexes to form floc and settle. This *cold break* process helps to remove most of the chill haze, thus improving the beer stability.

During the conditioning period the beer may also be treated with chill-proofing agents such as papain to improve the stability, which prevents the formation of haze and turbidity during low-temperature storage.

The chill stabilization may also be enhanced by the addition of tannic acid with subsequent precipitation and separation, or by adsorption of trouble-causing high-molecular-weight proteins present in the beer onto silica gel. In most instances, the haze is due to the formation of polyphenol-protein complexes. These polyphenols can be reduced by adsorption onto beads of polyvinylpyrrolidone (pvp) or Nylon 66.

The finished product is then filtered to appear sparkling clean and transferred to surge tanks for final distribution to the packaging area. It is a legally bonded area referred to as a government cellar where federal inspectors constantly monitor the volumes and strengths of the alcohol present in the beer for levying of appropriate taxes. During the final stages of packaging it is important that beer is kept free of air, the $CO_2$ equilibrium is not disturbed, and the beer is not contaminated with microorganisms. The increase in the sale of packaged beer over draft beer has made the whole subject of bottling and canning of great significance. Today's lager brewers in the United States package up to 80% of their beer production in bottles or in cans with 20% being racked into kegs. The beer packaged in kegs is known as draft beer. It is sold mostly in the local market and is unpasteurized. If the beer is to be delivered over long distances, it may be pasteurized by means of *bulk* or *flash* pasteurization before filling or it is kept refrigerated to prevent microbiological deterioration.

Bottled beer and canned beer are pasteurized after closing. In a typical beer pasteurization, using a tunnel pasteurizer, bottled and canned beer are passed through different heating and cooling zones on a moving platform. The beer is heated to the pasteurization temperature and held for a specific period of time before being cooled to the desired temperature. In a modern, fully automated pasteurizer, heating to pasteurization temperature (approximately 60°C) takes about 20 minutes; the holding time is about 10 minutes; and cooling to final temperature (approximately 30°C) also takes about 20 minutes. Thus, the total pasteurization time is about one hour.

## BREWERY CONTAMINANTS

### Wild Yeast

The performance of a yeast in a brewery fermentation is largely determined by its genetic constitution. Therefore, brewers consider the yeast strain used in their brews as a critical component in the process to achieve the unique character of the particular brand of beer. The term *wild yeast* is thus applied to any yeasts that occur in brewery fermentations other than the culture strain used as the pitch at the start of the brewery fermentation.

Accordingly, a strain of yeast *Saccharomyces cerevisiae* suitable for the production of top-fermenting beer will be considered a wild yeast when it is detected in the production of lager beer. Brewers who use more than a single strain of yeast to produce different brands of beer are constantly at risk of cross-contamination. Types of yeast known to cause adverse affects on beer quality are species of *Candida, Saccharomyces, Dekkera, Hanseniaspora, Kloeckera, Kluyveromyces, Hansenula, Rhodotorula, Torulaspora, Brettanomyces,* and *Pichia,* as shown below (Barnett, Payne, and Yarrow 1983).

| | |
|---|---|
| *Brettanomyces anomalus* | *Han. fabianii* |
| *B. claussenii* | *Kloeckera apiculata* |
| *B. custersii* | *Kluyveromyces marxianus* |
| *Candida beechii* | *Pichia fermentans* |
| *C. himilis* | *P. guilliermondii* |
| *C. intermedia* | *P. membranaefaciens* |
| *C. sake* | *P. orientalis* |
| *C. solani* | *Saccharomyces cerevisiae* |
| *C. tropicalis* | *S. exiguus* |
| *C. versatilis* | *S. unisporus* |
| *Dekkera bruxellensis* | *Schizosaccharomyces pombe* |
| *Dek. intermedia* | *Torulaspora delbrueckii* |
| *Hanseniaspora uvarum* | *Zygosaccharomyces bailii* |
| *Hsp. vineae* | *Z. rouxii* |
| *Hansenula anomala* | |

Wild yeasts may enter the brewery with the air, the water, and other raw materials, or they may be carried in by insects or in empty bottles. In general, the level of contamination is relatively small compared to the large number of culture yeasts in a fermentation. During the active stages of fermentation, the pure culture yeasts that are well adapted to wort as a nutrient generally outgrow the contaminating organism. However, a selective advantage for wild yeasts may occur after the completion of the primary fermentation when the supply of readily fermentable sugar has been almost exhausted. Some brewery manipulations may also result in a selective advantage for wild yeasts. For instance, differential flocculation may increase the percentage of wild yeasts in that portion of the culture yeast present in the suspension. Similarly, pasteurization may favor wild yeasts that are thermotolerant.

Routine microbiological examination of each batch of pitching yeast is necessary to check for the presence of wild yeast and other contaminants. Unfortunately, none of the currently available methods for determining wild yeasts is entirely satisfactory. Direct microscopic examination may be applicable if the level of contamination is already high, and it will not detect wild yeast if the appearance and dimensions of the contaminating cells are similar to those of the pure culture yeast.

A variety of inhibitory and selective media have been developed for the detection and enumeration of wild yeast. In the case of the former, certain inhibitors are incorporated in nonselective media to suppress the growth of the culture yeast but permit the growth of the contaminating wild yeast. Harris and Watson (1968) found that the presence of 5 $\mu$g/ml or less of actidione (cycloheximide) in a nonselective medium was inhibitory to a pure culture of brewing yeasts while it permitted the growth of the wild yeasts. Brewery microbiologists have now learned to modify the actidione concentration in the plating media to fully suppress growth of the culture yeast while providing the optimal conditions for the actidione-resistant wild yeast to form colonies for easy detection. Other inhibitors such as Crystal violet and fuchsin-sulfite have also been used in nonselective media for the detection of wild yeast. Kato (1967) used 20 $\mu$g/ml Crystal violet to suppress the culture yeast and Brenner et al. (1970) used 0.3% to 0.35% fuchsin-sulfite to suppress the growth of *S. cerevisiae* and *S. uvarum* while *S. cerevisiae* var. *ellipsoideus*, *S. diastaticus*, and several species of *Candida* and *Hansenula* grew in the medium. Use of freshly prepared media is important since the effectiveness of fuchsin-sulfite is known to diminish in storage.

Lysine agar is a common selective medium used by most brewery microbiologists for detection and isolation of wild yeasts due to the fact that only wild yeasts are capable of growing in media containing lysine as the sole nitrogen source. Selective media containing dextrin or starch as the carbohydrate source can be used to detect the presence of starch-utilizing yeasts like *S. diastaticus*, which are important contaminants in the brewing industry.

In order to ensure that the test results for the lysine and dextrin agar are accurate, it is absolutely necessary that there is no nutrient carryover from the yeast culture to the selective medium. For this reason, the yeast that is to be tested must be thoroughly washed prior to plating.

More recently, highly sensitive immunological methods have been developed for the identification of wild yeast when present even at very low concentrations in the culture yeast. The most widely used method is immunofluorescence developed by Richard and Cowland (1967). This technique makes use of a specific antibody molecule coupled to a fluorescent dye (e.g., fluorescein or rhodamine) to identify those yeasts with which the antibody reacts. The resulting complex fluoresces under ultraviolet light and permits identification and location of the wild yeast. This method is generally practiced to detect wild strains in the fermentations. Although lager yeasts differ immunologically from wild yeasts, the differences are too minor to be used as a reliable test to detect wild yeasts in lager fermentations.

## Bacterial Contaminants

The nature and magnitude of the bacterial flora during a brewery fermentation depend on the degree of microbiological control practiced during the entire operation. Fortunately, the number of bacterial genera that could survive in a brewing process is relatively small. Several factors may account for the limited growth of bacteria during a brewery fermentation. For instance, such spoilage organisms may be sensitive to hop resins, a high concentration of ethanol, low pH, lack of oxygen during a major part of the process, limited sugar after the fermentation, early heating steps, and low temperature maintained during fermentation and processing.

Experience suggests that the most useful single character for the primary subdivision of bacteria is the gram reaction. Gram-positive bacteria that are of significance to brewing include the members of the lactic acid bacteria, primarily those belonging to two genera, *Lactobacillus* and *Pediococcus*. Certain endospore-forming bacteria belonging to the genus *Bacillus* and certain *Micrococcus* species (*M. kristinae*) have, on occasion, caused problems in breweries.

The important gram-negative bacteria that have been responsible for beer spoilage are acetic acid bacteria, *Zymomonas*, and a few members of the family *Enterobacteriaceae*, *Pectinatus cerevisiiphilus*, and *Megasphaera*.

### Gram-Positive Bacteria

*Lactobacillus.* These are catalase-negative, nonsporulating, rod-shaped organisms that are either homofermentative or heterofermentative, based on temperature tolerance and the mode of fermentation. There are several species of lactobacilli isolated from beer, and these constitute the predominant spoilage organisms in the beer industry. They are resistant to hop-bittering compounds, and some of these lactobacilli have been implicated in the production of diacetyl that is responsible for the "buttery" flavor. Although all lactobacilli produce lactic acid, the level of the acid accumulated does not reach the high threshold levels needed to make a significant flavor impact on the final beer. However, the spoilage can be observed as "silky" turbidity.

*Pediococcus.* These are catalase-negative, homofermentative cocci that form characteristic pairs or tetrads due to their cell division in two planes.

*Pediococcus damnosus* is the most common spoilage organism of the genus found in breweries that produce lager beer. These organisms are seldom found in the pitching yeast, but are found at times during the late fermentation or in the final beer. The spoilage by pediococci is somewhat similar to that caused by lactobacilli. Pediococci are responsible for the sarcina sickness and give rise to high acidity and buttery aroma due to the production of diacetyl.

*Other Gram-Positive Microorganisms.* Several gram-positive microorganisms, although less important as spoilage organisms, have been detected in brewery fermentations. Although heterofermentative cocci like *Leuconostoc mesenteroides* have been detected in breweries, they are not known to produce any beer spoilage. On the contrary, *Streptococcus lactis* and *Micrococcus kristinae*, which are relatively acid-tolerant and hop-resistant, have been responsible for beer spoilage. Likewise, thermophillic endospore-forming bacilli like *B. coagulans* and *B. stearothermophilus* have been isolated from breweries. These organisms are known to produce high levels of lactic acid when sweet wort is left for an extended period of time at elevated temperatures.

## Gram-Negative Bacteria

*Acetic Acid Bacteria.* These are gram-negative, rod-shaped bacteria capable of producing acetic acid from ethanol. *Acetobacter* and *Gluconobacter* are two important genera under the group acetic acid bacteria that are traditionally associated with brewery fermentations. Acetobacter is known to oxidize ethanol to $CO_2$ and water via the hexose monophosphate pathway and TCA cycle (Rainbow 1981). In the case of *Gluconobacter*, the hexose monophosphate shunt constitutes the most important route for sugar metabolism. The entire glycolytic and TCA cycles are not functional in *Gluconobacter* (Williams and Rainbow 1964). Beer is not expected to contain oxygen in the final form, and these organisms cannot thrive in beer under highly anaerobic conditions. Problems with *Acetobacter* and *Gluconobacter* can only be seen in beers that have low oxygen tension due to process defects.

*Obesumbacterium proteus.* The best-known brewery contaminant in the family *Enterobacteriaceae* is *Obesumbacterium proteus*. It is a gram-negative, nonacid-fast straight rod that is sometimes found in the pitching yeast. It can grow in unhopped wort and is able to tolerate pH values ranging from 4.4 to 9.0. This organism is known to suppress the fermentation process and is also responsible for the increased level of dimethyl sulfide, dimethyl disulfide, diacetyl and fusel oils (Priest, Cowbourne, and Hough 1974). Beer contaminated with *Obegumbacterium proteus* has a characteristic parsniplike or fruity odor.

*Zymomonas.* These are gram-negative rods that occur mostly as single cells, in pairs, chains, or filaments, and most strains are nonmotile. Motile strains have one to four flagella. *Zymomonas mobilis* is the most common brewery contaminant, and its most distinctive characteristic is its ability to convert glucose or fructose to ethanol and $CO_2$ via the Entner-Doudoroff pathway. However, its growth is inhibited by 8% ethanol. It can also be completely killed by a few minutes exposure to 60°C. *Zymomonas mobilis* is known to produce unacceptable levels of acetaldehyde and hydrogen sulfide in lager beer.

*Other Gram-Negative Microorganisms.* Several other less important gram-negative microorganisms have been reported in brewery fermentations. One such gram-negative contaminant is *Pectinatus cerevisiiphilus*, which is known to produce acetic acid, propionic acid, acetoin, and hydrogen sulfide, either in fermenting wort or packaged beer (Lee et al. 1980). Beers contaminated with this organism are generally turbid and have a rotten egg odor. The other organism of minor significance is the gram-negative coccus, *Megasphaera*, which imparts a cloudiness and an unpleasant odor to the beer. These characteristics are primarily due to the production of butyric acid during the fermentation of wort.

## BIOCHEMISTRY OF BREWING

The flavor and aroma so characteristic of beer are influenced only in minor part by the ethanol content of the beverage. Raw materials such as malt and hops have long been recognized as important factors in defining the character of the beer. Nevertheless, the components that exert the greatest impact on the palate and on the sense of smell are often present in the beer in very low concentrations. They are mostly produced by the yeast during the brewery fermentation (Table 3-5).

Over the years, a brewer has been able to identify yeast strains that are capable of producing beer of superior quality. As one would expect, these strains of yeast discovered quite empirically have the ability to excrete certain organic compounds in sufficient quantities to the medium to make the beer acceptable in flavor and aroma. Those strains that produce these organic compounds disproportionately would not give the beer a balanced flavor and aroma profile and are considered unfit for use in brewery fermentations.

The following section will provide a limited introduction to the subject of yeast metabolism pertinent to brewing. The reader desiring more exten-

Table 3-5. Major Volatile and Nonvolatile Constituents of Beer

| Constituents | Concentration (ppm) | Flavor Threshold (ppm) | Aroma or Taste |
|---|---|---|---|
| *Alcohols* | | | |
| Ethyl alcohol (%) | 2.7-3.8 (%) | 1.5-2.0 (%) | Harsh, sweet |
| 2-phenyl ethanol | 25-30 | 45-50 | Rosey, sweet |
| 2,3-butanediol | 10-128 | 400-500 | Sweet aroma and taste |
| Isoamyl alcohol | 50-60 | 50-60 | Fusely, throat-catching |
| Active amyl alcohol | 8-15 | 50-60 | Fusely, pungent |
| n-propanol | 10-15 | 50-60 | Slightly pungent, harsh |
| Isobutanol | 9-12 | 80-100 | Fusely, penetrating |
| *Esters* | | | |
| Ethyl acetate | 20-25 | 50-60 | Fruity, solventy |
| Isoamyl acetate | 2-15 | 2-3 | Fruity, banana |
| Phenyl ethyl acetate | 0.5-1.2 | 3.0 | Honey, sweet, fruity |
| *Aldehydes* | | | |
| Acetaldehyde | 5-10 | 12-15 | Pungent, green house |
| 2-norenal | 0.0005 | 0.0003-0.0005 | Cardboardy |
| 2-hexenal | 4.5 | 0.5-0.7 | Spicy, cinnamonlike |
| 5-OH-methyl furfural | 0.1-3.0 | 1000 | Caramel, sweet |
| *Diketones* | | | |
| Diacetyl | 0.05-0.10 | 0.10-0.15 | Buttery |
| 2,3-pentanedione | 0.01-0.03 | 0.09 | Butterscotch |
| Acetoin | 1-4 | 50 | Buttery, smooth |
| *Sulfur Compounds* | | | |
| Dimethyl sulfide (ppb) | 5-100 | 35-60 | Cabbage, malty |
| Diethyl disulfide (ppb) | 0-10 | | |
| $H_2S$ (ppb) | 0-5 | 5 | Rotten eggs |
| 3-methyl butene thiol (ppb) | 2 | 1-3 | Sun-struck |

sive information is encouraged to refer to Briggs et al. (1982) and Rose and Harrison (1969, 1970, 1971).

## Alcoholic Fermentations

Yeasts have the metabolic mechanism to derive the carbon skeleton needed to synthesize new cell material as well as to generate the energy so vital for their survival. Metabolically, yeasts are predominantly facultative aerobes, capable of growing either in the absence of air (fermentative) or in its presence (oxidative). Hence, in the presence of aeration, yeast has the ability to instantaneously change its energy-yielding metabolism from fermentative to oxidative. Brewery fermentation that occurs predominantly under anaerobic conditions is a process by which the fermentable sugars in the malt wort are converted primarily to ethanol and $CO_2$ by the action of a specially selected strain of yeast.

In beer fermentations, fermentable sugars present in the wort enter the cell rapidly with the help of the stereospecific transport system that is many times faster than transport by simple diffusion. This mode of entry is termed *facilitated diffusion*. Under anaerobic conditions, yeast is known to perform the alcoholic fermentation, producing two moles each of ethanol and $CO_2$ from each mole of glucose. Yeast multiplication under such conditions is minimal. Pasteur demonstrated that under highly aerobic conditions and at a low concentration of sugar (<1.0%) yeast can utilize sugar exclusively by the aerobic oxidative pathway to produce biomass. This phenomenon is termed *Pasteur effect*. At high glucose concentrations, the aeration does not shift the metabolism of yeast from fermentative to a complete oxidative mode. Instead, the yeast continues to ferment the glucose and perform an aerobic fermentation. This phenomenon is referred to as the *glucose effect*, *Crabtree effect* or *reverse Pasteur effect*. The net results of oxidative and fermentative pathways are shown in the following equations:

Oxidative: $C_6H_{12}O_6 + 6\,O_2 \rightarrow CO_2 + 6\,H_2O \qquad F = 686\,\text{kcal}.$

Fermentative: $C_6H_{12}O_6 \rightarrow 2CO_2 + 2C_2H_5OH \qquad F = 54\,\text{kcal}.$

where $F$ = free energy.

According to the above equation, the fermentative pathway releases approximately 8% of the total energy trapped in the sugar. A more complete oxidative breakdown results in the release of the entire energy by the oxidative breakdown of the sugar in the form of ATP. There is a net production of 34 ATP molecules by the oxidative pathway. These are ultimately utilized, mainly promoting yeast growth. The biochemical pathway leading to the production of ethanol in an alcoholic fermentation is well documented (Lehninger 1975) and will not be discussed at length in this chapter.

At the early stages of the brewery fermentation, there is an adequate level of oxygen charged into the wort medium to make the yeast undergo two different overlapping metabolic phases. One is the initial aerobic growth phase that prevails until all the dissolved oxygen in the wort is utilized. The other is the anaerobic fermentative phase where the biochemical reactions lead to the production of ethanol, $CO_2$, and other flavor components characteristic of beer. Under these conditions, both the aerobic and anaerobic systems compete for the pyruvic acid that is formed as an intermediate by the classical glycolytic pathway. The aerobic system, which includes the TCA cycle, and the electron transport chain play a vital role in generating the energy and in providing the metabolic intermediates necessary for cell growth. Accordingly, a high proportion of budding cells become apparent with a simultaneous increase in the biomass at the early phase of the brewery fermentation. Molecular oxygen is utilized by these respiring cells as a hydrogen acceptor, generating the maximum energy by completely oxidizing the sugars taken in by the cell. The mitochondria are the primary organelles associated with energy metabolism. Due to the functioning of the anaerobic system, part of the pyruvic acid is converted to acetaldehyde and $CO_2$. The acetaldehyde formed as an intermediate is further reduced to ethanol, which together with $CO_2$ and the yeast produced, represents the major products at the early phase of a brewery fermentation.

As the oxygen disappears from the wort, the mitochondria begin to degenerate and the levels of the enzymes associated with the TCA cycle disappear so that the carbon flow through the oxidative system becomes practically insignificant. In the absence of the enzymes of the TCA cycle, the yeast becomes entirely dependent on the fermentative pathway to generate the energy required for its sustenance. The end products of this metabolic pathway are predominantly ethanol and carbon dioxide. The latter phase dominates in a brewery fermentation until most of the fermentable sugar in the wort is utilized.

At the beginning of the brewery fermentation, when cell multiplication takes place, an additional pathway is known to exist. It is termed the hexose monophosphate pathway or more commonly the pentose cycle. This pathway provides the ability for the yeast to synthesize the pentose sugars that the yeast cannot take in from the wort. These sugars are essential for the biosynthesis of nucleic acid as well as for many other enzyme cofactors, such as $NAD^+$ and $NADP^+$, needed for cell growth. The overall reaction is represented as follows:

$$G6 + 12NADP^+ \rightarrow 6CO_2 + 12NADPH + 12H^+ + P^+ + 6H_2O$$

This cycle together with the TCA cycle and the ancillary glyoxalate pathway are known to influence the cell propagation that exists at the early phase of the brewery fermentation. In brewer's yeast grown aerobically, 10-20% of the glucose is thought to be respired via the pentose cycle. The information presented here is a very limited introduction to the regulation of metabolism in an alcoholic fermentation.

During the primary fermentation, most of the important flavor and aroma compounds such as higher alcohols, esters, sulfur compounds, and organic acids are formed. An attempt will thus be made in the following section to review the biochemistry involved in the production of these important components in a brewery fermentation.

## Fusel Alcohols

In brewery fermentations yeast is primarily responsible for the conversion of fermentable sugar to ethanol and $CO_2$ as well as for the generation of a variety of by-products that contribute to the flavor and aroma of the beer. Among these by-products, the bulk of the volatile constituents are the higher alcohols that are less volatile than ethanol. During distillation, as practiced in the production of distilled spirits, the higher alcohols, based on their degree of volatility, are concentrated and separated at some specific level of the rectifying column. This complex mixture of higher alcohols is sometimes referred to as fusel oil. The principal components of fusel oil consist of isoamyl alcohol, active amyl alcohol, n-propanol, and isobutanol.

The levels of higher alcohols and other important alcohols formed in top and bottom fermentations are shown in Table 3-6. Approximately 60% of the higher alcohols consist of isoamyl alcohol with the proportions of active amyl alcohol, isobutyl alcohol, and n-propanol following in a descending order of magnitude. None of the above-mentioned individual alcohols are present in the beer in quantities higher than their threshold values. However, the combined level of these congeners may approach the threshold concentration, and it is more than likely that the total higher alcohols contribute to the final beer flavor. It has been noted that threshold values give, at best, an approximation of the concentrations that can be detected in beer or in degassed beer. Values given by Harrison (1970) represent the concentration of these alcohols that can be perceived in beer if they are added to the concentrations already present.

The fusel alcohol content of lager beer is in the 40-100 ppm range, and

Table 3-6. Concentrations and Threshold Levels of Higher Alcohols in Bottom- and Top-Fermented Beer

| Beer Constituent | Fermentation | | Flavor Threshold (ppm) |
| --- | --- | --- | --- |
| | Bottom (ppm) | Top (ppm) | |
| n-propanol | 9-18 | 10-38 | 50-60 |
| Isobutanol | 8-14 | 25-46 | 80-100 |
| Isoamyl alcohol | 36-52 | 50-84 | 50-60 |
| Active amyl alcohol | 11-17 | 18-25 | 50-60 |
| Tyrosol | 6-9 | 8-12 | 100 |
| Phenyl ethyl alcohol | 16-25 | 38-50 | 45-50 |

BIOCHEMISTRY OF BREWING 131

that of ale and high-gravity stouts is generally higher. The formation of higher alcohols occurs during the active primary fermentation and generally closely follows the disappearance of fermentable sugar. Fusel oil production by the fermenting yeast is influenced by the fermentation conditions such as temperature, aeration, agitation; by wort composition, especially with reference to the amino acid profile; and, most important, by the genetic makeup of the yeast. By appropriate adjustment of the fermentation variables, the brewer manages to have control over the production of higher alcohols. However, since yeast strains differ in their ability to produce fusel alcohols, the choice of yeast strain is probably the most important factor in controlling the level of fusel alcohol in a brewery fermentation.

The basic understanding that is currently available on the production of fusel alcohols is primarily derived from the pioneering studies of Ehrlich (1906). His original work was based on the concept that deamination of $\alpha$-amino acids by transamination, followed by a decarboxylation and reduction, results in the production of alcohols with one less carbon atom than that present in the original amino acid. This mechanism involves the participation of three enzymes, namely, transaminase, carboxylase, and alcohol dehydrogenase, to bring about the conversion of amino acid to the higher alcohol. The mechanism of the production of isoamyl alcohol from leucine is shown in Figure 3-6.

Although the amino acid uptake from wort in a brewery fermentation is more than adequate to account for the total fusel alcohol production by Ehrlich's mechanism, there are certain alcoholic fermentations where the higher alcohol levels produced during the fermentation far exceed what could be accounted for by the amino acid level present in the medium. Furthermore, production of n-propanol and n-butanol cannot be explained by Ehrlich's pathway due to the fact that the corresponding amino acids, $\alpha$-aminobutyrate and norvaline, do not exist naturally in wort for the yeast

$$\text{L-leucine} + \underset{\text{2 ketoglutaric acid}}{\overset{\overset{\text{COCOOH}}{|}}{CH_2CH_2COOH}} \xrightarrow{\text{transaminase}} \underset{\text{2 ketoisocaproic acid}}{(CH_3)_2CH\ CH_2COCOOH} + \text{glutamic acid}$$

$$(CH_3)_2CH\ CH_2COCOOH \xrightarrow{\text{carboxylase}} \underset{\text{isovaleraldehyde}}{(CH_3)_2CHCH_2CHO} + CO_2$$

$$(CH_3)_2CHCH_2CHO + NADH_2 \xrightarrow{\overset{\text{alcohol}}{\text{dehydrogenase}}} (CH_3)_2CHCH_2CH_2OH + NAD$$

Figure 3-6. Mechanism by which isoamyl alcohol is produced from leucine.

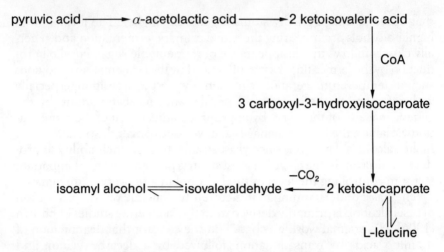

**Figure 3-7.** Mechanism by which isoamyl alcohol is produced from pyruvic acid.

to assimilate like other amino acids. Subsequent studies by Ayrapaa (1965) revealed that Ehrlich's mechanism is not the only pathway for the production of fusel alcohols. These studies showed an additional pathway via the ketoacids produced by carbohydrate catabolism. For example, the ketoisocaproic acid, which is an intermediate in the pathway that produces isoamyl alcohol from L-leucine, can also be generated from pyruvic acid or 2-ketoisovaleric acid, which are the intermediates of carbohydrate metabolism (Webb and Ingraham 1963). This process is shown schematically in Figure 3-7.

Genetic manipulation of yeast has been attempted in the past in the wine industry not only to develop strains with the ability to produce low fusel alcohols but also to elucidate pathways that lead to the production of different higher alcohols. Direct mutagenesis of the brewing yeast was shown to be successful. Certain leucine-requiring yeast mutants have already been developed that produce only trace quantities of isoamyl alcohol with all other characteristics associated with wine production unaffected. It is also known that some valine-requiring mutants produce markedly less isobutyl alcohol than do their parent strains. Although there are several publications related to this subject (Kunkee and Snow 1983; Ingraham and Guymon 1960), there is no evidence to show that these strains are of commercial significance.

## Diacetyl and 2,3-Pentanedione

Diacetyl is of considerable interest to the brewer because at very low concentrations it is known to affect the beer flavor. This product, also

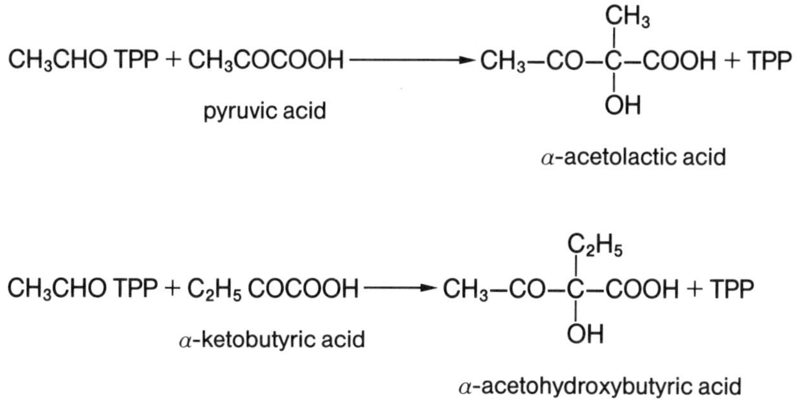

$$\text{CH}_3\text{CHO TPP} + \text{CH}_3\text{COCOOH} \longrightarrow \text{CH}_3-\text{CO}-\underset{\underset{\text{OH}}{|}}{\overset{\overset{\text{CH}_3}{|}}{\text{C}}}-\text{COOH} + \text{TPP}$$

pyruvic acid                  α-acetolactic acid

$$\text{CH}_3\text{CHO TPP} + \text{C}_2\text{H}_5\text{COCOOH} \longrightarrow \text{CH}_3-\text{CO}-\underset{\underset{\text{OH}}{|}}{\overset{\overset{\text{C}_2\text{H}_5}{|}}{\text{C}}}-\text{COOH} + \text{TPP}$$

α-ketobutyric acid          α-acetohydroxybutyric acid

**Figure 3-8.** Conversion of pyruvic acid to α-acetohydroxybutyric acid.

referred to as 2,3-butanedione, imparts a sweet buttery off-flavor to beer at concentrations exceeding 0.1 ppm in American lager beer. Diacetyl is often determined together with a related compound 2,3-pentanedione as vicinal diketone (VDK). Although these can be detected as low as 0.02 ppm in certain beers, their threshold or tolerance varies according to the beer because heavily flavored beers can mask higher concentrations of diacetyl so that it is not detected as an off-flavor.

The effect of valine in suppressing diacetyl production was first demonstrated by Owades, Maresca, and Rubin (1959) and elaborated by Portno (1966a, 1966b). Presently, there is sufficient data to indicate that diacetyl production is ultimately related to valine biosynthesis. Likewise, it has been shown that the production of 2,3-pentanedione is related to the biosynthesis of the amino acid isoleucine. Biosynthesis of valine and isoleucine are very similar in that each pathway begins with an active acetaldehyde group, which is an intermediate in the pathway that leads to ethanol production.

In amino acid biosynthesis, pyruvate serves as the initial substrate, and in the presence of the coenzyme thiamine pyrophosphate it first forms a complex that then undergoes decarboxylation to form a compound called α-hydroxyethyl-2-thiamine pyrophosphate, otherwise referred to as active acetaldehyde. The active acetaldehyde thus formed can react with pyruvic acid to form α-acetolactic acid and with α-ketobutyric acid to form α-acetohydroxybutyric acid (Fig. 3-8).

Although these two intermediates eventually form valine and isoleucine during yeast propagation as it occurs at the early phase of the brewery fermentation, these intermediates can also serve as a substrate for the production of diacetyl and 2,3-pentanedione by an oxidative process. The addition of valine to wort is known to reduce the final level of diacetyl by reducing the level of α-acetolactic acid by a feedback mechanism (Fig. 3-9). Likewise, suppression of α-acetohydroxybutrate by a high level of isoleucine suppression in the wort accounts for the low 2,3-pentanedione in beer. These pathways are shown schematically in Figure 3-10.

There is considerable variation in the production of diacetyl by different yeast strains. Hence, at the very outset, proper selection of the yeast strain is obviously the most important criterion in controlling the level of diacetyl in a brewery fermentation. Conditions like high aeration, high temperature, or pH values below 4.6 have a tendency to promote the

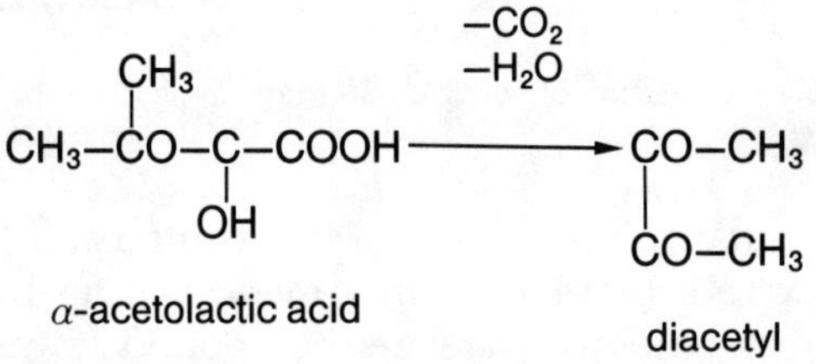

$$CH_3-CO-\underset{\underset{OH}{|}}{\overset{\overset{CH_3}{|}}{C}}-COOH \xrightarrow[-H_2O]{-CO_2} \underset{CO-CH_3}{CO-CH_3}$$

α-acetolactic acid    diacetyl

**Figure 3-9.** Conversion of α-acetolactic acid to diacetyl.

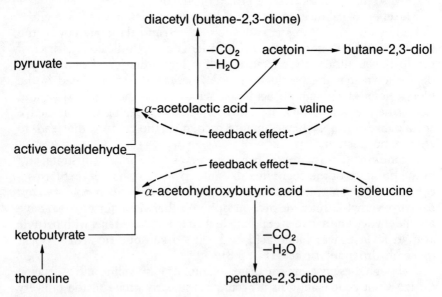

**Figure 3-10.** Reduction of diacetyl level in wort through reduction of α-acetolactic acid by feedback mechanism.

production of diacetyl. It is known that yeast in suspension reduces the diacetyl level if it is kept in contact with the wort and if access of air is prevented. Yeast cells in suspension reduce diacetyl to acetoin and 2,3 butanediol. Although both acetoin and 2,3 butanediol are present in appreciable quantities, these are not potent flavoring agents. Strains of yeast that are highly flocculant do not get an opportunity to assimilate the preformed VDK due to their early settling, and such strains are generally avoided for use in brewing. The respiratory-deficient or petit mutants produce high levels of VDK and therefore pitching yeast should be examined periodically for the presence of such mutants. Several of the contaminants, such as *Pediococcus* and *Lactobacillus*, are also capable of producing diacetyl. Hence, a high degree of microbiological control is necessary to keep these spoilage organisms out of the brewing process.

Practical suggestions for the production of beer with low diacetyl have been made by several authors (Lewis 1968b; Latimer et al. 1969; Brenner 1970), and these are as follows: (1) Fermentation should be carried out at low temperature; (2) use yeast strains that generally produce low diacetyl; (3) ensure that the wort is not contaminated with wild yeast, Pediococci, or Lactobacilli; (4) provide adequate nutrition, particularly valine and isoleucine; (5) avoid stimulation of yeast growth by aeration; and (6) avoid access of air during the final stages of fermentation and during transfer of fermented media to lager cellars and so forth.

## Esters

Esters produced during the fermentation undoubtedly contribute to the final beer flavor. Nevertheless, when these esters are produced at concentrations that are abnormally high, they contribute off-flavors, thus making the beer unpalatable.

The total amount of esters is generally expressed as ethyl acetate. The ester content in American lager beer is between 25 and 50 ppm. However, ethyl acetate is the predominant ester, accounting for less than half the total esters in the beer. A large number of esters have been identified and Table 3-5 shows the concentration of some of the known esters as well as their threshold values. It is thought that ethyl acetate, isoamyl acetate, and phenyl ethyl acetate contribute maximally to the estery or fruity flavor of the beer, but the contribution by other esters cannot be disregarded. Such estery flavors are desirable for top-fermented beer and for dark beer, but are judged undesirable for lager beer.

Ester formation occurs during the active phase of growth in yeast, and to a much lesser degree during the lagering period. Thus, it is unlikely that esters are formed by chemical reaction between the acids and the alcohols in the beer. According to Nordstrom (1965), ester production by yeast initially involves an activation of the acids to their acyl coenzyme A through the mediation of energy and the subsequent condensation of the active compound with an alcohol present in the medium to form the ester.

$$R.CO \sim SCoA + R.'OH \rightarrow R\text{-}COOR' + CoA \sim SH$$

The alcohol moieties for the esterification reaction are generally provided by the fusel alcohols formed during the early phase of the brewery fermentation. The acid moieties may arise as a result of small losses of acetyl CoA compounds that may occur during fatty acid synthesis.

According to the work of Nordstrom (1965) and Rainbow (1970), four major factors influence ester formation: (1) genetic constitution of the strain of yeast, (2) fermentation temperature, (3) extent of yeast growth, and (4) composition of the medium. Of these four factors, the genetic makeup of the yeast and to a lesser degree the temperature exert by far the most unequivocal and readily quantifiable effects.

Brewery fermentations are generally conducted at low temperature using a suitable strain to yield beer with optimum volatile ester profile. Control of the production of esters also requires the ability to control the extent of yeast growth in relation to the level of assimilable nitrogen in the medium. In general, oxygen is used as a growth regulator to achieve these objectives.

One disadvantage encountered in high-gravity brewing is the excessive production of volatile compounds such as esters and fusel alcohols. However, esterification can be controlled by providing appropriate aeration of the fermenting wort. As mentioned previously, acetyl CoA enters the TCA cycle and gets oxidized if the sugar concentration is low with easy access to oxygen. Sugar concentrations are high at the early phase of the brewery fermentation and, hence, the TCA cycle is inoperative as a result of the Crabtree effect. In the absence of oxygen in the fermenting wort the acetyl CoA present in yeast begins to acetylate the ethanol and higher alcohols to generate the corresponding esters. It is thus apparent that the production of acetate esters occurs when the condition in the medium is not conducive for cell growth.

## Sulfur-Containing Compounds

Certain sulfur-containing volatile compounds like $H_2S$, $SO_2$, and mercaptans in beer are of great interest because of their potential effect on beer flavor. The taste threshold values of many of these compounds are low. For example, the threshold value for $H_2S$ is 5 to 10 ppb and that for dimethyl sulfide is 35-60 ppb. Brenner, Owades, and Golyzniak (1953) showed that even though 5 ppm $H_2S$ modified beer flavor, more than 100 ppm was necessary before the unpleasant odor of $H_2S$ was detectable in the beer. Fortunately, the concentration of these sulfur compounds in beer are below the threshold values. However, production of any of these sulfur compounds in excess of threshold values should make these beers unpalatable.

Although sulfur is absorbed by the yeast at very low concentrations, it is considered an essential element for yeast growth. It can be supplied to the yeast as sulfate, as sulfur amino acid, or as other sulfur-containing organic compounds. Brewer's yeast contains sulfur-containing compounds like methionine, cysteine, cystine, biotin, thiamine, and certain inorganic sulfates that can be assimilated by yeast for the synthesis of proteins, coenzymes, and certain vitamins. When sulfate is used as the sole source of sulfur, it undergoes a series of reductive changes through sulfite and sulfide before it is incorporated into the organic molecule.

Yeast produces $H_2S$ mainly during the early phase of the brewery fermentation when the growth rate is at its maximum. At this stage, metabolism of the yeast, which includes both anabolism and catabolism, is at its peak level, and the energy for these processes is derived from a variety of compounds, although the predominant source is sugar. Although amino acids can serve as precursors for protein synthesis, these are sometimes used as a source of energy by oxidative degradation via the TCA cycle. There are several pathways by which carbon from amino acids enter the TCA cycle. In the case of cysteine, degradation may proceed by at least three different pathways, and one of these is through the action of the enzyme cysteine desulfhydrase. Wainwright (1971) suggested that a significant proportion of the hydrogen sulfide is probably formed by the action of the enzyme cysteine desulfhydrase on the amino acid cysteine.

$$HSCH_2CH(NH_2)COOH \xrightarrow{\text{cysteine desulfhydrase}}$$
cysteine

$$CH_2 = C(NH_2)COOH + H_2S$$
$\alpha$-amino acrylic acid

A high concentration of $H_2S$ was also detected in beer contaminated by the gram-negative bacterium *Zymomonas*. It is a more common contaminant in ale breweries than in lager breweries (Dadds, MacPherson, and Sinclair 1971). It is assumed that the low temperature at which the lager fermentations are conducted does not permit the development of *Zymomonas*. Beer contaminated with this microbe generally has a rotten apple odor that makes it unacceptable for consumption. The so-called sunstruck flavor, a well-known beer defect, is due to the formation of 3-methyl-butene-1-thiol by a reaction that involves the side chain of isohumulone and an $H_2S$ or -SH compound.

Very low concentrations of other organic sulfur compounds have also been found to show great potential in enhancing off-flavor in beer. However, many early attempts to identify these compounds failed because of inadequate methodology. Recent progress in isolation techniques and the development of analytical instrumentation have increased the interest to determine the impact of sulfur compounds in beer flavor. A major outcome as a result of these studies has been the establishment of dimethyl sulfide (DMS) as a key flavor compound that has the potential of imparting off-flavor (Williams and Gracey 1982a). Recently, much attention has been focused on hops as a source of sulfury off-odors in beer. Besides the familiar DMS, several other higher sulfides, like dimethyl disulfide (DMDS), dimethyl trisulfide (DMTS), and dimethyl tetrasulfide (DMTetS), though at much lower concentrations, have been shown to exhibit a high potential of enhancing sulfury character in the beer (Williams and Gracey 1982b).

The formation of $SO_2$ during wine fermentation has been mentioned in Chapter 4. This compound is also formed during the fermentation of wort. In contrast to $H_2S$, its concentration is not lowered during the lagering period. Most of the $SO_2$ is in the bound form. Sulfur dioxide receives less attention in the brewing industry than in wine making because it is not widely used as an additive in the brewing process. Its concentration in beer is generally well below levels of 50-100 ppm where it cannot be recognized by taste.

## RECENT DEVELOPMENTS

Despite the numerous technical advances that have taken place in recent years in the field of biotechnology, the progress the brewing industry has made by taking advantage of these new technologies has been surprisingly minor. This backwardness could be due in part to the innate conservatism on the part of brewers. Besides, when the brewers are generally happy with the existing conventional brewing process, such a reaction to new innovations is understandable. Unlike other industries, radical proposals to modify the conventional process have resulted in the production of beers of inferior quality that are often unpalatable. Such an outcome may have, perhaps, resulted in the exploration of less drastic and less debatable methods to modify and improve the brewing process.

### Low-Calorie Beer

During recent years, the general public has become increasingly calorie conscious. In response to this sudden change in lifestyle the food industry began to market their low-calorie product line, and the beer industry reacted by the introduction of beers containing a significantly lower calorie content. With the increasing demand for low-calorie beers, more and more brewers began to brew low-calorie beer, and before long this product became an important market segment in the brewing industry.

Although the introduction of low-calorie light beer is recent, the underlying technology for its production has been known to the brewer for several decades. A number of methods have been described in the literature, and almost all such approaches have involved the addition of fungal glucoamylase enzyme to reduce the dextrin content of the beer that would otherwise have contributed calories to the beverage.

The conventional type of brewing makes use of a wort that contains 60-70% fermentable sugar on the basis of solids. The sugars, which include glucose, fructose, maltose, and maltotriose, are all converted to ethanol and carbon dioxide by the yeast during the alcoholic fermentation. The maltotetroses, higher-molecular-weight dextrins, and some starches, which constitute the 30-40% unfermentable fraction of the wort, remain unchanged during the fermentation due to the inability of the yeast to utilize these high-molecular-weight compounds. As a result, these soluble carbohydrates contribute substantially to the final calorie content of the regular beer.

It is generally known that dextrins, ethanol, and protein constitute the major calorie-contributing components in a regular beer. The concentrations of lipids and amino acids in regular beer are so small that their impact on the final calorie content in relation to the three major components can be considered insignificant. The caloric values per gram of these beer components are fat, 9; ethanol, 7; protein, 4; and carbohydrate (expressed as monosaccharides), 3.75. The calories contributed by the ethanol, protein, and soluble unfermentable carbohydrates portion of the beer can be calculated from the following equation:

$$\text{calories}/100 \text{ ml} = [\text{alcohol}(g/100 \text{ ml}) \times 7] + [\text{total carbohydrate (as glucose)} \ g/100 \text{ ml} \times 3.75] + [\text{protein } (g/100 \text{ ml}) \times 4]$$

A regular 12-fluid-ounce bottle of beer generally contains 150-160 calories. It is, however, possible to reduce the overall caloric content by reducing the available carbohydrate content in a regular beer. In standard practice, use is made of glucoamylase enzymes to convert the dextrins into glucose monomers, which in turn results in an improvement of the efficiency of carbohydrate utilization by the brewing yeast. A low-calorie 12-fluid-ounce bottle of light beer produced by this procedure contains 90-110 calories. With further reduction of the alcohol content, it is possible to further reduce the calorie content of the beer. However, such drastic reduction in caloric content should obviously have to be made at the expense of the flavor and mouthfeel (Nagodawithana 1986).

Unlike the amylase enzymes present in malt, the glucoamylases have the capability of hydrolyzing $\alpha(1\text{-}6)$ linkages in the dextrin molecules in addition to hydrolyzing $\alpha(1\text{-}4)$ linkages, and this enzyme removes glucose units from the nonreducing end of the starch molecule. Commercially available preparations of glucoamylases are derived from strains of *Aspergillus* or *Rhizopus*. These enzymes have similar properties with pH optima between 4 and 6 and maximum activity at 55°-65° C. The enzymes are sometimes not destroyed by the normal beer pasteurization temperature of 60°-62°C. Such beers tend to get sweeter upon storage due to the release of glucose units by the continued activity of the enzyme on the dextrins.

Although the enzyme glucoamylase is not present in brewing strains, its presence as an extracellular enzyme has been reported in strains of *S. diastaticus*. Several attempts made in the past to improve the efficiency of the brewing process using *S. diastaticus* failed due to the quality problems detected in the final beer. These defects were primarily the heavy medicinal phenolic aroma due to the production of 4-vinyl guaicol as a result of enzymatic decarboxylation of ferulic acid derived from malt and hops. The genetics associated with this phenomenon are presented in Chapter 2.

One important application of genetic engineering to the brewing industry involves the improvement of the ability of brewing strains to degrade high-molecular-weight carbohydrates like dextrins and starches to produce well-attenuated or low-calorie light beer. The ability to produce the enzyme glucoamylase is controlled by the polymeric genes DEX 1, DEX 2, and STA 3 (Erratt and Stewart 1981). By use of the protoplast fusion technique, the ability to use dextrin has been introduced to *S. uvarum* and *S. cerevisiae* by fusing with *S. diastaticus*. Unfortunately, beer produced by such hybrids was of very poor quality due to the presence of phenolic off-flavor. Since the expression of traits as a result of protoplast fusion is random, this approach to selectively introduce a single trait like starch utilization ability into a yeast strain without introducing other undesirable characters would be difficult. For this reason the use of plasmid vectors for the transfer of DEX genes to brewing strains is presently being investigated in several laboratories. Despite the extensive research that has been carried out toward achieving this goal worldwide, there is no evidence to show the existence of a suitable dextrin-utilizing industrial strain of brewer's yeast for the production of low-calorie beer of acceptable quality.

### Killer Yeast

As we said in Chapter 2, the synthesis of proteins that impart the killer and resistance character to the yeast is directed by a double-stranded RNA (dsRNA) viruslike particle that is present in the killer strains. The yeast cells that do not contain the dsRNA are termed sensitive strains and are generally eliminated by the killer factor. The mechanism by which the killer phenomenon is expressed is not fully understood, although there are several models to explain some of the findings that have been made to date.

The genetic material that is associated with killer character can be introduced into sensitive industrial strains by hybridization (Hara, Timura, and Otsuka 1980) or by cytoduction (Ouchi et al. 1979). Inserting into brewing strains of yeast the ability to acquire the killer character would provide two significant advantages. First, these industrial strains will be immune to the toxin of their own and related wild killer strains, and second, the production of the toxin would prevent the proliferation of other wild yeasts in the brewery fermentation. Thus, protection against wild contaminants would be provided. There are several procedures to bring about the transfer of these genetic materials from laboratory strains of *S. cerevisiae* to brewing strains as described in Chapter 2. In addition to improving the biological stability of the beer produced, such genetically modified brewing strains can successfully be used in breweries with the cost of sterilization and pasteurization drastically reduced.

The genetic material that confers the killer character when inserted into the yeast cell remains in the cytoplasm of the yeast cell without integrating into the yeast chromosomes. There is thus a high probability for the newly acquired character to disappear during cell multiplication. Fortunately, with the rapid development in the field of genetic manipulation, new techniques are becoming increasingly available for stabilizing plasmids within a host cell, and such technology should provide genetic stability to those genetic materials like the killer factor newly inserted into brewing strains.

The presence of wild strains of yeast having the killer property can be a potential cause of faulty brewery fermentations. In general, it is good practice in quality control to test for the presence of such undesirable yeasts throughout the brewing operation. Although there is no single selective medium that would promote the growth of only one contaminant yeast while suppressing the culture strain, lysine agar can be considered an effective selective medium for the detection of wild yeast. The ability to grow on lysine agar is commonly shared by most wild yeasts. The industrial strains like *S. uvarum* and *S. cerevisiae* are incapable of utilizing lysine as a nitrogen source and do not appear on lysine agar plates. The other types of media used for detection of wild yeasts are described in the wild yeast section.

The most dramatic effect of the presence of killer yeasts can be demonstrated by inoculating a killer strain of yeast onto a lawn of sensitive yeast on a petri plate. The toxin that is released by the killer strain inhibits the growth of sensitive yeast in the vicinity, thereby creating a partially clear zone in areas where the inoculation was made. The phenomenon is apparent after two to four days of incubation, and this test can be used to detect the presence of killer contaminants in culture yeast.

## High-Gravity Brewing

In recent years the brewing industry worldwide has constantly been attempting to develop a cost-effective process to increase beer production without sacrificing quality with minimal additional capital outlay at existing plants. An important development has been the introduction of adjuncts to partially replace the more expensive malt used in the brewing process. The introduction of adjuncts also provided the brewer with the ability to prepare worts that are high in fermentable sugars and, hence, high in specific gravity. These developments in turn stimulated active research in the field of high-gravity brewing.

In a typical brewery fermentation, the wort contains about 9-12% fermentable sugar in addition to all other nutrients important for yeast growth. The brewing strains are not affected by the osmotic pressure of the medium or the ethanol produced during the fermentation under these conventional brewery fermentation conditions. Fermentation at high specific gravity obviously requires the use of brewing strains that can withstand high alcohol concentrations. Although the process changes associated with high-gravity brewing increase the productivity, these changes also create new environments that could perhaps not be suitable for average yeasts when the wort sugar exceeds a certain critical concentration. Under these new conditions brewer's yeast cannot be realistically expected to perform normally to produce the same flavor profile as in conventional brewing. It is clear that there are certain inherent limitations to high-gravity brewing if the only change made is the specific gravity of the wort used in the process.

Worts of fermentable sugar concentrations in the 12-18% range on a weight basis are normal for use in high-gravity brewing. The sugar concentration of worts higher than 18% by weight requires extraordinarily longer periods of fermentation time for achieving complete attenuation under typical brewing conditions and is not considered practical on a commercial scale.

The advantages and disadvantages of high-gravity brewing have been discussed by Pfisterer and Stewart (1976). The advantages as listed by these authors consist of the following: The novel process produces a beer with improved colloidal haze and flavor stability, provides more efficient use of existing plant facilities, reduces energy costs, yields more ethanol per unit of fermentable sugar primarily due to no apparent increase in yeast crop, permits high adjunct usage level, and produces a beer that is generally rated smoother in taste. The major disadvantages of high-gravity brewing include a substantial decrease in brewhouse extraction efficiency, a high level of $\alpha$-amino nitrogen in the final beer, and difficulties in matching flavor profile characteristics for regular beer made by conventional brewing. As presented previously, an increase in wort concentration does not necessarily result in a proportional increase in the yeast crop. Thus, with an increase in the wort concentration there is a corresponding increase in the $\alpha$-amino nitrogen that is more than adequate to meet the needs of the yeast crop. Hence, a high level of $\alpha$-amino nitrogen is not unusual for a beer brewed under high-gravity brewing conditions.

Abnormally high levels of acetate esters have also been detected in lager beer brewed according to the new process. However, recent studies have shown that the acetate levels could be controlled by the introduction of aerated water into the fermentor during the fermentation. High-gravity beer was also found proportionately low in certain congeners like amyl alcohol, phenylethanol, tyrosol, and tryptophol when diluted to the same alcohol level as that of a regular beer.

High-gravity brews carried out with 24% or more fermentable sugar have not reached limit attenuation even with prolonged fermentation (Nagodawithana 1986; Reed and Nagodawithana 1988). These studies indicated that the fermentation rate in high-gravity fermentations is greatly influenced by the combined inhibitory effect of ethanol and osmotic pressure. The effect of alcohol produced at any stage in a high-gravity fermentation on the fermentation rate was significantly higher than when the corresponding level of alcohol was produced in a typical conventional-type brewery fermentation (Nagodawithana and Steinkraus 1976). These studies have suggested the possibility that at high-gravity fermentation conditions yeast could be affected by the combined inhibitory effect of ethanol and the osmotic pressure caused by the residual soluble solids present in the fermenting medium (Fig. 3-11). The combined inhibitory effect has been mathematically represented by the product of the concentration of substrate (real extract) and the percentage of ethanol on a weight basis at that stage of the fermentation. The following equation has been useful to derive the kinetics at any stage of the high-gravity fermentation:

$$V_1 = K_1(RE_t - n) + K_2(RE_0)(E_t)$$

where

$V_1$ = specific rate at which the substrate is utilized at time $t$ ($\mu g$ of sugar utilized per minute);
$n$ = fraction of the dissolved solids unfermented
= 0.38 $RE_0$ (for worts with 38% unfermentables);
$K_1$ = proportionality constant for the forward reaction;
$K_2$ = proportionality constant for the combined inhibitory influence (see below);
$RE_t$ = real extract value of wort at time $t$;
$E_t$ = weight percent ethanol at time $t$;
$RE_0$ = real extract value of wort at time zero.

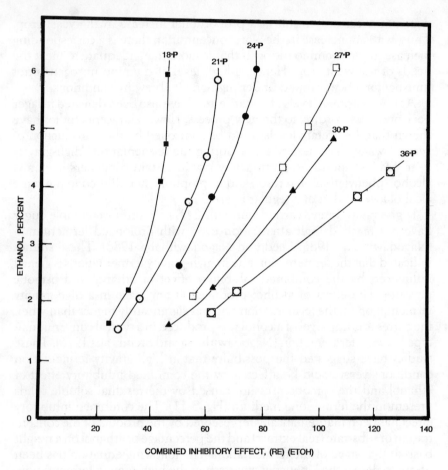

**Figure 3-11.** Variation of combined inhibitory effect with ethanol concentration for fermentation initiated with 18-36°Plato worts. RE = real extract.

The proportionality constants $K_1$ and $K_2$, which can be derived graphically, characterize a given set of fermentation conditions. Under high-gravity brewing conditions, the term $K_2$ assumes a negative value. Table 3-7 shows some of the data generated in the aforementioned study.

In the above equation, $V_1$ represents the net result of two opposing factors:

$$V_1 = \text{forward reaction (A)} - \text{combined inhibitory effect (B)}$$

The simple equation shown above predicts the kinetics of the brewery fermentation. In order to achieve high fermentation rates, a logical approach is to take maximum advantage of the factor A at the early phase of the

Table 3-7. End Real Extract Values for 18°-36°P Worts by Experimentation and as Predicted by Equation

| Initial °Plato | $K_1$ | $K_2$ | End Real Extract Based on Equation | Experimental |
|---|---|---|---|---|
| 18 | 4.8 | 0.239 | 5.44 | 5.97 |
| 21 | 4.73 | 0.120 | 6.90 | 7.20 |
| 24 | 4.18 | −0.120 | 11.82 | 11.10 |
| 27 | 3.55 | −0.170 | 16.72 | 15.1 |
| 30 | 3.02 | −0.170 | 16.72 | 15.1 |
| 36 | 3.29 | −0.380 | 27.59 | 27.65 |

Note: The corresponding K values were used in the calculation of the end RE value.

fermentation when the combined inhibitory effect, B, is relatively minimal. With the increase in ethanol concentration, there is a corresponding increase in the combined inhibitory effect (factor B) that would result in a decline in the overall fermentation rate. This effect is usually accompanied by a collapse of the foam head following the reduction in the generation of $CO_2$ within the fermentation medium. At certain wort concentrations, the combined inhibitory effect can reach values that are abnormally high to bring the fermentation virtually to a standstill. At this stage, one way to regain fermentative ability is to make an appropriate adjustment in the fermentation to minimize the combined inhibitory effect on the yeast. For example, this goal has been accomplished by diluting the fermenting medium at a critical stage of the fermentation as outlined in U.S. patent 4,140,799 (Nagodawithana and Cuzner 1979). The collapsing of the foam head creates a large enough head space in the fermentor to accommodate the volume of water necessary for the required dilution. Carbonated water is used for the dilution to ensure final product quality. This procedure provides not only a method to complete a brewery fermentation within an acceptable time but also makes it possible to increase the productivity by utilizing the head space of the fermentors without having to spend additional capital for new fermentors.

The beer brewed by high gravity brewing receives an additional dilution prior to packaging for final adjustment of the alcohol concentration of the beer. The underlying principle to account for good quality beer by the novel process stems from the fact that there is an immediate shift from high gravity brewing conditions to a more conventional type of condition after the critical dilution step has been completed at the early phase of the fermentation. Studies have shown that the beer produced by this procedure is similar in flavor, aroma and stability characteristics to that produced by conventional procedures.

# REFERENCES

Atkin, L. A., P. P. Gray, W. Moses, and M. Feinstein. 1962. Growth and fermentation for different brewery yeasts. *Wallerstein Lab. Commun.* 25:122-134.

Ayrapaa, T. 1965. Formation of phenyl ethyl alcohol from $C^{14}$ labeled phenylalanine, *J. Inst. Brew.* 71:341-347.

Barnett, J. A., R. W. Payne, and D. Yarrow. 1983. *Yeasts: Characterization and Identification.* Cambridge University Press, New York.

Beavan, M. J., D. M. Belk, G. G. Stewart, and A. H. Rose. 1979. Changes in electrophoretic mobility and lytic enzyme activity associated with development of flocculating ability of *S. cerevisiae. Can. J. Microbiol.* 25:885-895.

Bevan, E. A., and M. Makower. 1963. The physiological basis of the killer character in yeast. In *Genetics Today,* S. J. Goerts (ed.). Pergamon Press, Oxford, England, pp. 202-203.

Brady, B. L. 1965. Utilization of amino compounds by yeasts of the genus *Saccharomyces. Antonie van Leeuwenhoek* 31:95-102.

Brenner, M. W. 1970. A practical brewer's view of diacetyl. *Master Brewer's Assoc. Am. Tech. Quart.* 7:43-49.

Brenner, M. W., J. L. Owades, and R. Golyzniak. 1953. Determination of volatile sulfur compounds. I. Hydrogen sulfide. *Am. Soc. Brew. Chemists Proc.*, pp. 83-89.

Brenner, M. W., M. Karpiscak, H. Stern, and W. P. Hsu. 1970. Differential media for detection of wild yeasts in the brewery. *Am. Soc. Brew. Chemists Proc.*, pp. 79-88.

Briggs, D. E., J. S. Hough, R. Stevens, and T. W. Young. 1981. *Mating and Brewing Science,* vol. 1, Chapman and Hall, London.

Broderick, H. M. 1977. *A Manual for the Brewing Industry,* 2nd ed. Master Brewers Association of the Americas, Madison, Wis.

Coors, J. H. 1977. *The Practical Brewer,* 2nd ed. Master Brewers Association of the Americas, Madison, Wis.

Curtis, N. S., and A. G. Clark. 1960. New yeast culture plant. *J. Inst. Brew.* 66:287-292.

Dadds, M. J. S., A. L. MacPherson, and A. Sinclair. 1971. Zymomonas and acetaldehyde levels in beer. *J. Inst. Brew.* 77:453-456.

Damle, W. R., and R. S. W. Thorne. 1949. The growth and fermentation of yeast 6479 with simple peptides as nitrogen nutrients. *J. Inst. Brew.* 55:13-18.

Densky, H., P. J. Gray, and A. Buday. 1966. Further studies on the determination of zinc and its effect on various yeasts. *Am. Soc. Brew. Chemists Proc.*, pp. 93-100.

Ehrlich, F. 1906. Uber eine Methode zur Spaltung racemischer Aminosauren mittels Hefe. *Biochem Z.* 1:8.

Enari, T. M., M. Linko, M. Loisa, and V. Makinen. 1970. The effect of wort amino acids on fermentation. *Master Brewers Assoc. Am. Tech. Quart.* 7:237-240.

Erratt, J. A., and G. G. Stewart. 1981. Genetics and biochemical studies in glucoamylase from *Saccharomyces diastaticus.* In *Current Developments in Yeast Research,* G. G. Stewart and I. Russell (eds.), Pergamon Press, Toronto, pp. 177-183.

Frey, S. W., W. G. DeWitt, and B. R. Bellomy. 1967. Effect of several trace metals on fermentation. *Am. Soc. Brew Chemists Proc.*, pp. 199-205.

Gilliland, R. B. 1956. Maltotriose fermentation in the species differentiation of *Saccharomyces. C. R. Trav. Lab. Carlsberg, Ser. Physiol.* 26:139-148.

Hara, S., A. Timura, and K. Otsuka. 1980. Breeding of useful killer wine yeasts. *Am. J. Enol. Vitic.* 31:28-33.

Hardwick, W. A. 1983. *Beer In Biotechnology*, vol. 5, H.-J. Rehm and G. Reed (eds.). Verlag Chemie, Deerfield Beach, Fla., p. 191.
Harris, G., and D. J. Millin. 1963. Sequential induction of maltose permease and maltase system in *Saccharomyces cerevisiae*. *Biochem. J.* **88:**89-95.
Harris, J. O., and W. Watson. 1968. The use of controlled levels of actidione for brewing and non-brewing yeast strain differentiation. *J. Inst. Brew.* **74:**286-290.
Harrison, G. A. F. 1970. The flavor of beer: A review. *J. Inst. Brew.* **76:**486-495.
Hough, J. S., E. D. Briggs, and R. S. Stevens. 1971. *Malting and Brewing Science*. Chapman and Hall Ltd., London, pp. 255-288.
Hudson, J. R. 1967. Factors affecting yeast performance. *Europ. Brew. Conv. Proc.*, pp. 187-195.
Ingraham, J. L., and J. F. Guymon. 1960. The formation of higher diaphetic alcohols by mutant strains of *S. cerevisiae*. *Arch. Biochem. Biophys.* **88:**157-166.
Jayatissa, P. M., and A. H. Rose. 1976. Role of phosphomannan in flocculation of *S. cerevisiae*. *J. Gen. Microbiol.* **96:**165-174.
Johnston, J. P., and H. Oberman. 1979. Yeast genetics in industry. In *Progress in Industrial Microbiology*, vol. 15, M. J. Bull (ed.). Elsevier Scientific Co., New York.
Johnston, J. R., and H. P. Reader. 1983. Genetic control of flocculation. In *Yeast Genetics*, J. F. P. Spencer, D. M. Spencer, and A. R. W. Smith (eds.). Springer-Verlag, New York.
Jones, M., and J. S. Pierce. 1964. Absorption of amino acids from wort by yeasts. *J. Inst. Brew.* **70:**307-315.
Jones, M., M. J. W. Pragrell, and J. S. Pierce. 1969. Absorption of amino acids by yeast from a semidefined medium simulating wort. *J. Inst. Brew.* **75:**520-536.
Kato, S. 1967. A new measurement of infectious wild yeast in beer by means of crystal violet media. *Bull. Brew Sci.* **13:**19-24.
Kreger-van Rij, N. J. W. 1984. General classification of the yeasts. In *The Yeasts, A Taxonomic Study*. N. J. W. Kreger-van Rij (ed.) Elsevier Science Publication, Amsterdam, pp. 1-44.
Kunkee, R. E., and S. R. Snow. 1983. Method of reducing fusel oil in alcohol beverages and yeast strains useful in that method. U.S. patent 4,374,859.
Latimer, R. A., P. R. Glemster, K. G. Koepple, and F. C. Dallos. 1969. A review of the diacetyl problem. *Master Brew. Assoc. Am. Tech. Quart.* **6:**24-29.
Lee, S. Y., M. S. Mabee, N. O. Jangaard, and E. K. Horiuchi. 1980. Pectinatus, a new genus of bacteria capable of growth in hopped beer. *J. Inst. Brew.* **86:**28-30.
Lehninger, A. L. 1975. *Biochemistry*, 2nd ed. Worth Publishers, New York.
Lewis, M. J. 1968a. American lager beer. *Proc. Biochem.* **3**(8):47-62.
Lewis, M. J. 1968b. Recent research on diacetyl. *Brewers Digest* **43:**9-81.
Lodder, J. 1970. *The Yeasts: A Taxonomic Study*, 2nd ed. North Holland Publishing Co., Amsterdam.
MacWilliam, I. C. 1968. Wort composition: A review. *J. Inst. Brew.* **74:**38-54.
MacWilliam, I. C., and J. F. Chapperton. 1969. Dynamic aspects of nitrogen metabolism in yeast. *Eur. Brew. Conv. Proc.*, pp. 271-279.
Montreuil, J., S. Mullet, and R. Scriban. 1961. A study of fermentation of wort sugar in the brewery. *Wallerstein Lab. Commun.* **24:**304-315.
Mortimer, R. K., and D. Schild. 1985. Genetic map of *Saccharomyces cerevisiae*, edition 9. *Microbiol Rev.* **49:**181-212.
Nagodawithana, T. W. 1986. Yeasts: Their role in modified cereal fermentations. In *Advances in Cereal Science and Technology*, vol. 8, Y. Pomeranz (ed.). American Association of Cereal Chemists, St. Paul, Minn., pp. 15-104.

Nagodawithana, T. W., and K. H. Steinkraus 1974. The effect of dissolved oxygen, temperature, initial cell count and sugar concentration on the viability of *Saccharomyces cerevisiae* in rapid fermentations. *Appl. Microbiol.* **28**:383-391.

Nagodawithana, T. W., and J. M. Cuzner. 1979. Method of fermenting brewer's yeast. U.S. patent 4,140,799.

Nordstrom, K. 1965. Possible control of volatile ester formation in brewing. *Proc. Eur. Brew. Conv. 10th,* Stockholm, pp. 195-208.

Ouchi, K., R. B. Wickner, E. A. Toh-e, and H. Akiyama. 1979. Breeding of killer yeasts for sake brewing by cytoduction. *J. Ferment. Tech.* **57**:483-487.

Owades, J. L., L. Maresca, and G. Rubin. 1959. Nitrogen metabolism during fermentation in the brewery process. II. Metabolism of diacetyl formation. *Am. Soc. Brew. Chemists Proc.,* pp. 22-26.

Pfisterer, E., and G. G. Stewart. 1976. High gravity brewing. *Brew. Digest* **6**:34-42.

Portno, A. D. 1966a. Some factors affecting the concentration of diacetyl in beer. *J. Inst. Brew.* **72**:193-196.

Portno, A. D. 1966b. The influence of oxygen on the production of diacetyl during fermentation and conditioning. *J. Inst. Brew.* **72**:458-461.

Priest, F. G., M. A. Cowbourne, and J. S. Hough. 1974. Wort enterobacteria – A review. *J. Inst. Brew.* **80**:342-356.

Rainbow, C. 1968. Measurement of yeast concentration. *J. Inst. Brew.* **74**:427-429.

Rainbow, C. 1970. Brewers's yeast. In *The Yeasts,* vol. 3, A. H. Rose and J. S. Harrison (eds.). Academic Press, London. pp. 147-224.

Rainbow, C. 1981. Beer spoilage microorganisms. In *Brewing Science,* vol. 2, J. R. A. Pollock (ed.). Academic Press, New York and London, pp. 491-550.

Reed, G., and T. W. Nagodawithana. 1988. Technology of yeast usage in wine making. *Am. J. Enol. Vitic.* **39**(1):83-90.

Richard, M., and T. W. Cowland. 1967. The rapid detection of brewery contaminants belonging to the genus *Saccharomyces* by a serological technique. *J. Inst. Brew.* **73**:552-558.

Rose, A. H. 1980. Recent research on industrially important strains of *S. cerevisiae.* In *Biology and Activities of Yeasts,* F. A. Skinner, S. M. Passmore, and R. R. Davenport (eds.). Academic Press, London, pp. 101-121.

Rose, A. H., and J. S. Harrison (eds.) 1969. *The Yeasts,* vol. 1, *Biology of Yeasts.* Academic Press, London and New York.

Rose, A. H., and J. S. Harrison (eds.) 1970. *The Yeasts,* vol. 3, *The Yeast Technology.* Academic Press, London and New York.

Rose, A. H., and J. S. Harrison (eds.) 1971. *The Yeasts,* vol. 2, *Physiology and Biochemistry of Yeasts.* Academic Press, London and New York.

Stewart, G. G., and I. Russell. 1977. The identification, characterization and mapping of a gene for flocculation in *Saccharomyces* sp. *Can. J. Microbiol.* **23**:441-447.

Stewart, G. G., and I. Russell. 1981. Yeast flocculation. In *Brewing Science,* vol. 2, J. R. A. Pollock (ed.). Academic Press, London, pp. 61-92.

Stewart, G. G., I. Russell, and E. F. Garrison. 1975. Some considerations of the flocculation characteristics of ale and lager yeast strains. *J. Inst. Brew.* **81**:248-257.

Thorne, R. S. W. 1949. Mechanism of nitrogen assimilation from amino acids by yeast. *Nature* **164**:369-370.

Wainwright, T. 1971. Production of hydrogen sulfide by yeasts. *J. Appl. Biotechnol.* **34**(1):161-171.
Webb, A. D., and J. L. Ingraham. 1963. Fusel oil. *Adv. Appl. Microbiol.* **5**:317-353.
Williams, P. J. Le. B., and C. Rainbow. 1964. Enzymes of the tricarboxylic acid cycle in acetic acid bacteria. *J. Gen. Microbiol.* **35**:237-247.
Williams, R. S., and D. E. F. Gracey. 1982a. Beyond dimethyl sulfide: The significant flavor of this ester and polysulfides in Canadian beer. *J. Am. Soc. Brew. Chem.* **40**(2):68-71.
Williams, R. S., and D. E. F. Gracey. 1982b. Factors influencing the level of polysulfides in beer. *J. Am. Soc. Brew. Chem.* **40**(2):71-74.

CHAPTER

# 4

# WINE YEASTS

## ECOLOGY

Fermentations of fruits and fruit juices occur naturally and spontaneously without human intervention. Ripe berries can undergo spontaneous fermentation in the field if they are overripe or injured. Birds that eat such fermented berries can show all of the signs of alcohol intoxication and they may fly erratically.

These fermentations are initiated by yeasts found on the grape skins usually near the pedicels or stomate and on spots where the skin of the fruit has been broken. It has been long believed incorrectly that the yeasts survive the winter in the soil of vineyards. Yeasts survive from year to year in the intestines of bees and wasps (Mooser 1958) and they are readily transferred during the crushing season by fruit flies (*Drosophila melanogaster*) during the transport of grapes to the winery and within the winery (Wolf and Benda 1967). Once the crushing season is underway the major source of yeasts is likely to be found on winery equipment such as crushers, destemmers, and tanks.

## TERMINOLOGY

The terms *brewer's yeast* and *baker's yeast* are restricted to certain specific strains of *Saccharomyces cerevisiae*. These yeasts are always or almost always produced from selected pure strains of this species. In contrast, wine yeasts are strains of various species participating in the must fermentation.

In a broad sense, any yeast species occurring on grapes and participating

Table 4-1.  Wine Yeast Species—Synonyms

| | |
|---|---|
| Saccharomyces cerevisiae | (S. elipsoideus, S. bayanus, S. uvarum, S. carlsbergensis, S. chevalieri, S. vini, S. oviformis, S. cheresiensis, S. italicus, S. beticus) |
| Saccharomyces exiguus | (imperfect form Candida holmii) |
| Torulaspora delbrueckii | (Saccharomyces delbrueckii, S. rosei, S. fermentati, Torulaspora rosei) |
| Zygosaccharomyces rouxii | (Saccharomuces rouxii, S. mellis) |
| Zygosaccharomyces bailii | (Saccharomyces bailii) |
| Candida stellata | (Torulopsis stellata, T. bacillaris) |
| Candida vini | (Saccharomyces vini, Mycoderma vini) |
| Candida krusei | (Saccharomyces krusei) |
| Metschnikowia pulcherrima | (Torula pulcherrima, Torulopsis pulcherrima imperfect form Candida pulcherrima) |

in the must fermentation, no matter how transient, may be called a wine yeast. In a narrow sense, strains grown in pure culture for use in winery fermentations are wine yeasts. Such strains almost always belong to the species *Saccharomyces cerevisiae* (at least according to the recently revised classification). *Wine yeast* will be used here in the narrow sense.

The term *wild yeasts* will not be used. It often leads to confusion because it implies undesirable characteristics. However, a film-forming yeast, which may become a spoilage organism in a white wine, may be a desirable yeast for the secondary sherry fermentation. Because of this ambiguity, the term *natural yeasts* will be used for yeasts that occur naturally on grapes and in musts and that play a role in wine fermentations.

Taxonomists classify yeast species on the basis of several characteristics such as morphology, sexuality, physiology, and genetics (Kockova-Kratochvilova 1981). On the basis of such findings, yeasts are not infrequently reclassified. Kreger van Rij (1984) has recently reclassified numerous species of wine yeasts, often by including them in *Saccharomyces cerevisiae*. Table 4-1 shows the current designation of these yeasts as well as the synonyms generally found in the literature dealing with wine yeasts. This terminology creates some problems because species now included in *S. cerevisiae* may show important differences in the technology of wine making, yet these differences are of no concern to the taxonomist. In this chapter the new revised classification will be used except when reference is made to the work of earlier authors. It should be kept in mind that older synonyms for yeast strains may be used for many years after a reclassification.

A traditional method of distinction between species is based on their patterns of fermentation and assimilation of various sugars. The method is still quite useful. Table 4-2 shows such patterns based on the taxonomy of Kreger van Rij (1984).

Table 4-2. Fermentation and Assimilation of Various Sugars by Yeasts

| | Fermented | | | | | Assimilated | | | | | |
|---|---|---|---|---|---|---|---|---|---|---|---|
| | Glucose | Galactose | Sucrose | Maltose | Lactose | Glucose | Galactose | Sucrose | Maltose | Lactose | Raffinose |
| Saccharomyces cerevisiae | + | v | v | v | – | + | v | v | v | – | v |
| Saccharomyces exiguus | + | + | + | – | – | + | + | + | – | – | +/s |
| Kloeckera apiculata | + | – | – | – | – | + | – | – | – | – | – |
| Torulaspora delbrueckii | + | v | v | v | – | + | v | v | v | – | v |
| Hansenula anomala | + | v | +/w | v | – | + | v | + | + | – | v |
| Zygosaccharomyces rouxii | + | – | +/– | v | – | + | v | v | v | – | – |
| Zygosaccharomyces bailii | + | – | v | – | – | + | v | v | – | – | –(+) |
| Candida stellata | + | – | + | – | – | + | – | + | – | – | + |
| Candida vini | – | – | – | – | – | + | – | – | – | – | – |
| Candida krusei | – | – | – | – | – | + | – | – | – | – | – |
| Pichia membranefaciens | –/w | – | – | – | – | + | – | – | – | – | – |
| Hanseniaspora guilliermondii | + | – | – | – | – | + | – | – | – | – | – |
| Metschnikowia pulcherrima | + | +/– | – | – | – | + | + | + | + | – | – |
| Saccharomyces ludwigii | + | – | + | – | – | + | – | + | – | – | + |
| Schizosaccharomyces pombe | + | – | + | + | – | + | – | + | + | – | + |

Source: Based on Kreger van Rij 1984.
Notes: v = variable; w = weak; s = strong.

## NATURAL YEASTS AND THEIR OCCURRENCE ON GRAPES AND IN MUSTS

Yeast cell counts on grapes differ by several orders of magnitude as reported by various authors. For example, Peynaud and Domercq (1953) report from 1 to 160 cells per milliliter of must of aseptically harvested grapes. Others have reported 50,000 to 100,000 cells per milliliter of must (or per berry). Such differences may be due, at least in part, to differences in grape maturity, climate, viticultural practices, and so on. However, there is a great deal of similarity with respect to the species that are found on grapes and in musts in different parts of the world. *Saccharomyces cerevisiae*, the wine yeast par excellence, is often absent. By far the largest number of yeast cells found on grapes belong to the genera *Kloeckera*, *Hanseniaspora*, and to a lesser extent, *Metschnikowia* and *Candida*. The proportions of these naturally occurring yeasts vary greatly.

There is a considerable increase in cell counts as the grapes are transported and the must is prepared in the winery. The fermentation is initiated by weakly fermenting yeasts, mainly *Kloeckera apiculata*. The

Table 4-3. Succession of Yeast Species During the Fermentation of White Wines

| | Frequency of Occurrence of Species as Percentage of the Number of Wines Tested | | |
|---|---|---|---|
| | Before Fermentation (213 samples) | Half Fermented (208 samples) | End of Fermentation (165 samples) |
| Kloeckera apiculata | 36.1 | 0 | 0 |
| Torulopsis bacillaris | 23.4 | 1.0 | 0 |
| Saccharomyces cerevisiae var. ellipsoideus | 31.4 | 87.5 | 50.3 |
| Saccharomyces oviformis | 0.5 | 2.4 | 38.1 |
| Saccharomyces chevalieri | 0.5 | 3.8 | 3.6 |
| Saccharomyces fructuum | 0.9 | 0 | 0.6 |
| Saccharomyces carlsbergensis | 2.3 | 1.0 | 0 |
| Saccharomyces uvarum | 0.5 | 0 | 0 |
| Saccharomyces steineri | 2.3 | 1.9 | 1.2 |
| Saccharomyces heterogenicus | 0 | 0.5 | 1.8 |
| Saccharomyces acidifaciens | 0.5 | 0.5 | 1.2 |
| Saccharomyces elegans | 0 | 0.5 | 0 |
| Saccharomyces veronae | 0 | 0.5 | 0 |
| Saccharomyces rosei | 0.9 | 0.5 | 2.4 |
| Torulopsis delbrueckii | 0.5 | 0 | 0 |

*Source:* Data from Domercq 1957.

early fermenters are replaced by more alcohol-tolerant yeasts, and during the main part of the fermentation and toward the end the true wine yeasts (*S. cerevisiae*) dominate. Table 4-3 shows the succession of yeast species before, during, and at the end of the fermentation (Domercq 1957). Similar studies have been carried out in Italy (Castelli 1954), Czechoslovakia (Vojketova and Minarik 1985), the United States (Mrak and McClung 1940; Amerine et al. 1982), Germany (Benda 1982), and in most other wine regions of the world. In each case, the genera and species of yeasts were similar, although their proportions varied; and in each case there was a succession of species as described above. Table 4-3 does not permit a quantitative estimate of the actual yeast counts. This estimate is supplied in Figure 4-1, which shows cell counts for four yeast species over a 28-day fermentation of a white wine (Dittrich 1987). Note the rapid and immediate drop of the *Kloeckera apiculata* cell count, the rapid increase and subsequent fast decline of the *Candida* species, as well as the dominance of *Saccharomyces cerevisiae* after the second day of the fermentation.

The occurrence of yeasts on grapes of *Vitis rotundifolia* (muscadine grapes) is quite similar to that reported above, except for the absence of *Kloeckera apiculata*. During the early stages of the fermentation, *Hanseniaspora osmophilia* and *Pichia membranefaciens* were dominant, and after the eighth day, *Saccharomyces cerevisiae* was dominant (Parrish and Carroll 1985). This report as well as the previously mentioned ones have to be considered as

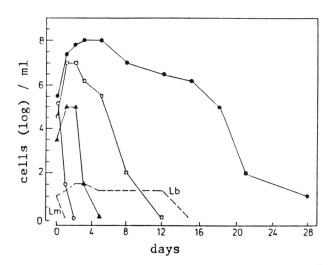

**Figure 4-1.** Occurrence of yeasts and lactic acid bacteria during the spontaneous fermentation of a must of white grapes: ● = *Saccharomyces cerevisiae*; □ = *Candida stellata*; Lb, *Lactobacillus*; ○ = *Kloeckera apiculata*; ▲ = *Candida krusei*; Lm, *Leuconostoc mesenteroides*. (*From Dittrich 1987*)

*qualitative* guides to the occurrence of yeasts on grapes and to the succession of yeast species during the fermentation. Obviously there are tremendous variations from vineyard to vineyard, from season to season, and at different maturities of the grapes.

## NATURAL FERMENTATIONS

Spontaneous or natural fermentation has been practiced for several thousand years. It is still the practice in the majority of less industrialized countries. It is also practiced in many wineries in highly industrialized countries, mainly in wineries in which wine making by traditional methods is highly prized. As previously mentioned, there is a succession of species of yeasts beginning with the genera *Kloekera, Hansenula, Hanseniaspora, Candida,* and *Pichia. Torulaspora delbrueckeii* is often encountered. In the most active stages of the fermentation and toward the end these yeasts are replaced by *Saccharomyces cerevisiae.*

There has been some controversy over the relative merits of spontaneous fermentations with the natural flora of the winery and fermentations carried out with inoculations with selected yeast strains. The latter can be carried out by growing pure cultures of a selected yeast in the winery or with an active dry wine yeast. Traditional wine makers have strongly suggested that spontaneous fermentations affect the quality, particularly the organoleptic quality, of the wine. Specifically it is claimed that such wines lead to a more complex — that is, better — aroma. Some of the extensive literature on this question has been reviewed by Kraus, Reed, and Villettaz (1983/1984). Benda (1982) found that spontaneous fermentations produce a better rounded and more complex aromatic quality. Schmitt, Curshmann, and Koehler (1979) and Schmitt et al. (1984) find a significant preference for wines produced with selected yeasts. Still others find significant differences in the aroma or flavor of wines made by spontaneous fermentations versus wines made with selected yeasts, but cannot find significant preferences for wines made by either method (Bidan and Maugenet 1981; Rossini, Bertoluccioli, and Pasquale 1981). Finally, some authors have found no significant differences at all.

The noted discrepancies are not surprising. Indeed they should be expected. Spontaneous fermentations by their very nature show a great deal of variability, and hence the reproducibility when compared with selected yeast fermentations must be poor. There is an additional complicating factor. It seems to have been tacitly assumed that inoculation with a selected *Saccharomyces cerevisiae* strain leads to the immediate dominance of that strain and to the rapid disappearance of the natural flora. However, Figure 4-2 shows prolonged survival of *Kloeckera apiculata* and *Candida*

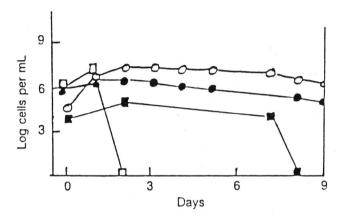

**Figure 4-2.** Growth of yeast during the fermentation of a white wine inoculated with *Saccharomyces cerevisiae*: ○ = *Saccharomyces cerevisiae*; □ = *Candida stellata*; ● = *Kloeckera apiculata*; ■ = *Candida pulcherrima*. (From Heard and Fleet 1985)

*stellata* in a must inoculated with $10^5$ to $10^7$ cells of *S. cerevisiae* per milliliter (Heard and Fleet 1985). Twenty-two years earlier Rankine and Lloyd (1963) had inoculated white musts with 2% of a starter culture of *S. cerevisiae* containing $10^8$ cells per milliliter. The must had an original cell count of $10^4$ yeasts per milliliter. After 7-8 days of vigorous fermentation, 62 of 66 isolates from the white wine could not be distinguished from the inoculated wine yeast, indicating survival of some of the original yeast flora at the end of the fermentation. In these experiments, fermentation was visible 2 days after inoculation when cell counts had reached $10^7$ cells per milliliter. The unyeasted white must started to ferment 4 days later than the inoculated must.

A very brief description of the more important yeast species follows. More detailed descriptions have been provided by Kunkee and Amerine (1970), Benda (1982), Lafon-Lafourcade (1983), Minarik and Navara (1986), and Dittrich (1987).

## DESCRIPTION OF SPECIES

The *Saccharomyces* yeasts are the strongest fermenters. Cells propagate vegetatively by multilateral budding. Sporulation can be induced on acetate agar. Asci carry from one to four spores.

*Saccharomyces cerevisiae* is the species par excellence for wine fermentations. The cells are usually spheroid, ovoid, ellipsoid, or elongated with a cell size often 3-7 × 4-12 μm. The species may produce up to 18-20% ethanol by volume. *S. bayanus* also produces high concentrations of ethanol. It is often a more flocculent yeast. *S. uvarum*, generally used in the production of lager beer, produces lesser concentrations of ethanol. *S. exiguus* forms somewhat smaller, ellipsoid cells, 2.5-5.0 × 3.5-6.5 µm. The yeast tolerates low pH levels, which accounts for its use in sour doughs.

*Torulaspora delbrueckii* cells are globular to oval, 2.5-6.5 × 2.5-7.0 µm. The yeast formerly classified as *Saccharomyces rosei* produces from 6-10% ethanol by volume, which accounts for its absence at the end of the fermentation. It can produce one, two, and rarely up to four ascospores.

*Hansenula anomala* is often found on grapes. Its shape may be spherical or elongate. Cell size is small, 1.9-4.1 × 2.1-6.1 µm. It liberates considerable amounts of ethyl acetate and some acetic acid. Diploid cells produce one to four hat-shaped spores. The yeast produces little alcohol (0.7%) and is a spoilage yeast.

*Zygosaccharomyces bailii* and *Zygosaccharomyces rouxii* are osmotolerant yeasts. The haploid cells are globular, ellipsoid, or ovoid. Cell size of *Z. bailii* is 3.5-7.0 × 5.5-14.0 µm; *Z. rouxii* cells are somewhat smaller. The species produce about 7.5% ethanol by volume. They form ascospores and can be considered haploid forms of *Saccharomyces* species. The alcohol tolerance of the species is greater than their ability to form ethanol. They tolerate high concentrations of sugars, ethanol, $SO_2$, and acetic acid. *Z. bailii*, forms two to four ascospores, and *Z. rouxii* forms one to four.

*Hanseniaspora guilliermondii* are apiculate, oval, or elongate with a cell size of 2.2-5.8 × 4.5-10.2 µm. The species forms little alcohol and its growth is inhibited by 3-4% ethanol by volume. It forms four hat-shaped ascospores.

*Metschnikowia pulcherrima* are globular to ellipsoid and usually 2.5-6 × 4-7 µm. The yeast produces needlelike spores. The yeast produces little alcohol and may produce higher concentrations of glycerol and five carbon alcohols.

*Saccharomycodes ludwigii* are lemon-shaped, sausage-shaped, oval, or elongated. The cells are large, 4-7 × 8-23 µm. The yeast produces four spores per ascus in two groups of two. The yeasts may produce 8% ethanol by volume and tolerate higher concentrations of ethanol and $SO_2$. Consequently, they may occur as spoilage organisms in highly sulfited wines.

*Schizosaccharomyces pombe* are usually cylindrical. Vegetative growth occurs by fission after a wall has been formed across the middle of the elongated cells. The cell size is 3-4 × 6-16 µm. The yeast may produce as much ethanol as *Saccharomyces cerevisiae*. The yeast ferments malic acid to ethanol and has been used for the reduction of acidity in highly acidic musts. Ascus formation follows conjugation of two vegetative cells, and four to eight spores are formed per ascus.

*Pichia membranefaciens* are ovoid or elongate. The cell size is 1.8-4.5 ×

2.4-17 μm. The cells may form long chains or clusters. It is a film-forming yeast. One to four spores are formed per ascus. The yeast produces very little ethanol (0.2% by volume), but it readily survives higher alcohol concentrations. The yeast grows readily on the surface of young wines where it forms a continuous film.

Kloeckera apiculata are apiculate and sometimes round to oval and elongate. The cell size is quite small, 1.5-5.0 × 2.5-11.5 μm. The yeast is the imperfect form of Hanseniaspora uvarum. It produces 5-6% ethanol but tolerates higher concentrations. It occurs abundantly in musts and generally dominates the very early phase of the wine fermentation. Kloeckera and Hanseniaspora species produce relatively large concentrations of acetic acid, ethyl acetate, mannitol, sorbitol, and ribitol.

Candida stellata are ovoid cells that are rather small, 2-3 × 4-6 μm. They are often arranged in starlike clusters; hence the name stellata. Candida stellata is an asporogenous yeast. It is a weaker fermenter than Saccharomyces cerevisiae (up to 10% ethanol by volume). The yeast is osmophilic and tolerates fairly high temperatures (30°-35°C). It ferments fructose faster than glucose.

Candida vini or Mycoderma vini are medium-size anascosporogenous cells, 2.9-4.3 × 5.8-10.1 μm, which are ovoid or cylindrical. They occur singly, in pairs, chains, and clusters. The yeast forms a pseudomycelium. Candida vini as well as Candida krusei form little ethanol. The yeasts grow as film formers on the surface of low-alcohol wines. The species can grow on media containing ethanol as the sole carbon source.

The very brief descriptions of the above-mentioned species do not mean that all strains belonging to a particular species behave similarly. Not all strains of Saccharomyces bayanus (now S. cerevisiae) are flocculent; not all strains of S. cerevisiae produce low levels of volatile acidity, and so forth.

## SELECTED PURE CULTURE YEASTS AND ACTIVE DRY WINE YEASTS

The use of selected yeast cultures for inoculation of brewer's worts and for bread doughs was widely practiced during the nineteenth century. These substrates do not ferment spontaneously; the pH is higher (about 5.2), and hence spoilage organisms grow more readily. In the wine industry inoculation with selected strains of wine yeasts was not practiced to any extent until the twentieth century. In the 1950s, the practice of using selected wine yeasts for inoculation was very common in the United States, Australia, New Zealand, and South Africa. It was also practiced in Europe, although the practice was not always admitted. In the 1960s, active dry wine yeast (WADY) was introduced in the United States and its use spread quickly to

Australia, New Zealand, and South Africa. Since the late 1970s, WADY has been used in Europe. Its use in Germany, Italy, and France is now common, and it is being introduced in other wine-producing countries of Europe and South America (Reed and Nagodawithana 1988).

Pure yeast cultures have been isolated by wine research institutes and are kept with frequent transfers in their collections. The number of strains so preserved may number several hundred each in several institutes. There may be duplications within and between institutes. The origin of the strains is often not known. For instance, one of the most widely used strains (Montrachet, University of California no. 422) was brought to the United States by Professor W. Cruess in the 1930s, presumably from the Montrachet region in France. But there is no specific information on the wine maker or the scientist who first isolated and cultured the strain.

Yeast cultures are available from research institutes in the form of agar slants, and from many commercial organizations in the form of slants or liquid cultures. They are usually designated as Epernay, California Champagne, Prise de Mousse, Steinberg, Beaujolais, Tokay, and so on. Such names are not particularly informative, since different strains may be marketed under the same designation. It is advisable to identify the strains by the name of the institute and the number of the institute's collection. Pure yeast cultures may also be isolated by wine makers and kept as proprietary strains.

Several wine yeast strains had been grown by Ulbrich and Saller (1951) under both aerobic and anaerobic conditions. They found no difference in the fermenting power of the yeasts. In California, Castor (1953) proposed the idea of producing wine yeasts by aerobic methods commonly used for the production of baker's yeast. At the same time, in Ontario, Adams (1953/1954) produced wine yeast aerobically in 1-liter fermenters and collected the yeast in the form of wet press cakes. The yeast could be stored for several weeks at $-29°C$. Production of wine yeast in continuous culture in 10-liter fermenters has also been described (Fiechter 1963). None of these attempts led to actual commercial uses.

In the early 1960s, the U.S. wine industry became interested in a commercial source of bulk wine yeasts. Several strains were successfully produced at that time and used in the production of table wines (Thoukis, Reed, and Bouthilet 1963). The methods used for the large-scale commercial production of wine yeasts need not be described here, because the production methods follow closely those used for the aerobic, fed-batch production of baker's yeast (Reed 1974; Kraus, Reed, and Villettaz 1983/1984). However, wine yeasts are generally grown in the presence of bisulfite to acclimatize the yeasts to the presence of $SO_2$ in musts. For distribution and commercial use the yeasts are dried to WADY with a moisture content of 5-7.5%. WADY requires protective packaging under vacuum or in a nitrogen or carbon dioxide atmosphere. Such yeasts are stable for at least one year unless they have been exposed to air. Prior to use, WADY must be rehydrated in water or grape juice at a temperature of about 40°C (Kraus, Scopp, and Chen 1981).

The suitability of WADYs is, of course, a function of the strain. More than ten commercially available strains have been listed by Kraus, Reed, and Villettaz (1983/1984). Quality depends further on viability, fermentation activity, and absence of gross contamination with other microorganisms. Radler, Dietrich, and Schonig (1985) reported the presence of 11 to 39 × $10^9$ yeast cells (colony-forming units) per gram for three brands of commercial WADYs.

The fermentation activity can be determined by well-known gasometric methods. For purposes of quality control a short time method (2.5 hours) with a large yeast inoculum is preferred. Longer fermentation times with small inocula that imitate commercial fermentations do not reflect the innate gassing power of the yeast but a combination of gassing power and cell multiplication. The gassing power of three yeast strains determined by the short time method of Reed and Chen (1978) is shown in Table 4-4. The table also shows the inhibiting effect of increased $SO_2$ levels for these yeasts.

Results are shown in mMole $CO_2$ per hour per gram of yeast solids. In commercial fermentations such values are usually expressed as decrease of °Brix per hour because the amount of yeast solids is not known. Such values reflect both the gassing power of the yeast and the growth of the

Table 4-4. Determination of Fermentation Activities of Active Dry Wine Yeasts by a Gasometric Method (activity expressed in m Mole $CO_2$/g yeast solids/hr)

| Strain | Free $SO_2$ Concentration (ppm)* | | | |
|---|---|---|---|---|
| | 0 | 50 | 100 | 125 |
| Montrachet | 11.2 | 10.8 | 10.2 | |
| | 12.3 | 12.0 | 11.0 | 7.9 |
| | 11.0 | | 9.2 | |
| Champagne | 8.8 | | 5.6 | |
| Sherry | 8.6 | | 3.1 | |

Source: Data from Reed and Chen 1978.
*Containing 0 ml, 1.50 ml, 3.0 ml, and 3.75 ml of the stock sodium bisulfite solution (1,000 ppm $SO_2$) in each reaction vessel. The total substrate volume in the vessel was made up to 30 ml with distilled water.

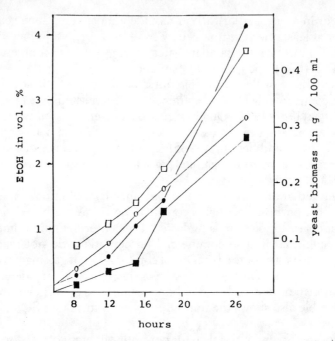

**Figure 4-3.** Yeast multiplication and ethanol formation of two active dry wine yeasts during the start of a wine fermentation: ○ = ethanol formation of yeast A; ● = ethanol formation of yeast B; □ = biomass formation of yeast A; ■ = biomass formation of yeast B. (*After Malik, Minarik, and Kutlik 1984*)

yeast throughout the fermentation period. Figure 4-3 shows ethanol production and yeast biomass production in a grape must medium. It is apparent that the rate of fermentation increases greatly after 15-18 hours (Malik, Minarik, and Kutlik 1984). It can also be seen that the increase is due to an increase in yeast biomass. Calculations from the authors' data show a fermentation rate of the yeast in the range of 10-20 mM sugar per hour per gram of yeast solids.

The level of contaminants in commercial WADYs varies greatly. Radler, Dietrich, and Schonig (1985) reported the presence of less than $10^4$ to $6 \times 10^5$ non-*Saccharomyces* cells per gram for three commercial WADYs, and a total aerobic bacterial count of 0.3 to $15 \times 10^6$ per gram. For some uses a low level of malolactic organisms is essential.

## PREPARATION OF STARTER CULTURES

Natural wine fermentations require no preparation except for crushing of the grapes. Fermentations usually start slowly at the beginning of the season and faster thereafter because of the increase in yeast counts in

PREPARATION OF STARTER CULTURES    163

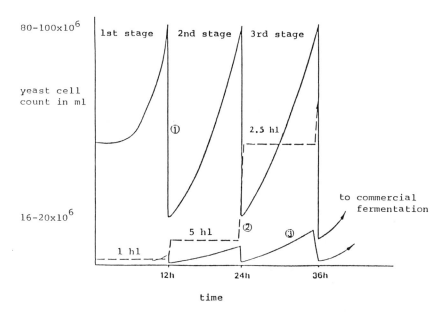

**Figure 4-4.**   Preparation of yeast inoculum in three stages (schematic):

(1) ——— wine yeast cells/ml
(2) ——— wild yeast cells/ml
(3) ----- relative volume of ferment

(*Courtesy of Prof. Rouanet*)

winery equipment. For the *maceration carbonique* the grapes may merely be covered with a blanket of $CO_2$.

The use of selected pure culture yeasts requires the growth of the yeast through successively larger vessels. Starting with a pure culture slant the yeast is inoculated into grape must in laboratory flasks. The content is then transferred into larger and larger vessels, generally with a 5-15-fold volume increase. One such procedure is illustrated in Figure 4-4. The three-step procedure shows 5-fold scale-ups of the starter as well as the numbers of yeast cells and contaminating cells throughout the 36-hour growth period. Figure 4-5 shows a schematic drawing of a typical installation consisting of two 50-gallon, two 350-gallon, and one 5,340-gallon tanks. In such anaerobic installations the time required between transfers is usually between 24 and 48 hours (De Soto 1955).

It is desirable to aerate the must in order to stimulate yeast growth. An installation for the aerobic propagation of selected yeast cultures in a 2,000-gallon propagation tank has been described (Knappstein and Rankine 1970). The methods for yeast starter production from slants begin with a pure yeast culture. However, other microorganisms can and do enter the

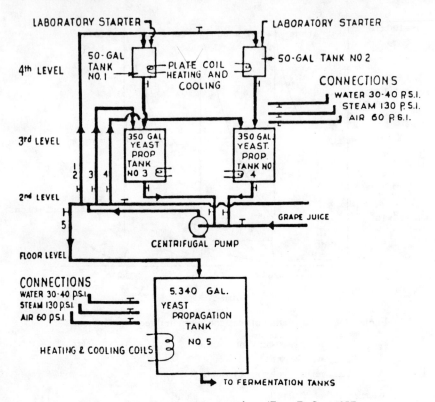

**Figure 4-5.** Yeast propagation plant. (*From De Soto 1955*)

propagation throughout the many transfers. The use of selected pure cultures offers the opportunity of using a very large number of different strains as they are available from wine research institutes. It has the disadvantage of requiring a considerable expenditure of effort and time during the crushing season, as well as during champagne fermentations throughout the year.

WADY requires rehydration in warm water (40°C) or grape juice. The rate of inoculation is usually between 5 and 20 g/hl. This rate corresponds to a starting yeast population of $1.5\text{-}6 \times 10^6$ cells-ml. The lower levels of inoculation are usually quite satisfactory, but delay the onset of fermentation.

The identification of wine yeast species is often important to avoid contaminants during fermentation and spoilage yeasts in the finished wine. Several attempts have been made to shorten the time period of conventional methods of identification. Subden, Cornell, and Noble (1980) based their method of identification on a 72-hour growth response test with a variety of media components. Tredoux et al. (1987) based their method on the determination of the methyl esters of fatty acids (C14, C16, C18) extracted from yeasts. It is frequently more important to distinguish between strains of yeasts within a species. Van Vuuren and Van der Meer (1987) employed gel electrophoresis of the total soluble proteins of *S. cerevisiae* yeasts in a so-called fingerprint method. They were able to distinguish closely related wine yeast strains of *S. cerevisiae*. A method for the identification of spoilage yeasts in bottled wine based on growth response in pantothenate-free agar and lysine agar plates has been reported by Rodriguez (1987). All of the mentioned reports have been tested with a large number of wine yeast species and strains.

## BIOCHEMISTRY OF WINE FERMENTATION

The formation of ethanol and carbon dioxide from fermentable sugars, glucose, and fructose by the well-known glycolytic pathway need not be described again. The overall reaction leads theoretically to 51.1% ethanol and 48.9% carbon dioxide as a percentage of the weight of monosaccharides. A number of by-products of the fermentation—glycerol, acetaldehyde, organic acids—lower the yield of ethanol. In addition, some ethanol is lost in the evolving $CO_2$ gas. At a temperature of 20°C this loss may amount to 0.50% ethanol. Some sugar is used for the growth and maintenance of the yeast. Hence, in winery practice the yield of ethanol is about 47% of the weight of fermentable sugars.

In the wine industry ethanol concentrations are usually expressed as volume %; 10% ethanol by volume correspond to about 8% by weight. Normal ethanol concentrations in wines may vary from 10% to 12.5% by volume depending on the original sugar concentration of the must. In northern climates or with immature grapes the sugar concentration may be too low, and the addition of sucrose to the must is required. The other major product of the fermentation, $CO_2$, is lost to the atmosphere except in the secondary champagne fermentation. Recovery of the gas is rarely economical. The volume of evolved $CO_2$ is considerable: 1 liter of must containing 180 grams of sugar (1 mole) produces about 45 liters of $CO_2$ (2 moles).

The overall relationship of concentration of sugar and ethanol and the number of yeast cells is shown in Figure 4-6. This graph is just one of the several hundred in the literature showing the same general picture. In contrast to Figure 4-3, which shows ethanol production during the first day, Figure 4-6 shows the first five days of a wine fermentation. It covers the most active fermentation period. It permits some insight into the role of yeast growth that precedes the period of rapid ethanol formation by almost a day.

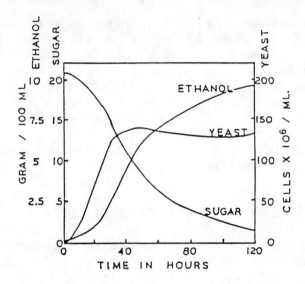

**Figure 4-6.** Number of yeast cells and concentration of sugars and ethanol during a wine fermentation. (*From Wick 1968*)

## Sugars and Fermentation Rate

The fermentable sugars of grape must are glucose and fructose. The concentration of the sugars in ripe grapes usually expressed in Brix varies from 20-24° Brix. The fructose/glucose ratio varies greatly. It often approaches 1:1 at full maturity (Amerine and Thoukis 1958). Almost all strains of *Saccharomyces cerevisiae* are glucophilic, that is, they ferment glucose at a slightly faster rate than fructose. Residual (unfermented) sugar in wine contains a higher percentage of fructose. Some sauterne yeasts are fructophilic (Sols 1956). Many of the osmotolerant species of yeasts are fructophilic. Trehalose, a sugar characteristically present in yeast cells, has been reported in wine (Bertrand, Dubernet, and Ribéreau-Gayon 1975).

The addition of sugar (sucrose) to high-acidic/low-sugar musts merely increases the concentrations of fructose and glucose because of the rapid conversion of the disaccharide by yeast invertase. This process of amelioration is sometimes practiced in the eastern wine regions of the United States. It is not legal in California, where the sugar concentration of musts is sufficiently high to permit production of table wines with more than 11% ethanol by volume without the addition of cane or beet sugar.

The rate of fermentation of *Saccharomyces cerevisiae* wine yeasts is in the range of 15-30 mM $CO_2$ per gram of yeast solids per hour. This figure is reasonable for initial rates at pH values above 3.5 and at 25°-30°C; and, of course, in the absence of inhibitors. There are some differences between strains, but these are rarely decisive for performance of the strain. The industry sometimes refers to a yeast strain as a "strong" fermenter, but what is meant is generally a good rate of fermentation under adverse conditions or completeness of the wine fermentation.

An accurate assessment of the rate of fermentation in relation to yeast solids is usually not determined because the amount of yeast solids is unknown. Even if the amount of yeast solids is known at the start of the fermentation, it is not known after yeast growth has occurred. Cell counts may, of course, be performed, but cell size varies, and hence, an estimate of yeast solids is difficult. With an original inoculum of $2\text{-}6 \times 10^6$ per milliliter, the final cell count may reach $50\text{-}150 \times 10^6$ per milliliter. In industrial fermentations the rate is usually expressed as the drop in Brix per hour, a value that depends not only on the percent of yeast solids or the number of cells but on all other environmental variables. It must now be shown how these variables affect the rate of yeast fermentation.

## Temperature

The rate of fermentation increases with increasing temperature up to about 30° to 33°C. Table 4-5 shows such changes in the range of 10° to 32°C for white California wines (Ough 1964). The rate triples approximately between 10° and 20°C and it doubles approximately between 20° and 30°C. These data span the range of usual temperatures for white wines (15°-22°C) and red wines (25°-30°C).

Table 4-5. Rate of Fermentation in 10 White California Wines

| | | Loss of °Brix per hr | | | | |
|---|---|---|---|---|---|---|
| | | 10°C | 15.5°C | 21.1°C | 26.6°C | 32.1°C |
| Rate of fermentation 20° to 0° Brix | lowest rate | 0.034 | 0.071 | 0.128 | 0.188 | 0.219 |
| | highest rate | 0.064 | 0.192 | 0.318 | 0.487 | 0.540 |
| | average rate of 10 wines | 0.089 | 0.182 | 0.270 | 0.374 | 0.455 |
| Maximum rate of fermentation | lowest rate | 0.050 | 0.090 | 0.145 | 0.215 | 0.289 |
| | highest rate | 0.086 | 0.235 | 0.422 | 0.738 | 0.739 |
| | average rate of 10 wines | 0.130 | 0.247 | 0.371 | 0.519 | 0.642 |

Source: Data from Ough 1964.

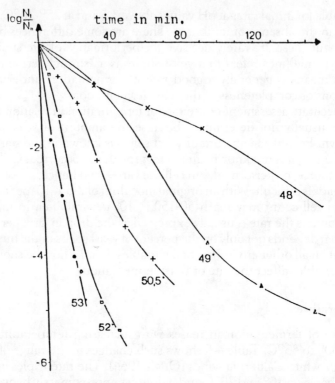

**Figure 4-7.** Survival of a strain of *Saccharomyces cerevisiae* at different temperatures: $N_o$, original yeast cell count (about $2\text{-}10 \times 10^6$/ml); $N_t$, cell counts at times $t$/ml. (*From Barillère, Bidan, and Dubois 1983*)

The temperature of the fermentation also affects ethanol yields, by-products of the fermentation, and aroma of the wine. The slightly lower yield of ethanol at the higher temperatures is at least partly due to a greater ethanol loss in the escaping $CO_2$. At lower temperatures there is less yeast growth. The aroma of white wines fermented at lower temperatures is said to be cleaner (Bisson, Daulny, and Bertrand 1980). Some wineries have carried out white wine fermentations at 8° to 10°C, but the rate of fermentation is extremely slow.

The anaerobic fermentation of sugars by yeast is an exothermic reaction that liberates about 23 to 24 kCal per mole of fermented glucose. Similar increases in temperature are also experienced in fermenting bread doughs and beer worts. The increase in temperature during a fermentation can, therefore, be calculated but requires adjustment for heat losses from the

BIOCHEMISTRY OF WINE FERMENTATION 169

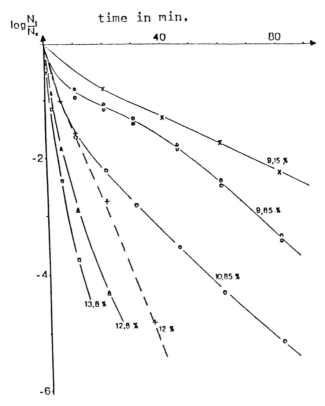

**Figure 4-8.** Survival of a strain of *Saccharomyces cerevisiae* at 45°C and at varying ethanol concentrations: $N_o$, original yeast cell count (about 2-10 × $10^6$-ml; $N_t$, cell counts at times $t$/ml. (*From Barillère, Bidan, and Dubois 1983*)

walls of the tanks (greatly dependent on volume) and for losses in escaping $CO_2$. For large-volume fermentations, for fermentations in warm climates, and for very rapid fermentations cooling is obviously required. Boulton (1979) has dealt very well with calculations for cooling requirements.

Increases of the temperature above 30° to 35°C will ultimately kill yeast cells, particularly at higher ethanol concentrations, and lead to "stuck" fermentations. Barillère, Bidan, and Dubois (1983) have reviewed the earlier literature on thermal resistance of wine yeasts and reported their results. Figure 4-7 shows survival curves for a wine strain of *S. cerevisiae* at temperatures between 48° and 53°C. Figure 4-8 shows the considerable lowering of thermal resistance in the presence of higher concentrations of ethanol. The values were obtained at the relatively low temperature of

45°C. The effect of $SO_2$ in lowering thermal resistance is equally striking. At 49°C and in the absence of $SO_2$ the time required for a 10-fold reduction in viability was about 28 minutes. In the presence of 210 ppm of $SO_2$ a thousandfold reduction in viability occurred within about 4 minutes.

## Ethanol

Ethanol inhibits yeast viability, the specific yeast growth rate, and the specific rate of fermentation. But different inhibition constants apply for each of these effects. Ethanol inhibition is of crucial importance to the wine industry. However, our knowledge of the phenomenon is confused because of the number of dependent and independent variables such as sugar concentration, pH, and temperature that affect this inhibition. For instance, yeast growth stopped at 11% by weight of ethanol in the temperature range of 15°-28°C, but at 8°C and at 32°C it already stopped at 8% ethanol by volume (Sa-Correia and Van Uden 1983). Of course, yeast growth is greatly inhibited at lower ethanol concentrations, and little growth seems to occur above 9% ethanol. Stated simply, yeast growth and viability are inhibited at lower ethanol concentrations than fermentation.

There are considerable differences between yeast species and between strains within a species in the tolerance to ethanol. The reasons for variability of ethanol tolerance have been the subject of many investigators. Nagodawithana and Steinkraus (1976) observed the maximum rate of ethanol production per cell unit at the early stages of the fermentation and under conditions of rapid fermentation. Under these conditions the concentrations of ethanol within the cells were higher than in the environment, which suggests that accumulation of ethanol in the cells was faster than diffusion out of the cells. That is, ethanol tolerance depends on the ability of yeast cells to export ethanol into the surrounding medium, a process known to depend greatly on cell membrane composition and fluidity. The finding that the intracellular concentration of ethanol is greater than the extracellular concentration by a factor of 2 to 20 has been confirmed by many authors (as listed by D'Amore and Stewart 1987), but it has also been contested, for instance, by Guijarro and Lagunas' (1984) publication "*S. cerevisiae* Does Not Accumulate Ethanol Against a Concentration Gradient." Obviously, it depends also on the rate of ethanol formation within the cell whether the internal concentration will exceed concentration in the medium, which may explain the contrary findings.

Nagodawithana, Whitt, and Cutaia (1977) have identified the noncompetitive feedback inhibition of hexokinase in the presence of ethanol. At a critical ethanol concentration an inactive hexokinase-ethanol complex was formed that stops the reaction of glucose to glucose-6-phosphate. Complex formation is reversible, and upon dilution of the reaction mixture the active hexokinase is again released. This hypothesis gains credence from the fact that under rapid fermentation conditions and at high ethanol concentrations a simple dilution will restore fermentation activity (Nagodawithana and Cuzner 1979).

Pascual et al. (1988) investigated the effect of ethanol on specific enzymes and found the strongest inhibition for hexokinase and pyruvate kinase. Novak et al. (1981) suggest that permeability of ethanol from the outside to the inside of the cell is very low, which explains the fact that ethanol added to the medium is less inhibitory than ethanol produced by the cells. Regulation of ethanol formation under rapid fermentation conditions has been considered only briefly because it requires future elucidation.

## Sterols

In the late 1960s, Bréchot found that grapes crushed without removal of the grape stems fermented faster than crushed destemmed grapes (Bréchot et al. 1971). He found the cause in the presence of ergosterol on the grape stems. Most likely he was aided in this recognition by the earlier pioneering work of Andreasen and Stier (1953) on the requirement of ergosterol for yeast growth in a refined, unaerobic medium.

The role of lipids and specifically the role of sterols in yeast cell wall membranes as they affect solute transport and ethanol tolerance have been reported by Thomas and Rose (1979) and Prasad and Rose (1986). These effects on wine yeasts have been reviewed by Ribéreau-Gayon in 1985. If one inoculates a must with yeasts grown under aerobic conditions, such as commercial active dry yeasts, the addition of sterols to the must does not increase yeast growth or fermenting power during the first 10 days of the fermentation. But in the declining phase the yeast cells retain their viability better and ferment more actively (Lafon-Lafourcade 1983). This effect has been called quite appropriately *facteur de survie* (survival factor) by Lafon-Lafourcade et al. (1977).

Figure 4-9 shows the effect of ergosterol addition on the live cell count at two temperatures. Viability is improved by a factor of about 8 at 25°C after 20 days of fermentation. And Table 4-6 shows the sterol concentrations in yeasts during wine fermentations in strict anaerobicity and with constant (light) aeration (Lafon-Lafourcade et al. 1977). This table demonstrates the effect of aeration on retention of a fair concentration of sterols. Under these circumstances the addition of 25 milligrams ergosterol per liter has no effect on the sterol concentration or fermentation rate. In contrast,

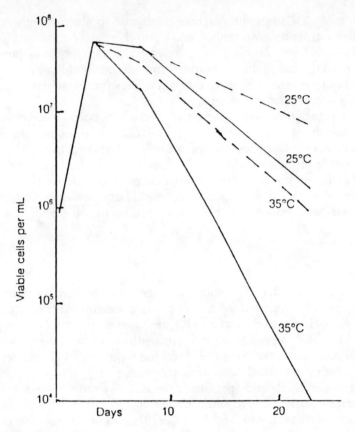

**Figure 4-9.** Effect of ergosterol addition on viable yeast cell counts during a fermentation of grape must at 25°C and 35°C. (*From Ribéreau-Gayon 1985*).

under anaerobic conditions the sterol content of the yeast drops rapidly and fermentation activity lags. The addition of ergosterol makes a contribution to cell survival, fermentation rate, and sterol concentration of the yeast, but does not match the performance under aerobic conditions.

Intermittent aeration for as little as 5 minutes as it is often practiced in the wine industry induces rapid synthesis of sterols. Actual data will not be reproduced here because the literature does not define aeration in terms of actual oxygen transfer. Traverso-Rueda and Kunkee (1982) have dealt with the problems of oxidation during the aeration of musts or fermenting wines. Commercial active dry yeasts are produced under highly aerobic conditions in a fed-batch fermentation. Pichova et al. (1985) have described ergosterol synthesis under such conditions.

Table 4-6. Variation in Sterol Concentrations in Yeasts During the Alcoholic Fermentation of Grape Juice

|  | Constant Aeration | | Strict Anaerobiosis | |
| --- | --- | --- | --- | --- |
|  | C | +E | C | +E |
| Sugar fermented (g/L) | | | | |
| 2nd day | 30 | 27 | 37 | 36 |
| 5th day | 116 | 101 | 113 | 111 |
| 9th day | 187 | 175 | 164 | 169 |
| End of fermentation | 256 | 234 | 170 | 199 |
| Sterols (% of dry weight) | | | | |
| 2nd day | 2.70 | 2.80 | 1.60 | 1.40 |
| 5th day | 1.90 | 1.90 | 0.60 | 1.10 |
| 9th day | 1.20 | 1.10 | 0.40 | 1.00 |
| End of fermentation | 1.00 | 0.80 | 0.30 | 0.60 |
| Viable cells ($10^6$/ml) | | | | |
| 2nd day | | | 22 | 20 |
| 5th day | | | 13 | 10 |
| 9th day | | | 5 | 7 |
| End of fermentation | | | 0.05 | 0.5 |

*Source:* Data from Lafon-Lafourcade et al. 1977.
*Notes:* C = Control must; +E = plus 25 mg ergosterol/L.
Must after sugar enrichment: 250 g/L; yeast: active dry *S. cerevisiae*; initial sterol conc.: 1.5%; initial cell population: $2.2 \times 10^6$/ml.

## Alcohol Yield

The yield of ethanol in wine fermentations does not reach the theoretical value that can be calculated from the well-known Gay-Lussac equation that predicts the formation of 2 moles of ethanol and 2 moles of $CO_2$ from each mole of glucose; that is, an ethanol yield of 51.1% by weight based on fermented glucose. This yield is due to the formation of by-products of the alcoholic fermentation and to the assimilation of sugar by the yeast. Ethanol yield is affected by several variables including temperature, pH, amount of yeast inoculum, and presence or absence of $O_2$.

Laranidis and Lafon-Lafourcade (1982/1983) determined ethanol yields by fermenting a muscat must (sugar level adjusted with added sucrose) at pH 3.2 with a yeast inoculum of 100 mg per 100 ml. Table 4-7 shows the results. Ethanol yields were lower at 30°C than at 19°C, and lower under aerobic conditions and at higher initial sugar concentrations. Less sugar was fermented under anaerobic conditions, particularly at high sugar

Table 4-7.  Alcohol Yield as a Function of Sugar Concentration, Aerobiosis, and Temperature

|  |  | Sugar | | Ethanol | | Ethanol Yield |
|---|---|---|---|---|---|---|
|  |  | initial, g/L | fermented, g/L | % by vol. | g/L | % (w/w) |
| 19°C | an  | 156.3 | 151.4 | 9.75  | 77.1  | 50.9 |
|      | aer | 156.3 | 155.1 | 9.43  | 74.6  | 48.0 |
|      | an  | 198.8 | 185.1 | 11.7  | 92.5  | 46.6 |
|      | aer | 198.8 | 197.9 | 11.79 | 93.3  | 47.1 |
|      | an  | 233.2 | 173.9 | 10.96 | 86.7  | 49.8 |
|      | aer | 233.2 | 231.9 | 13.75 | 108   | 46.9 |
|      | an  | 248.2 | 181.1 | 10.52 | 83.2  | 45.9 |
|      | aer | 248.2 | 247.2 | 14.16 | 112   | 45.3 |
| 30°C | an  | 156.3 | 113.0 | 6.98  | 55.2  | 48.9 |
|      | aer | 156.3 | 143.6 | 8.11  | 64.2  | 44.6 |
|      | an  | 198.8 | 107.8 | 6.42  | 50.8  | 47.1 |
|      | aer | 198.8 | 125.4 | 7.0   | 55.4  | 44.2 |
|      | an  | 233.2 | 113.2 | 6.73  | 53.2  | 47.0 |
|      | aer | 233.2 | 131.2 | 7.14  | 56.5  | 43.0 |
|      | an  | 248.2 | 127.8 | 7.0   | 55.4  | 43.3 |
|      | aer | 248.2 | 142.2 | 7.5   | 59.3  | 41.7 |

Source: Calculated from Lanaridis and Lafon-Lafourcade 1982/1983.
Notes: Fermented sugar = initial sugar − residual sugar; an = anaerobic; aer = aerobic.

concentrations, but ethanol yields based on fermented sugar were generally higher.

The effect of aeration in improving survival of yeast cells during the later stages of the fermentation has already been mentioned, and intermittent aeration is sometimes practiced in wineries. In winery practice, 47-47.5 grams ethanol are usually obtained for each 100 grams of fermented sugar. Yields are often expressed as ethanol by volume per °Brix. Such values are generally in the range of 0.57-0.60 and vary somewhat from region to region and with the grape variety (Jones and Ough 1985).

## pH

The pH of grape must of California grapes at maturity normally varies between 3.1 and 3.9. The rate of fermentation is greater at the higher pH values. For instance, musts with an initial Brix of 25 and at a temperature of 21°C had the following rates of fermentation (expressed as drop of Brix per hour): 0.12 at pH 3.0; 0.16 at pH 3.5; and 0.19 at pH 4. At the higher pH values yeast fermentation starts sooner and yeast growth rates are faster (Ough 1966a, 1966b). Baker's yeast strains of S. cerevisiae ferment only very slowly at pH values below 4.2. They are not suitable for wine fermentations except for less acidic fruit juices such as apple juice, and their use for wine fermentations in general is not advisable. The effect of pH on the dissociation of sulfurous acid will be discussed below.

## Osmotic Pressure and $CO_2$

Yeast growth and fermentation activity of industrial strains are inhibited by increasing osmotic pressures. In bread doughs the source of osmotic pressure is first sodium chloride and second the sugar of sweet goods. In wine making the primary source of high osmotic pressure is a high sugar concentration, essentially glucose and fructose. At very high sugar concentrations, as in jams or honey, yeast fermentation does not occur or occurs only rarely with highly osmotolerant yeasts such as *Saccharomyces mellis*.

Even within a fairly normal range of sugar concentrations the effect of increased osmotic pressure on growth and fermentation activity of the yeast can be demonstrated, as shown in Table 4-8. However, the effect of some variables cannot always be clearly quantified. For instance, at increasing sugar concentrations the final ethanol concentration may account for a good deal of the inhibition. As is well known and expected, cell volume of yeast cells shrinks with increasing osmotic pressure. For a wine yeast strain

Table 4-8. Effect of the Initial Sugar Concentration on Yeast Growth and Fermentation of Grape Must

| | Initial Sugar Concentration (w/v) | | |
| --- | --- | --- | --- |
| | 20% | 23.5% | 26% |
| Yeast growth | | | |
| Lag phase, hr | 20 | 28 | 40 |
| Max. viable cell counts (cells/ml) | $8.1 \times 10^7$ | $7.8 \times 10^7$ | $6.9 \times 10^7$ |
| Percent dead cells during lag phase | 9 | 20 | 40 |
| Fermentation | | | |
| Duration, days | 10 | 12 | 16 |
| Residual sugar in g/L | 3 | 6 | 28 |

Source: Data from Nishino, Miyazaki, and Tohio 1985.

of *S. cerevisiae*, Nishino, Miyazaki, and Tohio (1985) reported a shrinkage from a cell volume of 65 $\mu$m$^3$ to 45-50 $\mu$m$^3$.

A high concentration of $CO_2$ and a high pressure occurs in the secondary champagne fermentation. It is not clear at all whether the inhibition of yeast growth and fermentation at higher $CO_2$ concentrations is due to the increased osmotic pressure of the wine, to the exertion of pressure per se, or to a specific effect of the $CO_2$ molecule. Kunkee and Ough (1966) reported that yeast growth ceased at 15 g/L of $CO_2$ (corresponding to a pressure of 7.2 atmospheres at 15°C). Fermentation activity continued to a concentration of 30/gL, and yeast cells were killed at concentrations exceeding this amount. Fifteen g of $CO_2$ per liter has a 0.34 molar concentration, and hence it should exert the same osmotic pressure as a 6.1% glucose solution. This $CO_2$ concentration can be expected in a secondary champagne fermentation in which the pressure may well increase by about 5 to 6 atmospheres.

Haubs, Muller-Spath, and Loescher (1974) reported on the effect of even lower concentrations of 0.6-0.8 gram $CO_2$ per liter on growth inhibition. In still wines about 0.1-0.5 gram $CO_2$ are dissolved in 1 liter at atmospheric pressure. An indication of the atmospheric pressure is not informative unless the temperature is also given. Twelve grams per liter $CO_2$ give a pressure of 0.4 atm. at 0°C, 4.8 atm. at 5°C, 5.8 atm. at 10°C, 6.6 atm. at 15°C, and 7.5 atm. at 20°C.

The fermentation rate of white wines can be controlled by maintaining sufficient $CO_2$ pressure in bulk fermentation tanks. This procedure slows the rate of fermentation particularly at lower pH values. Ough and Berg (1969) concluded that the ability to regulate the fermentation rate does not warrant the cost of pressure tanks and the prolonged fermentation period.

## Sulfur Compounds

The use of $SO_2$ in wine making goes back to antiquity when sulfur was burned in wine barrels. Today $SO_2$ is still used almost universally in wine making where it exerts chemical and biological functions. Chemically it prevents oxidation that leads to browning reactions. It combines with several compounds, principally acetaldehyde, pyruvic acid, and 2-ketoglutaric acid, by forming sulfonates.

Biologically it limits the development of acetic acid bacteria and lactic acid bacteria. It also inhibits the growth of some yeasts, for example, *Kloeckera apiculata* and *Metschnikowia pulcherrima*, which may produce larger concentrations of volatile acids. Other "wild" yeasts, such as *Saccharomyces ludwigii* and *Saccharomyces bailii*, show a greater tolerance to $SO_2$.

At present, regulatory agencies try to reduce the use of $SO_2$ by obligatory labeling of wines because of the sensitivity of some consumers to $SO_2$. Obviously, it is not easy to replace $SO_2$ in wine making because of the diverse functions of the compound. In solution there is a pH-dependent equilibrium between three forms of the compound, $SO_2$, $HSO_3$, and $SO_3$. Molecular $SO_2$ is responsible for the toxic effect of the compound, and lower pH levels shift the equilibrium to this compound. *Saccharomyces cerevisiae* can be acclimatized to the presence of $SO_2$, at least within limits.

The addition of $SO_2$ at the crusher or to the must delays the onset of the fermentation, particularly at higher levels. Traditionally the level of addition of $SO_2$ is between 50 and 125 mg/L and will depend on many factors such as the maturity of the grapes, the pH, and the desired function of the compound. Table 4-4 gives some indication of the effect of added free $SO_2$ on the fermentation activity of an acclimatized active dry wine yeast. Wine yeasts for use in starter cultures are fermented in the presence of 50-200 mg/L $SO_2$ in order to acclimatize them to the microbicide. During the production of active dry wine yeasts $SO_2$ may be added directly to the molasses feed for the same reason (Reed 1974).

Wine yeasts produce various sulfur compounds during the fermentation, mainly $SO_2$ and $H_2S$. The sulfur may be derived from the inorganic sulfate of the must, from sulfur-containing amino acids of the must, or from elemental sulfur (used for dusting the crop) adhering to the grapes after harvest. The pathway of the formation from sulfate in the inductive environment of a yeast fermentation has been indicated by Dittrich (1987) as follows:

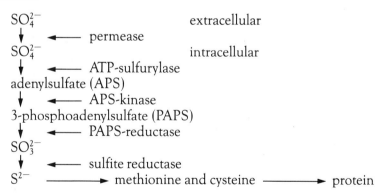

Almost all wine yeast strains produce measurable quantities of $SO_2$ generally in the range of 5-15 mg/L. Dittrich and Staudenmayer (1968) tested 162 strains in the collection of their institute. They found 6 strains that produced more than 50 mg/L $SO_2$. All of the other strains produced less than 20 mg/L. The sulfite formed by wine yeasts acts as a stabilizing factor in wine production (Suzzi, Romano, and Zambonelli 1985). The authors established a relationship between the stabilizing activity and $SO_2$ production.

The temperature during the fermentation is a major factor in $SO_2$ formation. At 10°C the level of $SO_2$ is five times higher than at 25°C, which explains why the formation of $SO_2$ is generally observed in white wines but not in red wines (Larue et al. 1985).

$H_2S$, that is, sulfur in its most reduced form, is a highly undesirable component of wines. In concentrations below the threshold of taste perception (50-90 $\mu g/L$) it may even contribute to the complexity of wines. But at concentrations exceeding this threshold it contributes the typical aroma of spoiled eggs. It may be formed during the fermentation by any sulfur-containing compounds of the must. For instance, in a synthetic medium a strain of *S. cerevisiae* formed the following amounts of $H_2S$ from the individually added compounds shown in parentheses: 20 mg/L (30 mg/L elemental sulfur); 6 mg/L (125 mg/L $(NH_4)_2SO_4$); 6 mg/L (115 mg/L cysteine); 2.3 mg/L (115 mg/L cystine); 4 mg/L (300 mg/L glutathione); 0 mg/L (150 mg/L methionine); and 11 mg/L (260 mg/L thiamine) (Hernandez 1964). All of the mentioned sulfur compounds had been added at equivalent sulfur concentrations. Hence, elemental sulfur was converted most efficiently to $H_2S$. Indeed, in commercial practice elemental sulfur remaining on the grapes is most likely to lead to excessive $H_2S$ production, but other factors are also important.

There is a positive correlation between free amino nitrogen of the must and $H_2S$ development, suggesting the presence of sulfur-containing amino acids as important sources for $H_2S$ formation (Vos and Gray 1979). Indeed, the addition of inorganic nitrogen sources to musts that are more readily assimilated by yeasts lessens $H_2S$ formation. Pasteurization of musts and the presence of larger amounts of insoluble solids increase $H_2S$ formation.

The presence of $H_2S$ in wines can be minimized or prevented by one of the following, alone or in suitable combination: separation of any elemental sulfur or of all insoluble solids prior to fermentation; bentonite treatment and removal of the precipitate before fermentation; addition of an inorganic source of nitrogen to the must; fermentation at a low temperature; and choice of a yeast strain low in $H_2S$ formation. There is often little agreement whether a particular strain is a good producer of $H_2S$ or not. Finally, it is known that a considerable portion of the formed $H_2S$ is swept out of the wine by evolving $CO_2$. Aeration of the fermented wine or preferably flushing with nitrogen removes a major portion of the sulfur compound.

If $H_2S$ remains in the wine during storage it may form other sulfur compounds, such as ethyl mercaptan, diethyldisulfide, methyl mercaptan, and dimethylsulfide, presumably by reaction with ethanol or acetaldehyde. These compounds have a penetrating, often garliclike, odor (Rapp, Gunther, and Almy 1985; Schreier 1979). They cannot be removed by aeration because their gas pressure is too low at room temperature. They may sometimes be removed by precipitation with copper sulfate (blue fining).

## Nitrogenous Compounds

Nitrogenous compounds of the must supply required nutrients for yeast growth. The level of nitrogen in musts may vary greatly. Among other variables it depends on the availability of nitrogen in the soil and on fertilization. For instance, in clean-cultivated vines of Thompson Seedless grapes nitrogen levels in the must were doubled by application of 112 kg N/ha (Bell, Ough, and Kliewer 1979). Ranges from 100-1,000 mg $N_2$/L (Lafon-Lafourcade 1983) and from 100-1,100 mg $N_2$/L (Amerine, Berg, and Cruess 1967) have been cited with median values in the neighborhood of 600 mg/L.

The distribution over various nitrogenous compounds is also variable, but generally falls within the following percentages (as percent of total nitrogen): $NH_3$ 3-10%; amino acids 25-30%; peptides 25-40%; and proteins 5-10% (Lafon-Lafourcade 1983). *Saccharomyces* wine yeasts assimilate ammonium nitrogen and amino acids readily, and may assimilate peptides and proteins to the extent to which they are hydrolyzed to amino acids. Among the amino acids arginine, glutamic acid, proline, and threonine predominate, accounting together for about 85% of the amino acids.

Active dry wine yeasts contain about 7% $N_2$, and an inoculation of 10 g/hl yields about $3 \times 10^6$ cells per ml. A 10-20-fold increase in yeast cell counts and a corresponding increase in biomass would require the assimilation of 70-140 mg N/L. It is apparent that the nitrogen available in the must is in most cases sufficient to sustain this yeast growth. It is also clear that it may not be sufficient in musts with a low concentration of assimilable nitrogen, which is often borne out by fermentations that lag because of a deficiency in available nitrogen. In such instances an ammonium salt such as $(NH_4)_2HPO_4$ may be added in concentrations up to 30 g/hl. Wineries also use yeast extracts that in addition to nitrogen supply some trace minerals and vitamins (Ingledew and Kunkee 1985). Urea, which has sometimes been used as a nitrogen source, should be avoided because of the danger of urethane formation in the wine.

During the fermentation the concentration of amino acids in the must decreases because of uptake by the yeast. With the cessation of yeast growth the concentration of amino acids remains stable until it increases slightly due to slow autolysis of the yeast cells. This process is shown in Figure 4-10 for a Riesling must.

The same sequence is seen during the secondary champagne fermentation. But long storage periods on the yeast in bottle fermentations result

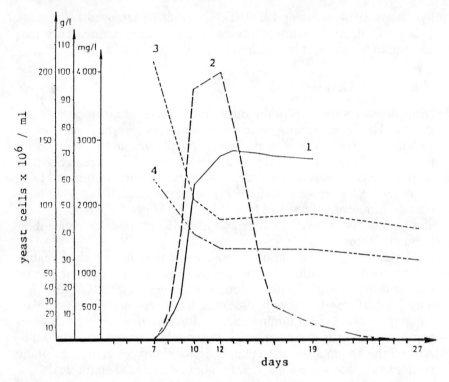

**Figure 4-10.** Uptake of amino acids during the active fermentation of a Riesling must:

(1) ──────── ethanol concentration, g/L scale
(2) ── ── ── yeast cell count
(3) ─ ─ ─ ─ ─ total amino acids, mg/L scale
(4) ─ · ─ · ─ glutamic acid family (arginine, proline, glutamine), mg/L scale

(*After Dittrich 1987*)

in extensive autolysis of the cells and release of considerable concentrations of amino acids (Bergner 1968). The relative concentrations of released amino acids do not usually correspond to the ratios found in the must.

Feuillat and Charpentier (1982) followed the release of amino acids from champagne yeast and the intracellular proteolytic activity of the yeasts during prolonged maturation. They concluded that the enrichment of champagne after completion of the bottle fermentation starts with the slow excretion of amino acids from the yeast during the first 6 months of storage. After 9 months there is a rapid release of amino acids due to yeast autolysis. Intracellular proteolytic activity of the yeasts increases after 9 months of storage to coincide with the rapid release of amino acids.

In apple juice and in the juice of others fruits and berries the concentration of assimilable nitrogen is generally lower than in grape must. Therefore, the addition of ammonium salts or yeast extracts is frequently required for rapid and complete fermentation.

## Fermentation Inhibitors and Stuck Fermentations

Stuck fermentations refer to fermentations that have not gone to completion for a variety of reasons and to fermentations in which complete fermentation of sugars is desired. The effect of higher concentrations of ethanol and the roles of temperature, $SO_2$, aeration, and sterol content of yeasts have already been discussed, but additional problems may be due to the presence of specific inhibitors or to some poorly defined variables.

It is well known that highly clarified white musts are more likely to get stuck than musts containing a normal concentration of insoluble particles of the grape pulp. In the presence of such insoluble particles fermentation proceeds faster (Groat and Ough 1987). The following reasons have been cited for the effect of insoluble solids: removal of dissolved $CO_2$, which serves as an inhibitor by gas release at the surface of insoluble particles. This phenomenon can also be observed by addition of other insoluble materials such as diatomaceous earth or bentonite. The release of $CO_2$ bubbles at the surface of insoluble solids induces a stirring action that keeps yeast cells from settling and accelerates diffusion of sugars to the surface of yeast cells (Dittrich 1987). It is obvious that the magnitude of these effects depends greatly on the size of the fermentation vessel, and that investigation of these phenomena on the bench scale is not likely to provide satisfactory explanations.

Earlier work has demonstrated the inhibitory effect of polyphenolic compounds such as caffeic acid, as well as the stimulatory effect of chlorogenic and isochlorogenic acid (Sikovec 1966). De Soto and Huber (1968) showed the inhibitory effect of some added tannin, but red wines, which contain a much higher content of polyphenols than white wines, are not more difficult to ferment. Goldman (1963) reported that the secondary fermentation of red-base wines (for making sparkling burgundy) is more difficult than that of white-base wines of the same ethanol content. The slow conversion of chlorogenic acid to caffeic acid during aging of wines may be responsible for this result.

Treatment of grapes with pesticides may introduce fermentation inhibitors into the must. Conner (1983) tested 14 fungicides, 7 insecticides, and 4 herbicides and listed the lowest concentrations at which they were inhibitory. Only one of the insecticides was inhibitory at a level of 5 mg/L. None of the herbicides was particularly inhibitory. Four of the fungicides were inhibitory at levels below 10 mg/L. In the European fall climate fungal infections are more likely to occur. The use of specific botrycides is common and their effect as fermentation inhibitors is of greater concern (Benda 1983).

The inhibition of the wine fermentation by fatty acids and their esters has been reported in recent years (Lafon-Lafourcade, Geneix, and Ribéreau-Gayon, 1984; Larue et al. 1985). The inhibition is due to the presence of octanoic and decanoic acids, the latter being much more inhibitory. These fatty acids are formed during the fermentation. Levels of decanoic acid in white wines ranged from 0.27 to 7.48 mg/L (Lafon-Lafourcade 1983).

The acids may be removed and the fermentation rate restored by adsorption on charcoal or yeast cell wall material. For practical reasons yeast cell wall material is used. Yeast cell walls are obtained by autolyzing yeast and separating the insoluble material by centrifuging. The French literature refers to the material as *écorces de levure*. In the United States the terms *yeast hulls* and *yeast ghosts* have been used. The inhibition can be removed by the addition of 1 gram of cell wall material (dry weight) per liter (Wahlstrom and Fugelsang 1988).

The fatty acids also affect the growth rate of wine yeasts. Figure 4-11 shows the synergistic effect of added octanoic and decanoic acids on growth rate in the presence of 8% added ethanol (Viegas, Sa-Correia, and Novais 1985). The fatty acids also inhibit the malolactic fermentation, and their inhibitory effect can be eliminated by the addition of yeast cell wall material (Lonvaud-Funel, Desens, and Joyeux 1985; Edwards and Beelman 1987). The use of ascorbyl-6-decanoate as a preservative for wines has been suggested (Bartley et al. 1987).

There is no simple answer to the question of lagging or stuck (incom-

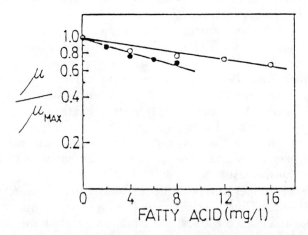

**Figure 4-11.** Effect of the addition of octanoic and decanoic acids on the maximum specific growth rate of *Saccharomyces bayanus* in the presence of 8% (v/v [volume per volume]) added ethanol: ○ = octanoic acid; ● = decanoic acid. (*From Viegas, Sa-Correia, and Novais 1985*)

plete) fermentations. Occasionally, the answer may be obvious. If the number of viable yeast cells is extremely low because of a very high temperature at a considerable ethanol concentration, it is possible to separate the dead yeast and to inoculate the cooled wine with fresh yeast. More likely stuck fermentations are due to nutritional deficiencies or to the presence of one or several of the known inhibitors.

Nutritional deficiencies occur because of a lack of assimilable nitrogen, and only rarely because of a lack of biotin or required trace metals. An insufficiency of yeast sterols and/or a lack of minimal concentrations of oxygen may also be considered a nutritional deficiency. The interplay of these factors is exemplified in Figure 4-12 (Ingledew and Kunkee 1985),

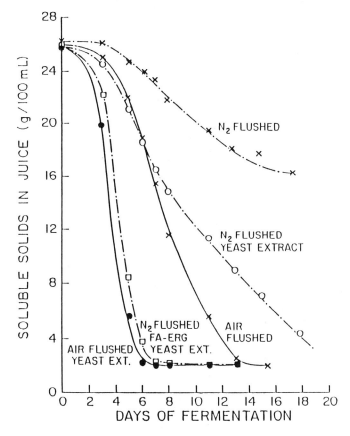

**Figure 4-12.** Fermentation of supplemented and unsupplemented Ruby Cabernet juices under conditions of nitrogen or air flushing of the headspace. (*From Ingledew and Kunkee 1985*)

which shows the effect of aeration and the addition of yeast extract on the fermentation of a Ruby Cabernet juice. The unsupplemented juice sticks at a concentration of about 16% soluble solids. The addition of yeast extract or flushing with air permits completion of the sluggish fermentation in 13 to 20 days. Only the provision of air plus nutritional supplementation permit a rapid fermentation that is completed in 7 days. The figure also shows that the addition of ergosterol plus some sorbitan oleate could be substituted for the aeration process.

Some fatty acids are strongly inhibitory. Acetic acid, which can be readily recognized, is a strong inhibitor of yeast growth and of the fermentation activity at 0.2%. The presence of decanoic acid can only be determined by special analytical techniques. Its adsorption on yeast cell wall material has already been mentioned. The inhibiting effect of *Botrytis* on the fermentability of infected grapes will be discussed separately.

## THE KILLER FACTOR

Antagonistic effects between yeast strains and between species have been found to be caused by specific killer factors. Yeasts called killer yeasts excrete a toxin that is a protein or a glycoprotein. Several such killer toxins have been called K1, K2, and so on. At least 10 types of toxins have now been identified. A strain of killer yeasts secretes only one specific toxin. Sensitive strains are killed by the toxin. However, most strains produce toxins against a narrow range of wine yeasts, and *Saccharomyces cerevisiae* killer strains are not effective against *Kloeckera apiculata*, *Pichia*, *Candida*, *Debaromyces*, and other genera. The utility of such killer strains for wine making is also limited because of the rather narrow limits of effectiveness, generally between pH values of 4.0 to 5.4 and at temperatures below 30°C (Jacobson 1985). However, Hara, Iimura, and Otsuka (1980) reported effective killer action of a hybrid *S. cerevisiae* culture at pH values between 3.0 and 4.5 and at temperatures between 15° and 35°C.

The killer factor can be introduced into suitable wine yeast strains by hybridization of haploids of a killer strain and a wine yeast (Hara, Iimura, and Otsuka 1980) or by protoplast fusion (Jacobson 1985). Figure 4-13 shows a test for the determination of killer activity. Killer yeast strains in the form of active dry yeasts have been made available commercially, but published reports on their successful use in wine making have been absent to date. A specific glycoprotein killer toxin from *S. cerevisiae* was adsorbed by bentonite and by a commercial yeast cell wall preparation. Fractionation of the yeast cell wall showed that the mannan fraction adsorbed the toxin (Radler and Schmitt 1987). The last-mentioned publication is also a good guide to the earlier literature.

BY-PRODUCTS AND AROMA COMPOUNDS 185

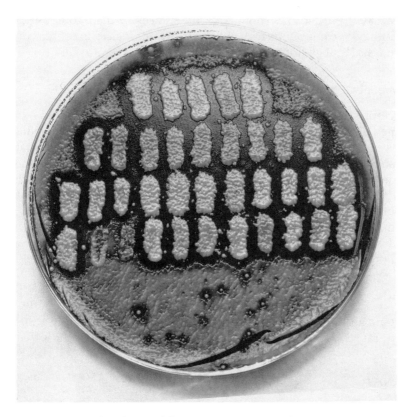

Figure 4-13. Test plate showing killer yeasts and two nonkiller strains. The second and third patches from the left of the bottom row are nonkiller strains. (From Jacobson 1985)

## BY-PRODUCTS OF THE ALCOHOLIC FERMENTATION AND AROMA COMPOUNDS

By-products of the fermentation may affect the flavor, aroma (odor), and body of wines. The effect of acids of the grape must, principally tartaric acid, malic acid, citric acid, and organic acids liberated during the fermentation, on acid perception is obvious. Equally obvious is the deleterious effect of higher concentrations of acetic acid that may be formed by bacterial action. However, the effect of the very large number of volatile by-products on the aroma is quite difficult to determine and evaluate. Such an evaluation depends on a knowledge of the aroma threshold of a given compound, on the combined effect of several compounds that singly may be present below threshold values, and on synergistic or antagonistic effects. Wines that may show large differences in their composition of volatile odor compounds may not show any differences in sensory difference tests (Ussegli-Tomasset and di Stefano 1981).

## Glycerol

Glycerol is not found in grape must except in botrytized grapes. It is always formed during the alcoholic fermentation, usually one-tenth to one-fifteenth the weight of ethanol. The amounts formed depend greatly on the strain of yeast. Radler and Schütz (1982) found concentrations between 2.5 to 12.5 grams of glycerol per liter for eight strains of *Saccharomyces* wine yeasts. The formation of glycerol was correlated with the concentration of glycerol-3-phosphate dehydrogenase. Glycerol formation during a 60-hour fermentation is shown in Figure 4-14. The concentration of the polyol is by far the largest of any fermentation by-product. It contributes body to the wine and probably adds a slight sweetness.

## Acids

Tartaric and malic acids account for almost all of the titratable acidity of grape musts. The concentration of tartaric acid in wines may be lower due

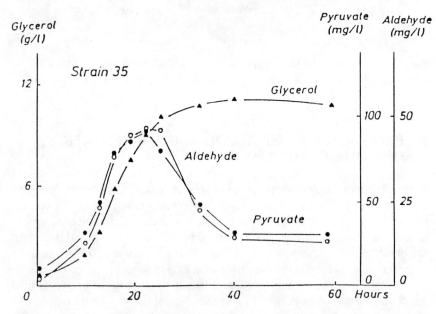

**Figure 4-14.** Formation of acetaldehyde, pyruvate, and glycerol during a fermentation by *Saccharomyces cerevisiae*: ○ = aldehyde; ● = pyruvate; ▲ = glycerol. (*From Radler and Schütz 1982*)

to tartrate precipitation. That of malic acid may be considerably lower for wines that have undergone the malolactic fermentation (see below). Titratable acidity of wines that includes both fixed and volatile acidity is usually expressed as tartaric acid. For wines of medium acidity it is close to 0.65%. A medium to high acidity is generally preferred for white table wines and champagne.

Succinic acid accounts for about 90% of the total organic nonvolatile acids formed during the fermentation (up to 0.15%) (Thoukis, Ueda, and Wright 1965). The amounts of pyruvic, alpha-ketoglutaric, and citric acids formed by yeasts are very slight. The concentration of lactic acid formed by yeasts is negligible, but concentrations as high as 0.5% may be formed by bacterial fermentation. Acetic acid may be formed by *Saccharomyces cerevisiae* up to 40 mg/L, but higher amounts may be formed by other *Saccharomyces* species and especially by other genera of yeasts. Incipient acetic acid formation by acetic acid bacteria will quickly spoil a wine.

### Acetaldehyde

Acetaldehyde, pyruvic acid, and alpha-ketoglutaric acid form sulfonates with $SO_2$. Acetaldehyde binds $SO_2$ more strongly than pyruvic acid and it binds pyruvic acid more strongly than alpha-ketoglutaric acid. The addition of thiamin to the must reduces the amounts of pyruvic and alpha-ketoglutaric acids formed, but not those of acetaldehyde. Actual concentrations of acetaldehyde in wines vary greatly. For 300 German wines Dittrich and Barth (1984) reported an average of 50 mg/L acetaldehyde, 27 mg/L pyruvate, and 48 mg/L alpha-ketoglutarate.

The formation of acetaldehyde parallels the production of ethanol during the early rapid stages of the fermentation. During the later stages of the fermentation acetaldehyde and pyruvate are metabolized by the yeast cells (Fig. 4-15, see also Fig. 4-14). However, for sluggish fermentations in which the yeast has settled to the bottom of the fermenter the reduction of acetaldehyde concentration may be absent or slight.

For the production of sherry under the customary oxidative conditions the formation of high concentrations of acetaldehyde is desired. Acetaldehyde accounts for more than 90% of the total aldehyde fraction of wines.

### Methanol

Methanol is not a by-product of the fermentation. It occurs in wines usually in concentrations of less than 500 mg/L. Methanol is formed from grape pectin by pectinmethyl esterase. The enzyme occurs naturally in must, but larger amounts of the enzyme are active in grape musts that have been treated with commercial pectic enzymes. High levels of methanol are present in wine or brandy made from apple pomace.

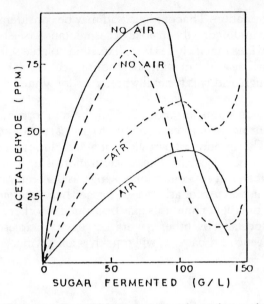

**Figure 4-15.** Formation of acetaldehyde during aerobic and anaerobic fermentation by two strains of *Saccharomyces cerevisiae* var. *ellipsoideus*. Solid lines represent strain no. 51; dashed lines, strain no. 41. (From Ribéreau-Gayon et al. 1976)

## Higher Alcohols

The fraction of higher alcohols, sometimes called fusel oils, is quantitatively the largest group of flavor compounds. According to the Ehrlich reaction the alcohols are formed by the oxidative deamination of amino acids so that valine yields isobutanol (2-methyl-1-propanol), leucine yields isoamyl alcohol (3-methyl-1-butanol), and isoleucine yields active amyl alcohol (2-methyl-1-butanol). The mechanism is certainly possible since addition of an excess of any of these amino acids to must leads to an increase in the appropriate fusel oil fraction. However, the mechanism cannot account for all of the fusel oil fractions formed since the quantity of amino acids in must is not sufficient. Thoukis (1958) suggested that the fusel oils could be derived from glucose, and various schemes for such transformations have been devised. The following simplified scheme of Äyrapää (1971) accounts for both pathways.

pyruvic acid ⟶ alpha-acetolactic acid ⟶ 2-ketoisovaleric acid ⟶

$\xrightarrow{\text{CoA}}$ 3 carboxyl-3-hydroxyisocaproate ⟶ 2-ketoisocaproate ⟶

$\xrightarrow{-CO_2}$ isovaleraldehyde ⇌ isoamyl alcohol ⇕ L-leucine

BY-PRODUCTS AND AROMA COMPOUNDS    189

The total amounts of fusel oils in wines is quite variable and so is the distribution of individual fractions. The major constituents are n-propanol, isobutanol, active amyl alcohol, isoamyl alcohol, and phenethyl alcohol. Together these alcohols account for 99% of the total, and isoamyl alcohol alone accounts for more than 50%. The earlier literature (Guymon and Heitz 1952) reports mean values of 250 mg/L for California white wines (range 162-366 mg/L) and mean values for red wines of 287 mg/L (range 156-900 mg/L). Values for French wines were somewhat higher but with equally wide ranges. Similar values are still reported in more recent publications (Dittrich 1987; Minarik and Navara 1986).

The wide variations can be attributed to the many variables that affect fusel oil formation during a fermentation. Webb and Kepner (1961) show wide variations with different strains of *S. cerevisiae*. Aerobic conditions may increase fusel oil levels several times, and higher temperatures and an increase of pH from 3.0 to 4.2 also raise their concentration. The effect of the fusel oil fractions on the aroma of wines is difficult to assess. Reported threshold values differ widely. For isobutyl alcohol the threshold values were 75 mg/L, 228 mg/L, and 500 mg/L and for isoamyl alcohol they were 7 mg/L, 14.5 mg/L, and 300 mg/L as reported in this order by Salo (1970), De Wet et al. (1978), and Rankine (1967). In spite of these discrepancies it may be concluded that isoamyl alcohol is more likely to affect the aroma than isobutyl alcohol or propyl alcohol. The last-mentioned alcohols are generally present in concentrations below the reported threshold values, which does not, however, exclude the possibility of a synergistic effect or that of an additive effect for compounds that are present below threshold values.

**Esters**

The pathways of ester formation during yeast fermentations have been explored. They will not be discussed here because they have not been established with sufficient clarity. As with other flavor compounds the concentration of various esters in wine depends on several variables. In general, ester concentrations are higher in wines from musts with higher sugar concentrations and in musts fermented at higher temperatures. Ester concentrations also depend on the yeast strain. They may also change during the storage of young wines. Table 4-9 shows the concentrations of various esters. The added data on taste threshold values show that the esters are present in concentrations above or near the threshold concentrations. These values merely confirm the perception of fruitiness in young wines, particularly of white wines fermented at low temperatures.

Soles, Ough, and Kunkee (1982) measured ester formation in a Chenin Blanc wine fermentation with 14 different yeast strains, most of them *S. cerevisiae*, but including one strain each of *Pichia fermentans*, *Schizosacchar-*

Table 4-9. Concentration of Esters in White Wines and Odor Thresholds of Esters

|  | Ester Concentration in mg/L | | | Odor Threshold in mg/L | |
| --- | --- | --- | --- | --- | --- |
|  | I | II | III | IV | V |
| Ethyl formate | 3.0 |  |  |  |  |
| Ethyl acetate | 61.9 |  | 50 | 12.3 | 17 |
| n-hexyl acetate |  | 0.4 | 0.46 | 0.67 |  |
| Ethyl hexanoate | 0.5 | 1.05 | 0.6 | 0.008 | 0.076 |
| Ethyl octanoate | 0.7 | 1.49 | 0.61 | 0.58 | 0.24 |
| Ethyl decanoate | 0.1 | 0.43 | 0.20 | 0.51 | 1.1 |
| 2-phenethyl acetate |  | 0.24 | 0.46 | 1.8 | 0.65 |
| Ethyl lactate | 141.5 |  |  |  | 14 |
| Isoamyl acetate | 0.7 | 3.78 |  | 0.16 | 0.20 |

I: Postel, Drawert, and Adam 1972; 25 German white wines.
II: Van der Merwe and van Wyck 1981; 35 South African Chenin Blanc wines.
III: Killian and Ough 1979; Chenin Blanc wine.
IV: De Wet et al. 1978.
V: Salo 1970.

*omyces pombe,* and *Saccharomyces bailli.* The concentrations of individual esters differed significantly with the yeast strain and spanned the following ranges: isoamyl acetate 0.86-2.94 mg/L; ethyl hexanoate 0.45-1.14 mg/L; ethyl octanoate 0.72-1.35 mg/L; hexyl acetate 0.07-0.20 mg/L; and 2-phenethyl acetate 0.04-0.18 mg/L. But there was no discernible pattern that would permit a sound judgment on the sensory effect of these variations.

The formation of esters during a fermentation and the effect of temperature on production and retention of ethyl hexanoate are shown in Figure 4-16. The higher temperatures showed increased formation of the ester but poorer retention in the wine during the latter stages of the fermentation. Killian and Ough (1979) concluded from their study of 10 esters that those with a fruity aroma (isoamyl acetate and n-hexyl acetate) are retained best at 10°C. The more "heady" aromatic esters (ethyl octanoate, ethyl decanoate, and 2-phenethyl acetate) are produced at higher concentrations and are retained better at 15°-20°C.

Van der Merwe and van Wyck (1981) undertook an intriguing study of the effect of various esters on the aroma of wine. The wines were stripped of esters by extraction with Freon 11. Different esters or combinations of esters were then added back to the wine, and odor quality and odor intensity were compared with the original wine. Ethyl acetate increased odor intensity and decreased odor quality only at concentrations exceeding 50 mg/L. A combination of 6 esters restored odor intensity and odor quality to the extracted wine (isoamyl acetate, n-hexyl acetate, 2-phenethyl

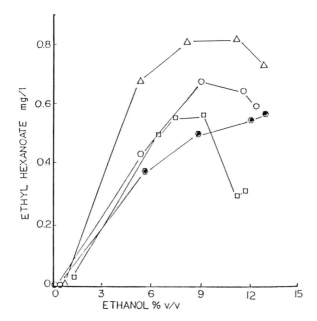

**Figure 4-16.** Effect of temperature and progress of the fermentation on the concentration of ethyl hexanoate: ⊙ = 10°C; ○ = 15°C; △ = 20°C; □ = 30°C. (*From Killian and Ough 1979*)

acetate, ethyl-n-hexanoate, ethyl-n-octanoate, ethyl-n-decanoate). However, the contribution of individual esters to the total odor cannot be easily isolated.

### Diacetyl, Acetoin, and Butanediol

Diacetyl (2-3-butanedione) and acetoin are probably formed by decomposition of acetolactate. These two compounds as well as 2-3-butanediol are part of a chain in which reduction leads from diacetyl to acetoin to butanediol as follows:

$$CH_3-CO-CO-CH_3 \xrightarrow[\text{(diacetyl reductase)}]{NAD(P)H_2 \longrightarrow NAD(P)}$$
diacetyl

$$CH_3-CO-CHOH-CH_3 \xrightarrow{NADH_2 \longrightarrow NAD}$$
acetoin

$$CH_3-CHOH-CHOH-CH_3$$
2,3 butanediol

Concentrations of butanediol may reach 300-900 mg/L in wines. Concentrations of acetoin reach 25-100 mg/L midway through the fermentation and drop to 5-20 mg/L at the end. Fermentation temperatures and aerobic conditions lead to higher acetoin concentrations.

Diacetyl is of particular interest because it imparts an undesirable buttery flavor and its threshold of perception is very low. Postel and Guvenc (1976) report the following ranges for diacetyl: white wines, 0.05-3.40 mg/L; rose wines 0.11-1.24 mg/L; red wines 0.02-4.05 mg/L. At concentrations of 0.2-0.4 mg/L, diacetyl may even make a positive contribution to the quality of the wine. The formation of diacetyl and its effect on beer flavor is highly important (see Chapter 3). For a detailed discussion of the role of vicinal diketones in wines see Dittrich (1987) and Nykanen (1986).

The flavor and aroma of a wine depends mainly on the type and quality of the must and on processing conditions. The strain of yeast used may produce subtle differences in flavor and aroma, but such differences cannot be derived simply from what has been described in this section. Gallander (1982) fermented a Vidal blanc (hybrid grape) must with five strains of commercially available WADYs (three *S. cerevisiae* and two *S. bayanus*) under identical conditions. The results of the sensory evaluation are shown in Table 4-10.

Ethanol concentration, pH, and total acidity of the wines were essentially the same. The volatile acidity of the *S. bayanus* fermented wines was considerably higher than that of the *S. cerevisiae* fermented wines. Three of the commercial yeasts produced wines that had identical ratings for taste. Two of these were also identical in the rating for aroma. The other two yeasts were rated lower but made acceptable wine. Gallander points out

Table 4-10. Aroma and Taste Ratings of Vidal Blanc Wines Fermented by Five Different Wine Yeasts

| Yeast | Sensory Rating* | |
|---|---|---|
| | Aroma | Taste |
| Montrachet no. 522, *S. cerevisiae* | 5.2 ab | 5.1 a |
| Epernay no. 2, *S. cerevisiae* | 5.5 a | 4.9 a |
| CU-2, *S. cerevisiae* | 3.9 d | 4.1 b |
| California Champagne, no. 505, *S. bayanus* | 4.5 c | 4.2 b |
| Pasteur Champagne, no. 585, *S. bayanus* | 5.0 b | 5.1 a |

Source: Data from Gallander 1982.
*Each attribute was scored on a 7-point hedonic scale, 7 being most acceptable. Values within each column having the same letter were not significantly different at the 5% level.

wisely that a change in fermentation conditions, for instance, from 18°C of the test to a different temperature, could well have changed the results.

## MICROBIOLOGICAL REDUCTION OF ACIDS

In many areas of the world with cool summer climates wines may be too acidic to be sufficiently palatable, for instance, Germany, France, Switzerland, and the northeastern and northwestern United States. For such wines excess acidity can be reduced chemically, but the use of microbiological methods is often preferred. Such methods include the well-known malolactic fermentation (MLF) and the rarely practiced use of malic acid metabolizing yeasts.

### Malolactic Fermentation

The MLF occurs spontaneously in wines, generally after completion of the alcoholic fermentation. It had already been recognized as a bacterial fermentation by Muller-Thurgau, Koch, and Pasteur toward the end of the nineteenth century. With wines that had been fermented in the fall the MLF usually took place in late spring, when warmer weather led to an increase in the temperature of the stored wine. Little was known at that time about the bacterial species or the particular conditions that favored or prevented the occurrence of the MLF.

The genera of malolactic bacteria occurring in wine are *Leuconostoc*, *Pediococcus*, and *Lactobacillus*. *Leuconostoc oenos* is the species most frequently responsible for the MLF, as well as the preferred species. *Leuconostoc mesenteroides*, *Pediococcus damnosus* (syn. *P. cerevisiae*), the homofermentative *Lactobacillus plantarum* and *L. casei*, and the heterofermentative *L. hilgardii* are often isolated.

The MLF consists in the conversion of the malic acid of the must or the wine to the weaker lactic acid, which results in an increase in pH and an amelioration of the acid taste of the wine. The overall formula for this conversion is

$COOH\text{-}CH_2\text{-}CHOH\text{-}COOH \longrightarrow COOH\text{-}CHOH\text{-}CH_3 + CO_2$
134 g malic acid                90 g lactic acid         44 g $CO_2$

There are three possible pathways for this transformation. Two of the pathways are based on pyruvic acid as an intermediate, which would suggest the formation of both $D(-)$ and $L(+)$ lactic acid. The third pathway is based on catalysis by the malolactic enzyme (Radler 1975). It is now considered the principal path for the MLF (Kunkee 1975; Lonvaud-Funel and Strasser de Saad 1982).

Wibowo et al. (1985) in an extensive review of the MLF suggest that it is not a true fermentation but rather an enzymatic reaction carried out by bacteria after they have grown and reached the stationary phase. Indeed the MLF can be carried out by inoculation with high numbers of bacteria ($10^6$ to $10^8$/ml) in the absence of active growth.

*Leuconostoc oenos* is the organism that usually carries out the MLF. It is rarely present on grapes and generally it cannot be isolated from must at the beginning of the alcoholic fermentation. Figure 4-17 shows a pattern that seems to be fairly characteristic of the sequence of microorganisms that carry out the alcoholic and the malolactic fermentation. During the alcoholic fermentation there is often some growth of a *Lactobacillus* species as shown in the figure for *Leuc. mesenteroides* or *P. damnosus*. The end of the alcoholic fermentation is indicated by a single arrow that signals the growth of *Leuc. oenos*. The double arrow indicates the end of the MLF.

The MLF occurs in wines to a variable extent. Dittrich and Barth (1984)

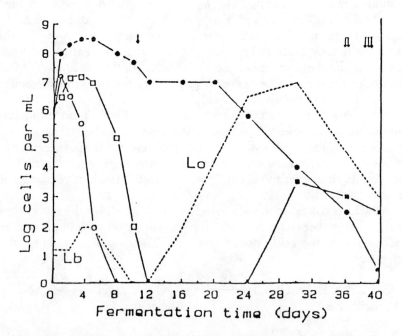

**Figure 4-17.** Evolution of yeasts and lactic acid bacteria during vinification of a red wine: ● = *Saccharomyces cerevisiae*; ○ = *Kloeckera apiculata*; ■ = *Pichia membranefaciens*; □ = *Torulopsis stellata*; Lb, *Lactobacillus* species; Lo, *Leuconostoc oenos*. (From Fleet, Lafon-Lafourcade, and Ribéreau-Gayon 1984)

report a reduction of malic acid to less than 1 g/L in 63% of red wines and 9% of white wines. Reduction to less than 0.5% malic acid occurred in 37% of the red wines and 2% of the white wines. The total acidity of wines is generally determined by titration of carboxyl groups and expression as tartaric acid. The MLF converts a dicarboxylic acid to a monocarboxylic acid, and hence a drop of the total acidity by 0.1% corresponds to a conversion of 0.2% malic acid to lactic acid. In wines with a malic acid content of 3-5 g/L the MLF will reduce it to 0.2-1.6 g/L, depending on the extent of the bacterial conversion (Dittrich 1980). Organic acids that occur in musts and wines at rather low concentrations such as citric, gluconic, and fumaric acids can also be degraded during the bacterial fermentation.

The growth of malolactic bacteria depends on several variables. Only the most important ones can be discussed here. The malolactic bacteria are more fastidious than yeasts in their nutritional requirements and for this reason alone their growth is more uncertain. Residual sugars in wine, often less than 0.1%, as well as xylose or arabinose can serve as carbon sources. The different species of bacteria require from 3 to 16 different amino acids for growth, and their requirements for various vitamins such as nicotinic acid, pantothenic acid, riboflavin, and folic acid are greater than for yeasts. It is, therefore, not surprising that spontaneous growth during the alcoholic fermentation is slight.

Growth usually follows the cessation of the alcoholic fermentation and the start of yeast autolysis, which supplies amino acids and other nutrients. Growth of all lactic acid bacteria is restricted at lower pH values. The limiting pH value differs with the species. Generally growth below a pH of 3.5 is restricted to *Leuc. oenos*, whereas *Pediococcus* and *Lactobacillus* grow well at pH values above 3.5. Growth of three *Leuc. oenos* strains was slight at pH 3.0, good at pH 3.3, and abundant at pH 3.7 (Silver and Leighton 1981).

Malolactic bacteria are highly sensitive to $SO_2$, particularly to free $SO_2$ in its molecular form, which is hence also reflected in the effect of pH. Figure 4-18 is instructive in demonstrating the effect of $SO_2$ for a red wine pH 3.5 (Lafon-Lafourcade 1983). Grapes without the addition of $SO_2$ developed a copious flora of lactic acid bacteria after cessation of the alcoholic fermentation ($10^7$ cells/ml), and the MLF was completed two weeks later. With the addition of 100 ppm of $SO^2$ to the grapes, growth of the bacteria and completion of the MLF was greatly delayed. It took longer at 14°C than at 19°C. Addition of 50 ppm of $SO_2$ to the free-run juice prevented significant growth and an effective MLF.

Bound $SO_2$ also inhibits the MLF but to a lesser extent. The temperature optimum for growth of the malolactic bacteria is in the neighborhood of 22°-25°C for most strains. Below 15°C it occurs rarely. Low ethanol concentrations have been reported to stimulate the MLF, but high ethanol

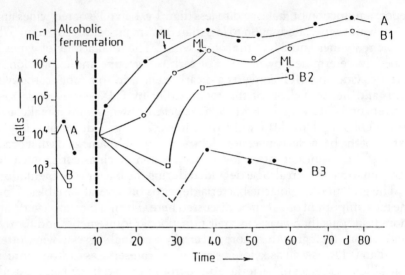

**Figure 4-18.** Development of populations of lactic acid bacteria during vinification: A, grapes not sulfited, storage at 19°C and 14°C; B, grapes sulfited (10 g/hl); B1, storage at 19°C; B2, storage at 14°C; B3, free-run juice, sulfited (5 g/hl), storage at 14°C; ML, end of the malolactic fermentation. (*From Lafon-Lafourcade 1983*)

concentrations are inhibitory. Growth is more strongly affected than the decarboxylation of malic acid. Increasing concentrations of insoluble solids favor the onset of the MLF, possibly by the reduction of $SO_2$ that has been observed (Liu and Gallander 1983). A requirement of $CO_2$ for *Leuc. oenos* has been reported (Mayer 1974).

As with yeasts, octanoic and decanoic acids are inhibitory to the growth of malolactic bacteria; and as with yeasts, the inhibition can be removed by addition of yeast cell wall material (Lonvaud-Funel, Desens, and Joyeux 1985; Edwards and Beelman 1987). *Leuc. oenos* can be lysed by phages, which may explain inconsistencies in the occurrence of the MLF (Gnaegi et al. 1984). The use of multiple strain starter cultures has been suggested to counteract phage infection (Henick-Kling, Lee, and Nicholas 1986). Phage infection has not received the attention it deserves.

Finally, an antagonistic interaction between *Saccharomyces cerevisiae* and *Leuc. oenos* has to be considered. Simultaneous inoculation of a model grape juice system with *S. cerevisiae* and *Leuc. oenos* decreased the growth of yeast and the malolactic bacterium compared to growth as individual inoculants. The inhibiting effect of yeast could not be ascribed merely to its formation of ethanol or acetaldehyde-bound $SO_2$ (King and Beelman 1986).

Liquid (frozen) cultures and freeze-dried cultures of lactic acid cultures are used routinely by the dairy industry. Such liquid cultures of malolactic bacteria are commercially available for use in the wine industry, but have not yet found widespread application. The cultures described as useful and those commercially available are all cultures of Leuc. oenos. Among these are the ML-34 strain of the University of California, the PSU-1 strain of Pennsylvania State University (Beelman, McArdle, and Duke 1980; King 1985; Silver and Leighton 1981), and additional strains listed by Davis et al. (1986). Such cultures can be used for the mass inoculation of musts and wines.

The immobilization of malolactic bacterial cells has also been attempted. Entrapment of Leuc. oenos in polyacrylamide gel gave a preparation that resembled the free cells in its pH and temperature profile (Rossi and Clementi 1984). Entrapment of Leuc. mesenteroides in calcium alginate gel resulted in a preparation with better resistance to low pH, higher temperatures, and higher ethanol concentration than the free cells (Totsuka and Hara 1983). Crapisi et al. (1987) used entrapment in K-carrageenan gels to obtain fairly stable and effective cell preparations.

The MLF is advantageous for wines of high acidity. If it does not occur naturally it may be achieved by inoculation with a strain of Leuc. oenos, preferably with a cell density of $10^6$/ml. It is not quite clear whether this inoculation should be done at the start of the alcoholic fermentation or at the end. It is likely that inoculation with a high count at the start of the alcoholic fermentation will be satisfactory as these cells can carry out the MLF in their stationary phase. California wine makers prefer a simultaneous MLF and alcoholic fermentation. In Bordeaux sequential alcoholic and malolactic fermentations are preferred (King 1985). A completed MLF before bottling is also advantageous since it prevents a later MLF in the wine. It cannot be assumed that it will stabilize the wine against the action of other spoilage bacteria (Davis et al. 1986).

A potential disadvantage of the MLF is the formation of higher concentrations of diacetyl, a tendency that was more pronounced in red wines than in whites (Mascarenhas 1984, Wibowo et al. 1985). The last-named authors also reviewed evidence for the formation of biogenic amines by some malolactic bacteria. For wines of low acidity the MLF must be prevented. Dittrich (1987) lists the following means toward that end: efficient clarification of the must, a fast alcoholic fermentation with selected wine yeast strains, adequate sulfiting, and cool storage.

**Malic Acid Degradation by Yeasts**

Malic acid is also metabolized by yeasts as follows:

$$COOH\text{-}CHOH\text{-}CH_2\text{-}COOH = CH_3\text{-}CH_2OH + 2\ CO_2$$
134 g malic acid       46 g ethanol    88 g $CO_2$

Wine yeasts with few exceptions are not as efficient converters of malic acid as the malolactic bacteria. Shimazu and Watanabe (1981) fermented a must with 2.05 g malic acid/L with nine strains of *S. cerevisiae* and single strains of eight other genera. The *S. cerevisiae* strains reduced the malic acid content by 3-51%; a strain of *Schizosaccharomyces* removed malic acid completely. More uniform results were found with a Muller-Thurgau must containing 6.7 g of malic acid/L (Wagner, Kreutzer, and Mahlmeister 1986). Fourteen commercial active dry yeasts, all strains of *Saccharomyces cerevisiae* or *S. bayanus*, reduced the relatively high concentration of 6.7 g malic acid/L by 13-22%, and only one strain reduced it by 54%.

The particular ability of *Schizosaccharomyces* species to degrade malic acid rapidly and usually completely has been known for a long time (Benda 1974; Dittrich 1963). The yeast species frequently tested is *Schiz. pombe*. The yeast produces 10-12% ethanol by volume during the alcoholic fermentation of musts, but it ferments more slowly than *Saccharomyces* species. The osmophilic species tolerates high $SO_2$ concentrations.

Most research workers have reported the development of off-flavors during fermentation with various strains of *Schiz. pombe*. For instance, Unterholzner, Aurich, and Platter (1980) isolated a strain of *Schiz. pombe* from a red wine with a dull, undesirable flavor. This flavor could be reproduced by fermentations with this isolated strain. Various attempts have been made to carry on fermentations with mixtures of a good wine yeast and *Schiz. pombe*. Snow and Gallander (1979) suggested initiation of the fermentation with *Schiz. pombe*, separation of the yeast, and reinoculation with a *S. cerevisiae* strain. However, none of the mentioned attempts have led to commercial use of the process.

## GENETIC MANIPULATION OF WINE YEASTS

The genetic manipulation of industrial yeast strains is difficult because many or most of the important characteristics for wine making are under multiple chromosomal gene control. However, some progress has been made in genetic strain improvement, and a few examples are given here.

Romano et al. (1985) determined the heritability of important winemaking characteristics by tetrad or random analysis of the spores of wine yeasts. The progeny showed the same flocculence and the same $H_2S$ formation as the parent strains, but their fermentation vigor and the percent ethanol formed were highly variable. By appropriate selection techniques the authors obtained a good flocculent champagne yeast with good fermentation vigor, high ethanol production, and lacking $H_2S$ formation.

A wine-making characteristic under single gene control can be manipulated by hybridization. Thornton (1985) mated the spores of a nonflocculent wine yeast with haploid cells of a heterothallic, flocculent laboratory strain. Five successive back breedings of the diploid hybrids with spores of the wine yeast resulted in a flocculent wine yeast with the same winemaking characteristics as the parent strain, except, of course, for the introduction of flocculence. Thornton (1982) found that the tolerance of wine yeast to $SO_2$ is under control of dominant polymeric genes.

Often the most difficult task is that of selecting a cell with a desired trait out of a very large number of cells that lack this trait. This task of selection is required whether one deals with variants of an industrial yeast, with hybrids, or with DNA recombinant cells. Rupela and Tauro (1984) reported an ingenious method for selecting a variant of a wine yeast that retains formed $H_2S$ within the cell and does not excrete it. Such cells are black in a nutrient agar medium containing 0.8% bismuth sulfite. Normal cells form colonies that excrete $H_2S$, and such colonies are white or brown. With this method they could find 4 colonies that remained black out of 69,000 colonies that excreted the sulfide. Wines produced with $H_2S$-retaining variants had only half the $H_2S$ concentration of controls.

It is sometimes desirable to lower the fusel oil content of a wine, and it is frequently desirable to lower it in distilled beverages. Rous and Snow (1983) mutagenized derivatives of a single spore isolate of a wine yeast. They obtained a leucine-requiring mutant that reduced the concentration of isoamyl alcohol in a wine by 50%. Pilot plant tests with this mutant showed that brandy produced with the yeast was equally lower in fusel oil content, and that the odor and flavor of the brandy were remarkably improved (Kunkee et al. 1987).

The introduction of the killer factor into a commercial strain of wine yeast (Montrachet) has already been mentioned. So far the powerful tools of genetic engineering and molecular biology have not made a substantial contribution to the improvement of wine yeast strains beyond traditional methods of hybridization and selection. "Undoubtedly one of the greatest difficulties facing those interested in improving wine yeasts, especially their flavor and aroma characteristics, is the exact specification of those metabolic changes which will result in a better wine" (Snow 1979). Additional information on methodologies for the genetic improvement of industrial wine yeast strains can be found in Vezinet (1981), Snow (1983), and Spencer and Spencer (1983).

## MICROBIAL SPOILAGE OF WINES

The microbial spoilage of wines will only be treated in summary form. It has been reviewed by Benda (1982), Lafon-Lafourcade (1983), Lafon-Lafourcade and Ribéreau-Gayon (1984), and comprehensively by Dittrich

(1987). Spoilage organisms produce a cloudiness that may vary from a slight haze to a heavy turbidity and to sediment formation in the bottle. Such microbial growth may not impair the palatability of the wine if it is due to fermentation in the bottle with species of wine yeasts, principally *Saccharomyces* or occasionally *Zygosaccharomyces bailii* or *Saccharomycodes ludwigii*.

The flavor of the wine is adversely affected by growth of film-forming yeasts that metabolize ethanol and produce a mat on the surface of the wine. Such yeasts belong to the genera *Pichia, Hansenula, Candida,* and others. The off-flavors, often described as oxidative, are due to the formation of acetic acid, acetaldehydes, and esters. (Film-forming yeasts of the genus *Saccharomyces* will be discussed below in connection with flor sherry production.)

In some instances the off-flavor produced by yeast spoilage can be traced to a specific substance. The mousy taint of wines is due to the growth of *Brettanomyces* species, but the characteristic compound, 2-acetyltetrahydropyridine, may also be formed by *Lactobacillus brevis* (Heresztyn 1986). In all of the mentioned instances of spoilage by yeasts the suggested countermeasures are a rapid and complete fermentation, that is, a sufficiently high ethanol content and absence of residual sugar, and in individual cases adequate sulfiting and absence of air, which can lead to the growth of film-forming yeasts. A system for identifying wine yeasts or spoilage yeasts in bottled wines may be based on colony formation in different growth media. Except for *Zygosaccharomyces bailii*, consistent growth responses can be obtained for 26 species of yeasts (Rodriguez 1987).

Wine may also be spoiled by the growth of bacteria that form polysaccharides and increase the viscosity of the wine. *Leuconostoc dextranicum* and *Pediococcus damnosus* as well as some other species have been found in such wines. Heterofermentative lactic acid bacteria may also produce mannitol, D-lactic acid, acetic acid, and glycerol. Lactic acid bacteria also are capable of forming higher concentrations of diacetyl as has been mentioned above. *Lactobacillus brevis* may metabolize glycerol to 1,3-propanediol. Acrolein, a bitter, acrid-tasting substance, can be formed from an intermediate of the glycerol degradation. The acetic acid bacteria are well-known wine spoilage organisms. It must merely be mentioned here that the organisms only grow in the presence of oxygen, resulting in the formation of acetic acid that may be converted later under anaerobic conditions to ethyl acetate.

Bacterial spoilage as well as yeast spoilage are generally contained by the following measures: elimination of spoiled or moldy grapes; rapid fermentation to dryness, particularly by inoculation with a heavy wine yeast population; fermentation to a sufficiently high ethanol concentration; adequate sulfiting; exclusion of air; and generally good sanitary practices.

# BOTRYTIS CINERIA

This fungus is a grape spoilage organism causing profound changes in the composition of musts. It is treated in this separate section because the changes caused by the mold are frequently beneficial and often eagerly desired. The mold develops on grapes during rainy or foggy periods, and particularly in the fall when rainy and warm periods alternate. The appearance of vineyards changes to a grayish color caused by the conidia of the fungus. Botrytic infections are infrequent in the United States because of the prevailing warm and dry weather in fall. They are quite frequent in northern Europe. Infection of unripe berries is very undesirable because the fungus metabolizes the sugars, but does not affect the acid composition to the same extent.

Infection of ripe berries also changes the composition of the must drastically. The hyphae of the mold penetrate the grape skin and make it porous. During dry weather water evaporates, and the must solids may be concentrated several fold. Sugar concentrations of 30-40% can often be reached. Only the most important changes of must composition due to growth of *Botrytis cineria* need be mentioned here. The concentration of glycerol may be increased from 0.1% to 2-3%.

The concentration of nitrogenous nutrients and of vitamins will be lowered. The must of botrytized grapes contains significant activities of extracellular enzymes of *B. cineria*. Two of these are most important for wine making. The activity of pectic enzymes is high, and the presence of pectinmethyl esterase leads to a somewhat higher concentration of methanol in the wine. A polyphenol-oxidizing enzyme, laccase (E.C. 1. 10. 3. 2.), is characteristic of botrytized grapes. It differs from the tyrosinase (E.C. 1. 10. 3. 1.) normally observed in musts by its broader spectrum of substrates, and hence botrytized musts are more sensitive to oxidation. The analysis of laccase in musts is a suitable means of determining the extent of *B. cineria* infection of grapes. Grassin (1987) has reported in detail on the extracellular enzymes of *B. cineria*.

The fermentation of grape musts from botrytized grapes is quite slow, largely because of the unusually high sugar concentration. Acetic acid concentrations may be high as a result of the metabolism of the mold and the metabolism of wine yeasts in high sugar media. The presence of a specific inhibitor of yeast fermentation has been reported by Ribéreau-Gayon et al. (1979) as well as that of a fermentation stimulant (Minarik 1983; Minarik, Kubalova, and Silharova 1986). These reports require further study.

Dittrich (1987) believes that the principal factor for slow and incomplete fermentations is the high osmotic pressure of the high-sugar musts. He recommends the addition of a nitrogen source, of thiamine, and adequate sulfiting to counteract the growth of the large populations of yeasts (particularly *Zygosaccharomyces bailii*, *Z. rouxii*, *Candida stellata*) and of lactic and acetic acid bacteria in musts from botrytized grapes.

However, excellent sweet white wines can be produced from botrytized grapes; the infection is known as noble rot (*pourriture noble*, *Edelfäule*). The characteristic flavor of the French Sauternes, produced in the Bordeaux area, is much sought after, and the wines are high-priced. It is believed with good reason that some of the best white German wines owe their slight sweetness and exceptional flavor to the occurrence of *Botrytis* infections.

Attempts have been made to produce the effects of noble rot in areas in which the infection is rare or absent. Nelson and Amerine (1957) and Nelson and Nightingale (1959) inoculated harvested grapes with *B. cineria* and could produce wines with a desirable Sauterne flavor. This process is very expensive and has not been used on a larger scale. Popper et al. (1964) and De Soto, Nightingale, and Huber (1966) attempted the production of natural sweet table wines with submerged cultures of *B. cineria*. De Jong, King, and Boyle (1968) achieved a similar effect with the addition of lyophilized *B. cineria* cells.

The mentioned processes have not been commercialized, possibly because of regulatory requirements. Little is known about the specific flavor compounds that are so characteristic of these wines. Watanabe and Shimazu (1981) point to the increased concentration of diethyl esters in wines from botrytized musts.

## BIOGENIC AMINES AND ETHYL CARBAMATE

Wines contain several substances that have been considered either harmful to humans or at least undesirable. Foremost among these substances is $SO_2$ to which some consumers are sensitive. The use of $SO_2$, which was practiced by the Romans, is almost universal in the wine industry. Even if no $SO_2$ is added to the must or the wine, wine yeast strains produce some $SO_2$ often in excess of 10 mg/L. Sulfites are principally used to prevent microbial spoilage in musts and wines. It is certainly not easy to strike a balance between the positive and negative effects of $SO_2$ on consumer health and safety.

Biogenic amines are produced in various foods by the decarboxylation of amino acids (Luthi and Schlater 1983). Tyramine is formed from tyrosine, histamine from histidine, putrescine from ornithine, cadaverine from lysine, phenylethylamine from phenylalanine, and so on. The reaction is catalyzed by some lactic acid bacteria, but biogenic amines are also found in very low concentrations in yeast extracts (Blackwell, Mabbit, and Marley 1969), and formation during incipient autolysis in wine cannot be excluded.

An extensive analysis of 230 European and North American wines showed the presence of histamine, putrescine, cadaverine, and tyramine, but 1,3-diaminopropane, agmaline, and tryptamine could not be detected (Zee et al. 1983). Table 4-11 shows the average concentration of the four detected amines in 102 red and 99 white wines. The concentrations of histamine and putrescine are significantly higher in red wines than in whites, which is in general agreement with other authors (for other tabulations see Lafon-Lafourcade 1983).

Some consumers experience migraine headaches and other seemingly allergic reactions upon drinking red wine, the so-called "red wine reaction" syndrome. It has been convenient to ascribe the reaction to the higher concentrations of biogenic amines in red wines. However, an objective evaluation with human subjects shed some doubt on the connection (Masyczerk and Ough, 1983). The question is still unresolved.

Ethyl carbamate (urethane) has been detected in table wines in concentrations often exceeding 20 $\mu$g/L, and in higher concentrations in fortified wines and distilled beverages. Ethyl carbamate is a known carcinogen; its toxicity is considerably less than that of some mycotoxins and nitrosamines. Ethyl carbamate was not formed during the fermentation, but developed if the wine was heated or stored for prolonged periods of time. For instance, addition of 0.3% urea as a yeast nutrient and heating of the final wine to 60°C for 20 minutes led to 176 $\mu$g/L ethyl carbamate (Ingledew, Magnus, and Sosulski 1987). Expected precursors of the formation of ethyl carbamate besides urea are arginine, ornithine, and citrulline.

The following suggestions have been made for limiting the amounts of ethyl carbamate in wine (Ough, Crowell, and Mooney 1988; Ough, Crowell, and Gutlove 1988): limitation of vineyard fertilization with nitrogen compounds plus addition of arginine-free yeast nutrients prior to the fermentation, or use of grape varieties low in arginine contents such as the White Riesling; the development of wine yeasts that will not metabolize arginine.

Table 4-11. Amine Contents in mg/L in Red and White Wines

| Amine | Red Wines | White Wines |
|---|---|---|
| Histamine | 5.73 ± 0.59 b | 3.35 ± 0.32 a |
| Putrescine | 5.13 ± 0.50 b | 1.94 ± 0.18 a |
| Cadaverine | 0.66 ± 0.11 a | 0.92 ± 0.07 a |
| Tyramine | 5.18 ± 0.43 a | 4.41 ± 0.48 a |

Source: Data from Zee et al. 1983.
Notes: Assays of 38 Canadian and 102 French wines.
Values within a given row without a common letter are significantly different ($P = 0.05$).

## TECHNOLOGY OF WINE MAKING

The processing technology of the wine industry will only be discussed in outline, since the microbiological aspects have already been treated above. However, two important secondary fermentations—the champagne and sherry fermentations—require special consideration.

### White and Red Table Wines

By far the largest production of wines falls into the category of white and red (and rose) table wines. In all wine-producing countries wines are made from varieties of the *Vitis vinifera* grape. In the United States, about 90% of the wine is produced with *Vitis vinifera* in California; the rest is produced mainly from the native *Vitis labrusca* strains (or hybrid strains) in the Northeast, Northwest, and a few other areas.

White wines are made from white (light) grapes. They may also be made from dark-skinned grapes if the juice is immediately separated after crushing, but it is not a common practice. Some common varieties of white wine grapes are French Colombard, Chenin Blanc, Chardonnay, Riesling, and Palomino. These are varieties of *Vitis vinifera*. Wines are also produced from varieties of raisin or table grapes.

White wines are designated either according to the wine type or according to the grape variety used. Wine type in this connection means such names as Rhine, Chablis, and Sauterne. The character of such white wines has been defined, but use of such designations remains vague. Wines named after grape varieties such as Chardonnay, Chenin blanc, and Gray Riesling must contain at least 75% of the must of the named grape. White California wine grapes yield a must of about 20°-22° Brix. A Brix of 20° results in a concentration of about 11% ethanol by volume in the dry table wine. The yield of wine per ton of grapes depends not only on the variety and maturity but also on processing conditions. It may vary from 150-180 gallons of wine per ton of grapes.

Figure 4-19 shows a flow diagram of white table wine production with an indication of the processing aids used (Long 1981). After crushing/destemming the juice is generally separated as quickly as possible from the skins and seeds. The free-run juice, that is, the juice that readily drains from the crushed grapes, is preferred for the production of high-quality wines.

After collection of the free-run juice, the grapes may be pressed. The juice is sulfited with 75-100 mg $SO_2$/L. The juice may then be clarified by settling and racking (decanting) or by use of centrifuges or vacuum filters. The juice is then inoculated with a selected yeast culture or active dry wine yeast as described above. White wines are usually fermented at low temperatures in the 10°-15°C range.

TECHNOLOGY OF WINE MAKING    205

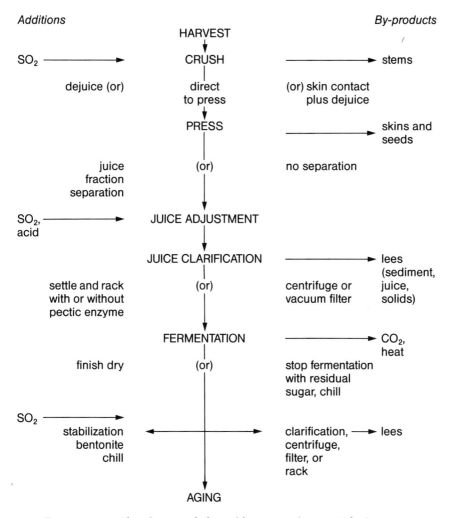

Figure 4-19.    Flow diagram of white table wine production. (*After Long 1981*)

The time required for the fermentation of white table wines varies inversely with the temperature. It may take 5 days at 15°C and 20 to 30 days at 10°C. After the wine has reached a low residual sugar level it may continue to ferment slowly until it is completely "dry." This point may be reached 4 to 6 weeks after cessation of the vigorous fermentation. White table wines should be fermented dry, that is, to a residual sugar concentration of no more than 0.2%, and preferably less than 0.12%. For table wines in which some sweetness is desired (e.g., Sauterne) one can start with must of high sugar content and arrest the fermentation when the desired degree of sweetness has been reached. Alternatively a dessert wine or grape juice concentrate may be added to sweeten the wine.

The white wine is then racked after most of the yeast and other insoluble materials have settled. It is then fined with bentonite or gelatin, and racked again and filtered. It may then be stored at a cool temperature until it is bottled.

The production of red table wines follows similar procedures except for several important differences. Obviously, they can only be made from dark-skinned grapes, and the coloring (anthocyanins) has to be extracted from the skins. For this reason the crushed grapes are fermented "on the skins." This fermentation is carried out at temperatures of $24°$-$29°C$, usually for 4 to 6 days until the Brix has dropped to 2 to 4. During fermentation on the skins the insoluble materials entrain carbon dioxide and rise to the top of the fermenter where they form a dense cap. The temperature inside the cap may well exceed $30°C$, and for this reason the cap must frequently be punched down into the wine or the wine must be pumped from the bottom of the tank and sprayed on top of the cap.

The fermentation on the skin serves to extract color, flavor, and tannins from the grape skins. The wine may then be drained off the pomace, and the pomace can be pressed to yield the wine that is high in color and astringency. The fermentation may then be completed to dryness. As described earlier, red wines frequently undergo a malolactic fermentation. For acid wines MLF is encouraged, often with the addition of cultures of malolactic bacteria.

Wineries of the coastal regions of California consider the malolactic fermentation desirable because it contributes to flavor complexity and smoothness. In the interior valley of California, MLF is usually avoided by means previously described. Such differences in treatment are dictated by the acidity and sugar-to-acid ratio of the musts in these different locations.

Red wines are designated by type, such as Claret, Burgundy, Chianti, or by a general term such as mountain red wine. Such terms are not particularly meaningful as applied to wines produced in the United States since there is no official specification for such different wine types. The wines may, of course, also be designated by the grape variety (at least 75%) used in their production, such as Cabernet Sauvignon or Pinot Noir. For a description of table wine production the following references should be consulted: Amerine et al. (1982), Long (1981), Martini (1981), Skofis (1981), and Cooke and Berg (1983). Specific processing conditions for native American grapes have been described by Klein (1981) and Wagner (1981).

### Fruit Wines

The production of fruit wines such as apple (cider), cherry, or berry wines generally follows procedures in use with grape table wines. The fruit juice

## TECHNOLOGY OF WINE MAKING

or fruit pulp often has a low Brix, which results in lower alcohol concentrations and may require amelioration with added sucrose. Fruit juices are often deficient in yeast nutrients and are more likely to require the addition of so-called yeast foods than grape wines (Beech and Carr 1977).

### Fortified Wines

These dessert wines are characterized by an alcohol content of 17-21% by volume. With the exception of "dry" sherry they are sweet wines with a Brix of 4 to 10. Sweetness can be obtained as previously described by fortification with brandy (only grape brandy may be used) before the fermentation has been completed. Neutral spirits of low fusel oil content are preferred for the fortification. The distilling material may be a wine, or more often the residual material at the bottom of a tank after the wine has been racked. Such dessert wines are port (red) and sherry (amber) as well as Angelica, Muscatel, and Tokay. Sherry, which may be produced by a secondary fermentation, will be considered separately because of the interesting aspects of an aerobic yeast fermentation (Goswell 1987).

### Sherry

Sherries are fortified dessert wines. They are distinguished from all other wines by the oxidative conditions during their production. This oxidation may be biological as in the traditional Spanish solera process with film (flor)-forming yeasts, or it may be chemical by application of heat and aeration as in the sherry baking process. The baked sherry, although readily identified by taste as sherry, differs in flavor from the flor sherry. A few related processes such as the submerged culture sherry will also be considered.

The Palomino grape is the grape of choice both in Spain and California. In Spain, the Pedro Ximenez grape is used to make concentrates for the production of sweeter sherries. In the United States, other white grapes with neutral flavors such as the Thompson Seedless may be used for sherry production. The grapes should be mature since high alcohol concentrations are desired, and the pH should be low, preferably 3.2. A lowering of the pH without an increase in titratable acidity can be achieved by the traditional plastering process. The addition of gypsum to the must leads to the following reaction (Ta is tartaric acid):

$$CaSO_4 + 2\ KHTa \longrightarrow CaTa + K_2SO_4 + H_2Ta$$

In the Spanish solera process the natural flora is responsible for the primary alcoholic fermentation and one finds the succession of species as previously described for table wines. At higher alcohol levels flor-forming strains of *Saccharomyces cerevisiae* predominate (including many earlier synonyms such as *S. beticus, S. bayanus, S. prostoserdovii, S. cheriensis,* at one time designated collectively as *Mycoderma vini). S. cerevisiae* strains are also responsible for the formation of a yeast film over the surface of the wine. This film or flor begins to form soon after completion of the alcoholic fermentation when the wine is exposed to air in partially filled barrels. The fermentation continues as an oxidative (respiratory) process with energy derived from the oxidation of ethanol to acetaldehyde.

A solera consists of a number of casks of 130-gallon capacity, arranged in three to five stages. Annually a portion (25-33%) of the wine of the final, or solera, stage is drawn from each cask, clarified, filtered, sulfited, and bottled. The wine is replaced with the same volume of wine drawn from the preceding stage, and this process is repeated until a portion of the wine has moved from each cask of each stage to a cask of the succeeding stage. During the transfer of wine from one stage of the solera to the next, care is taken to leave the film of flor yeast undisturbed. In essence, this process is a multistage, semicontinuous fermentation system with a dilution rate of about 40 to 130 per year. For a five-stage system the average age of the wine drawn from the last (solera) stage is 12 years. It represents a mixture of vintages of which the youngest component is at least 5 years old.

During the process of fermentation and maturation in the solera the wine has an ethanol concentration of 14-15% by volume. At higher ethanol concentrations the flor will not develop; at lower concentrations there is a danger of growth of acetic acid bacteria under the semiaerobic conditions of the process. The activity of the flor yeast results in a drastic reduction of the glycerol and malic acid content of the wine; acetic acid and amino acids are assimilated. There is a large increase in the concentration of acetaldehyde (to 200-400 mg/L) and acetal, and the typical flavor of flor sherry develops.

The identification and analysis of flavor compounds in sherry have been extensively studied by Webb and his colleagues at the University of California at Davis. He concludes that "while more than 400 volatiles have been identified in fermented beverages, recent wine aroma studies conclude that specific aromas result from unknown compounds or from complicated patterns of volatiles, not from the presence or absence of one or two *impact* compounds" (Webb and Noble 1976; Kung et al. 1980).

Martinez de la Ossa, Perez, and Caro (1987) have provided detailed analyses of each of the stages of the solera process. The development of acetaldehyde during the fermentation is generally taken as a measure of overall flavor development, but acetaldehyde by itself is not responsible for the aroma. The solera process that has been briefly described results in the production of *fino* sherries. If the maturation is achieved by longtime exposure of the fortified wine to air at elevated summer temperatures, and in the absence of yeast flor, the sherry is called *oloroso*.

A flor-type sherry may also be produced by submerged culture of a film-forming yeast in a sherry base wine of 14-15% ethanol by volume. This process was first developed in the 1950s in Australia, Canada, and California. More recently it has been described by Amerine et al. (1982) and Posson (1981). It is practiced on a limited scale in California, and some of the wines produced with the submerged culture process are used for blending with baked sherries to give a more complex flavor. The sherry-based wine, called *shermat*, is inoculated with a film-forming strain of *S. cerevisiae*, usually in the form of an active dry yeast. The yeast must first ferment grape juice in order to acclimatize itself to higher alcohol concentrations. Air or oxygen is supplied to the fermenters by sparging, which also serves to keep the yeast in suspension.

Figure 4-20 shows the development of acetaldehyde and yeast growth in submerged culture as a function of time. Final acetaldehyde concentrations that can be achieved with this process are quite high (800-1200 mg/L). The process permits good control over the degree of flavor

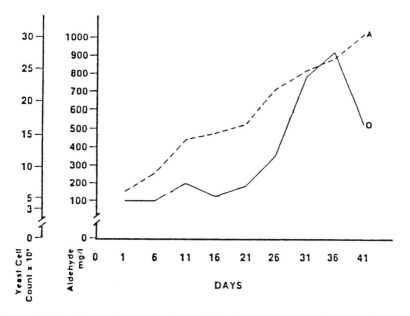

**Figure 4-20.** Yeast cell counts and acetaldehyde concentrations during a submerged sherry fermentation with a 50% transfer after 8 days:

(A)   _ _ _ _ _   aldehyde concentration
(O)   ⎯⎯⎯⎯   yeast cell counts

(*From Posson 1981*)

development, since it can be stopped by fortification of the wine to about 18% ethanol by volume.

The bulk of the sherry produced in the United States is so-called baked sherry. The shermat is fortified to 18-20% ethanol and "baked" at 50°-60°C for 10 to 20 weeks, the shorter time period applying to the higher temperature. In the eastern United States, the Tressler process calls for passage of finely dispersed oxygen through the shermat at 60°C. During baking the wine loses the fruity character of the *Vitis labruska* grapes and assumes a sherrylike character (Mattick and Robinson 1960). In many respects the baked sherry resembles Madeira, which is similarly heat processed.

Other variants of sherry processes exist. In Sardinia, a nonfortified sherrylike wine is produced with an ethanol content of 15% by volume. An authentic flor could be formed by film yeasts, with strains of *Saccharomyces bayanus* and *Saccharomyces prostoserdovii* giving the best results (Fatichenti, Farris, and Deiana 1983).

## Champagne

The production of sparkling wines in France dates from the middle of the sixteenth century. Sufficient carbon dioxide is trapped if the final stage of a table wine fermentation is conducted in a closed bottle. This "natural" method of fermentation is currently not important for the production of sparkling wines. The general availability of sugar made it possible in the early part of the nineteenth century to control the fermentation in the bottle by use of a still wine and addition of a predetermined amount of sugar. This *méthode champenoise* is the method of choice for the production of premium champagne. In Europe, the term *champagne* may be used only for wines produced by this method in the French Champagne region. The Germans refer to this type as *Sekt* or *Schaumwein;* the Italians refer to it as *Spumante* (foaming wine). In the United States and Australia, the term *champagne* may be used legally, a practice that is not appreciated in France.

By definition, a sparkling wine in the United States must have a carbon dioxide content exceeding 0.392 g/100 ml, which corresponds to a pressure of 15.9 psi at 15°C. Normally sparkling wines are produced to pressures of 75-90 psi at 10°C with a similar range for artificially carbonated wines. The artificially carbonated wines are made by the injection of $CO_2$ into a still wine and must be labeled to indicate artificial carbonation. Such wines do not undergo a secondary fermentation and require no further discussion here.

Sparkling wines are made from still table wines by a secondary fermentation. The table wine used is called *cuvée*. It should be made from free-run juice of no more than 19-20° Brix, and the wine should have a relatively low pH (3-3.2), a relatively high acidity (0.65-0.75% as tartaric acid), and an ethanol content not exceeding 10-11% by volume. At higher ethanol levels it is often more difficult to start the secondary fermentation.

Grapes for sparkling wines are harvested before they are fully mature in order to assure the preferred sugar-to-acid ratio. However, many wineries do not produce table wines specially for sparkling wine production, but choose cuvées of suitable composition after the white wines have been produced. Commonly used grape varieties are the French Colombard, the Chenin Blanc, and the Chardonnay and the white juice of the Pinot Noir for premium-quality sparkling wines. The fermentation of the wine is preferably carried out at the same low temperature as appropriate for dry white table wines.

Three types of secondary fermentation processes are in use. The traditional *méthode champenoise* requires the following steps:

1. Introduction of a dry white wine (cuvée) into bottles; addition of about 2.5% sugar or sugar syrup and yeast starter culture and tight closure of the bottles.
2. Secondary fermentation of the added sugar in the bottle to produce the desired pressure of 75-90 psi.
3. Maturing of the wine for a year or longer in the bottle and on the yeast, and inversion of the bottle to permit formation of a plug of sedimented yeast in the neck.
4. Freezing of the yeast plug in the neck of the bottle and removal (disgorging).
5. Addition of dry white wine or sugared white wine or brandy to bring the contents of the bottle up to full volume, and final closure of the bottle.

Sparkling wines produced by this method may be labeled "fermented in *this* bottle."

The second method is called the *transfer* method. It also involves fermentation in the bottle and follows steps 1 and 2 above. The time of maturation may vary considerably and is generally shorter than for the *méthode champenoise*. The yeast sediment is removed by transfer of the content of the bottles into a tank (under counterpressure to preserve the $CO_2$ content of the wine). The wine is then filtered to remove the yeast, and bottled again under counterpressure. Such wines may be labeled "bottle fermented."

The third method is called the *bulk fermented* or *Charmat* process. The cuvée is fermented in steel or other pressure-resistant tanks after the addition of sugar and yeast inoculum. After completion of the fermentation with generation of the desired pressure the wine is filtered and bottled. Such wines must be labeled "bulk fermented" or "Charmat

## 212 WINE YEASTS

process." Figure 4-21 shows a schematic diagram of the process for the *méthode champenoise* and the bulk fermentation (Berti 1981).

The microbiologist is greatly interested in the adaptation of yeast to the secondary fermentation, that is, its growth and fermentation rate in the cuvée and the flavor modification occasioned by yeast autolysis on longtime aging. For the secondary fermentation the use of flocculent strains is preferred. These are generally but not exclusively strains of *Saccharomyces bayanus* (now also classified as *S. cerevisiae*). Flocculence is needed to permit removal of a fairly compact yeast plug from the neck of the bottle, and it

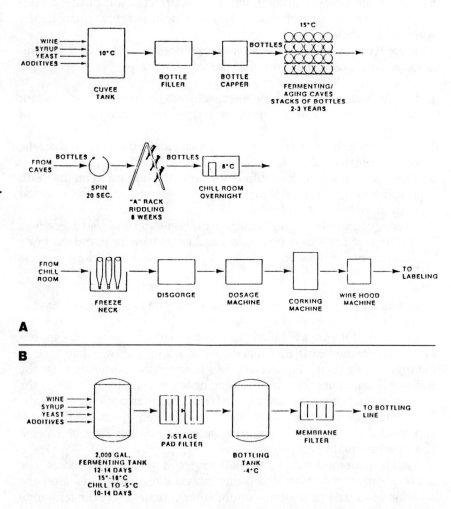

**Figure 4-21.** Flow diagram of the secondary sparkling wine fermentation: A, *méthode champenoise*; B, bulk fermentation. (*From Berti 1981*)

facilitates removal by filtration in the transfer and bulk processes. The importance of flocculence in yeast strains is treated in more detail in Chapter 3.

The effect of pressure on growth of yeast has already been mentioned. One of the more thorough investigations was undertaken early by Kunkee and Ough (1966). Under $CO_2$ pressure low pH (3.0, 3.3, and 3.7) was more inhibitory to growth, and so was a higher ethanol concentration (11%, 10%, and 9% by volume), but the inhibitory effect of higher pressures of $CO_2$ was overriding.

As expected, fermentation rate was not inhibited as strongly as yeast growth. Fairly rapid fermentation rates could be obtained by inoculation with a high level of yeast. The dependence of overall fermentation rate on yeast cell concentration, while obvious, is often overlooked. Figures 4-22A and 4-22B demonstrate this dependence (Cahill, Carroad, and Kunkee 1980). A slight overpressure of $CO_2$ slows yeast growth (Fig. 4-22A), but fermentation rates are equal once cell counts of 20-30 $\times$ $10^6$ cells/ml have been attained (Fig. 4-22B). In interpreting these figures, it must be remembered that these fermentations were carried out at relatively low pressures (0.3 and 0.6 atm) compared to practical secondary sparkling wine fermentations.

In practice, yeasts are usually grown in starter cultures to acclimatize to the high ethanol concentration encountered in the cuvée, which applies also to the use of active dry "champagne" yeasts. The level of inoculum is often 1-4 $\times$ $10^6$ cells/ml, which permits completion of the fermentation in the bottle in 4 to 6 weeks. Bulk fermentations with higher cell counts may be completed in 1 to 2 weeks.

Fermentation in the bottle is a labor-intensive process and in wineries that are more highly automated it requires expensive equipment. The *méthode champenoise* requires more labor than the transfer process for removal of the yeast plug from each individual bottle. Premium champagne is produced by fermentation in the bottle where the champagne is aged on the yeast for a year or longer. During fermentation in the bottle amino acids are taken up by the yeast. Amino acids are excreted after completion of the bottle fermentation.

However, a second phase of excretion of amino acids follows after several months when the wine is aged on the yeast. This phase corresponds to the autolysis of the yeast, with the greatest enrichment occurring between the sixth and the twelfth month (Feuillat and Charpentier 1982). Changes in the ultrastructure of the yeast cell wall and changes in lipid composition appear to be correlated with yeast autolysis (Piton, Charpentier, and Troton 1988).

The desirable development of flavor in the bottle-fermented champagne has long been attributed to yeast autolysis (Filipov 1963; Feuillat

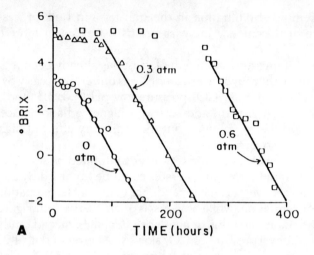

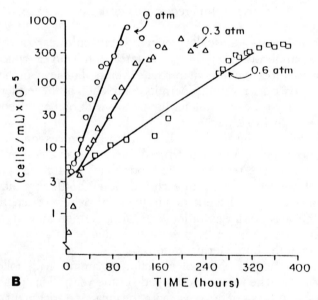

**Figure 4-22.** Growth of yeast and fermentation rate at varying pressures of $CO_2$: A, growth of yeast at ambient pressure and at 0.3 and 0.6 atm (gauge); B, fermentation rate expressed as drop in °Brix per time unit. The slopes of the regression curves are almost identical. (*From Cahill, Carroad, and Kunkee 1980*)

1980), and the preparation of various yeast autolysates for addition to sparkling wines has been suggested. So far it has not resulted in the commercial use of yeast autolysates for this purpose.

The ability to immobilize microorganisms has led to an interesting application of immobilized yeast in the secondary fermentation of champagne in the bottle. Immobilization of yeasts in alginate spheres of about 2 millimeters diameter greatly facilitated removal of the yeast after completion of the bottle fermentation (Coulon et al. 1983). Fumi et al. (1988) described the preparation of yeast Ca-alginate beads in more detail and found that fermentation rate and flavor development during the secondary bottle fermentation were equivalent to the use of free cells.

### Continuous Fermentation of Wines

In the process of Riddel and Nury (1958) grape must was fermented in a single-stage continuous reactor at a dilution rate of 0.5 per hour, at a Brix of 3, and at a temperature of 26.5°-29°C, but the fermentation had to be completed in a separate holding tank. The system was operated for several years and then abandoned. Wine fermentations are not suited to continuous, homogeneous operation because of the strong inhibition at the desired ethanol concentrations. Continuous heterogeneous operation, as it has been practiced for some years in the brewing industry, is of course possible. Egamberdiev (1967) described a 10-stage continuous system that required a 5-8-day dwell time to produce a wine with less than 0.2% residual sugar. A continuous system for sparkling wine production is operational in the Soviet Union (Brusilovskii et al. 1977).

## FURTHER READINGS

The following books and book chapters deal with the microbiology of wines: Ribéreau-Gayon et al. 1976; Kunkee and Goswell 1977; Troost 1980; Amerine et al. 1982; Benda 1982; Lafon-Lafourcade 1983; and Goswell 1987. The following two books are devoted *entirely* to the microbiology of wines: Dittrich (1987) and Minarik and Navara (1986).

## REFERENCES

Adams, A. M. 1953/1954. Studies on the storage of yeast. II. Wine yeast starter stored as moist yeast. *Rept. Ontario Hort. Expt. Sta. and Production Lab.* 98.

Amerine, M. A., H. W. Berg, and W. V. Cruess. 1967. *The Technology of Wine Making*. AVI Publishing Co., Westport, Conn.

Amerine, M. A., and G. Thoukis. 1958. The glucose-fructose ratio in California grapes. *Vitis* **1**:224-229.

Amerine, M. A., R. E. Kunkee, C. S. Ough, V. L. Singleton, and A. D. Webb. 1982. *The Technology of Wine Making*. AVI Publishing Co., Westport, Conn.

Andreasen, A. A., and T. J. B. Stier. 1953. Anaerobic nutrition of S. *cerevisiae*. I. Ergosterol requirement for growth in a defined medium. *J. Cell Comp. Physiol.* **41**:23-36.

Äyrapää, T. 1971. Biosynthetic formation of higher alcohols by yeasts. *J. Inst. Brewing* **77**:266:276.

Barillère, J. M., P. Bidan, and C. Dubois. 1983. Thermal resistance of yeasts and lactic acid bacteria isolated from wine (French). *Bull. de l'OIV* **56**(627):327-351.

Bartley, C., R. Beelman, K. Hicks, and G. Sapers. 1987. Ascorbyl decanoate—a potential new preservative to replace sulfur dioxide in wine (Abstract). *Am. J. Enol. Vitic.* **39**(1):99, 1988.

Beech, F. W., and J. G. Carr. 1977. Cider and perry. In *Alcoholic Beverages*, A. H. Rose (ed.). Academic Press, New York.

Beelman, R. B., F. F. McArdle, and G. R. Duke. 1980. Comparison of L. oenos strains ML-34 and PSU 1 to induce malo-lactic fermentation in Pennsylvania red table wine. *Am. J. Enol. Vitic.* **31**(3):269-276.

Bell, A. A., C. S. Ough, and W. M. Kliewer. 1979. Effects of must and wine composition, rates of fermentation, and wine quality of nitrogen fertilization of *Vitis vinifera* var. Thompson Seedless grapevines. *Am. J. Enol. Vitic.* **30**(2):124-129.

Benda, I. 1974. *Schizosaccharomyces* yeasts and their effect on acid reduction during vinification (French). *Vignes Vin, No. Spec., Internat. Symp. Oenol., Arc Senans, 31-36, 1973*.

Benda, I. 1982. Wine and brandy. In *Prescott and Dunn's Industrial Microbiology*, G. Reed (ed.). AVI Publishing Co., Westport, Conn.

Benda, I. 1983. Botrycides, their active ingredients and formulation, in microbiological tests (German). *Wein-Wiss.* **38**:41-50.

Bergner, K. G. 1968. Amino acids in sparkling wines and their variation as a function of manufacturing procedures (French). *Vigne Vin* **41**:450-467.

Berti, L. A. 1981. Sparkling wine production in California. In *Wine Production Technology in the United States*, M. A. Amerine (ed.). American Chemical Society, Washington, D.C.

Bertrand, A., M. O. Dubernet, and P. Ribéreau-Gayon. 1975. Trehalose, the principal disaccharide in wine (French). *C. R. Acad. Sci. Paris* **29**(3):220.

Bidan, P., and J. Maugenet. 1981. Recent information on use of active dry wine yeast (French). *Bull de l'OIV* **54**(601):241-254.

Bisson, J., J. Daulny, and A. Bertrand. 1980. Effect of fermentation temperature on the composition of white table wine (French). *Conn. Vigne Vin* **14**:195-202.

Blackwell, B., L. A. Mabbit, and W. Marley. 1969. Histamine and tyramine content of yeast products. *J. Food Sci.* **34**:47-51.

Boulton, R. 1979. The heat transfer characteristics of wine fermentations. *Am. J. Enol. Vitic.* **30**(2):152-156.

Bréchot, P., J. Chauvet, P. Dupuy, M. Groson, and A. Rabatu. 1971. Oleanoic acid as anaerobic growth factor of wine yeasts (French). *C. R. Acad. Sci.* **272**:890-893.

Brusilovskii, S. A., A. I. Mel'nikov, A. A. Merzhanian, and N. G. Sarishvili. 1977. *Production of Soviet Champagne in Continuous Culture* (Russian). Pishchevia Promy'shlennost, Moscow.

Cahill, J. T., P. A. Carroad, and R. E. Kunkee. 1980. Cultivation of yeast under carbon dioxide pressure for use in continuous sparkling wine production. *Am. J. Enol. Vitic.* **31**(1):46-52.

# REFERENCES 217

Castelli, T. 1954. The organisms responsible for wine fermentations (German). *Arch. Mikrobiol.* **20**:323-342.
Castor, J. G. 1953. Experimental development of compressed yeast fermentation starters. *Wines and Vines* **34**(8):27; **34**(9):33.
Conner, A. J. 1983. The comparative toxicity of vineyard pesticides to wine yeasts. *Am. J. Enol. Vitic.* **34**(4):278-279.
Cooke, G. M., and H. W. Berg. 1983. A re-examination of varietal table wine processing practices in California. I. Grape standards: grape and juice treatment and fermentation. *Am. J. Enol. Vitic.* **34**(4):249-259.
Coulon, P., B. Duteurtre, M. Charpentier, A. Parenthoen, C. Badour, and J. P. Moulin. 1983. New perspectives in the methode champenoise: Use of immobilized yeast (French). *Le Vigneron Champenois, France* 11, pp. 516-532.
Crapisi, A., M. P. Nuti, A. Zamorani, and P. Spettoli. 1987. Improved stability of immobilized *Lactobacillus* spec. cells for the control of malolactic fermentation in wine. *Am. J. Enol. Vitic.* **38**(4):310-312.
D'Amore, T., and G. G. Stewart. 1987. Ethanol tolerance of yeast. *Enzyme Microb. Technol.* **9**(6):332-330.
Davis, C. R., D. J. Wibowo, T. H. Lee, and G. H. Fleet. 1986. Growth and metabolism of lactic acid bacteria during and after malolactic fermentation of wines at different pH. *Appl. Environ. Microbiol.* **51**(3):539-545.
De Jong, D. W., A. King, Jr., and F. P. Boyle. 1968. Modification of white table wines with enzymes from *Botrytis cineria* Pers. *Am. J. Enol. Vitic.* **19**:228-237.
De Soto, R. T. 1955. Integrating yeast propagation with winery operation. *Am. J. Enol. Vitic.* **6**:26-30.
De Soto, R. T., and R. Huber. 1968. The effect of tannic acid on the secondary fermentation of champagne. *Am. J. Enol. Vitic.* **19**:246-253.
De Soto, R. T., M. S. Nightingale, and R. Huber. 1966. Production of natural sweet table wines with submerged culture of *Botrytis cineria* Pers. *Am. J. Enol. Vitic.* **17**:191-202.
De Wet, P., O. P. H. Augustyn, C. J. van Wyck, and W. A. Joubert. 1978. Odour thresholds and their application to wine flavour characteristics. *Proc. S. Afr. Soc. Enol. Vitic.*, pp. 28-42.
Dittrich, H. H. 1963. The alcoholic fermentation of L-malic acid by *S. pombe* var. *acidodevoratus* (German). *Zentralbl. Bakteriol. Parasitenkd. Infektionskr., Hyg. Abt. 2* **119**:406-421.
Dittrich, H. H. 1980. Effect of bacterial acid degradation on wine composition (German). *Wein-Wiss.* **35**:421-429.
Dittrich, H. H. 1987. *Microbiology of Wines* (German). Verl. Ulmer, Stuttgart.
Dittrich, H. H., and A. Barth. 1984. $SO_2$ content, $SO_2$ binding substances, and acid reduction in German wines (German). *Wein-Wiss.* **39**(3):184-200.
Dittrich, H. H., and T. Staudenmayer. 1968. $SO_2$ formation, formation of hydrogen sulfide odor and its removal (German). *Dtsche. Wein-Ztg.* **24**:707-709.
Domercq, S. 1957. Classification of wine grapes in the Gironde (French). *Ann. Technol. Agric.* **6**:5-58, 139-183.
Edwards, C. G., and R. B. Beelman. 1987. Inhibition of the malolactic bacterium, *L. oenos* (PSU-1), by decanoic acid and subsequent removal of the inhibition by yeast ghosts. *Am. J. Enol. Vitic.* **38**(3):239-242.
Egamberdiev, N. B. 1967. Study of the must fermentation by *Saccharomyces vini* yeast during continuous culture (Russian). *Prikl. Biokhim, Mikrobiol.* **3**:458-463.
Fatichenti, F., G. A. Farris, and P. Deiana. 1983. Improved production of Spanish-type sherry by using selected indigenous film-forming yeasts as starters. *Am. J. Enol. Vitic.* **34**(4):216-220.

Feuillat, M. 1980. Aging of champagne on the yeasts; effect on the enrichment and development of the aroma of the wine (French). *Rev. Franc. Oenol.* **16**(79):35-46.
Feuillat, M., and C. Charpentier. 1982. Autolysis of yeasts in champagne. *Am. J. Enol. Vitic.* **33**(1):6-13.
Fiechter, A. 1963. Pilot plant for the continuous production of microorganisms (German). *Chem. Ing.-Technik* **34**:696.
Filipov, B. A. 1963. Production of enzyme concentrates and their use (Russian). *Vinodelie Vinogradarstvo SSSR* **23**(2):11-14.
Fleet, G. H., S. Lafon-Lafourcade, and P. Ribéreau-Gayon. 1984. Evolution of yeasts and lactic acid bacteria during fermentation and storage of Bordeaux wines. *Appl. Environ. Microbiol.* **48**(5):1034-1038.
Fumi, M. D., G. Triloli, M. G. Colombi, and O. Colagrande. 1988. Immobilization of *Saccharomyces cerevisiae* in calcium alginate gel and its application to bottle-fermented sparkling wine production. *Am. J. Enol. Vitic.* **39**(4):267-272.
Gallander, J. F. 1982. Influence of different wine yeasts on the quality of Vidal blanc wines. *Dev. Ind. Microbiol.* **23**:123-129.
Gnaegi, F., et al. 1984. The bacteriophages of *Leuconostoc oenos* and progress in the mastery of the malolactic fermentation of wines (French). *Rev. Suisse Vitic., Arboric., Hortic.* **16**(2):59-65.
Goldman, M. 1963. Factors influencing the rate of carbon dioxide formation in fermented-in-the bottle champagne. *Am. J. Enol. Vitic.* **14**:36-42.
Goswell, R. W. 1987. Microbiology of fortified wines. In *Developments in Food Microbiology*, vol. 2, R. K. Robinson (ed.). Elsevier Appl. Sci. Publ., New York.
Grassin, C. 1987. Extracellular enzymes excreted by *Botrytis cineria* into must. Enological and phytopathological applications (French). Ph.D. thesis, Enological Institute, University of Bordeaux, France.
Groat, M. and C. S. Ough. 1987. Effects of insoluble solids added to clarified musts on fermentation rate, wine composition and wine quality. *Am. J. Enol. Vitic.* **29**(2):112-119.
Guijarro, J. M., and R. Lagunas. 1984. *Saccharomyces cerevisiae* does not accumulate ethanol against a concentration gradient. *J. Bacteriol.* **160**(3):874-878.
Guymon, J. F., and J. E. Heitz. 1952. The fusel oil content of California wines. *Food Technol.* **6**:359-362.
Hara, S., Y. Iimura, and K. Otsuka. 1980. Breeding of useful killer wine yeasts. *Am. J. Enol. Vitic.* **31**(1):28-33.
Haubs, H., H. Muller-Spath, and T. Loescher. 1974. The effect of carbon dioxide on wine (German). *Dt. Weinbau* **29**:930-934.
Heard, G. M., and G. H. Fleet. 1985. Growth of natural yeast flora during the fermentation of inoculated wines. *Appl. Environ. Microbiol.* **50**(3):727-728.
Henick-Kling, T., T. H. Lee, and D. J. D. Nicholas. 1986. Inhibition of bacterial growth and malolactic fermentation in wine by bacteriophage. *J. Appl. Bacteriol.* **61**:287-293.
Heresztyn, T. 1986. Formation of substituted tetrahydropyridines by species of *Brettanomyces* and *Lactobacillus* isolated from mousy wines. *Am. J. Enol. Vitic.* **37**(2):127-132.
Hernandez, M. R. 1964. Production of $H_2S$ by wine yeasts grown with several sulfur containing compounds (Spanish). *Semana Vitivinicola* **19**:2359-2360.
Ingledew, W. N., and R. E. Kunkee. 1985. Factors influencing sluggish fermentations of grape juice. *Am. J. Enol. Vitic.* **36**(1):65-76.
Ingledew, W. M., C. A. Magnus, and F. W. Sosulski. 1987. Influence of oxygen on proline utilization during the wine fermentation. *Am. J. Enol. Vitic.* **38**(3):246-248.
Jacobson, G. K. 1985. Eliminating undesirable yeast strains with killer yeasts. *Eastern Grape Grower and Winery News*, Aug./Sept. pp. 29-31.

Jones, R. S., and C. S. Ough. 1985. Variations in the percent ethanol (v/v) per °Brix conversions of wines from different climatic regions. *Am. J. Enol. Vitic.* 36(4):268-270.
Killian, E., and C. S. Ough. 1979. Fermentation esters—formation and retention as affected by fermentation temperature. *Am. J. Enol. Vitic.* 30(4):301-305.
King, S. W. 1985. Recent developments of industrial malolactic starter cultures for the wine industry. *Dev. Ind. Microbiol.* 26:311-321.
King, S. W., and R. B. Beelman. 1986. Metabolic interactions between *Saccharomyces cerevisiae* and *Leuconostoc oenos* in a model grape juice/wine system. *Am. J. Enol. Vitic.* 37(1):53-60.
Klein, J. K. 1981. Wine production in Washington state. In *Wine Production Technology in the United States*, M. A. Amerine (ed.). American Chemical Society, Washington, D.C.
Knappstein, A. T., and B. C. Rankine. 1970. Commercial application of pure yeast in wine making and its influence on wine quality. *Austral. Wine, Brew., Spirit Rev.* 89(3):52-54.
Kockova-Kratochvilova, A. 1981. Characteristics of industrial microorganisms. *Biotechnology,* vol. 5, H. J. Rehm and G. Reed (eds.). VCH Publishing Company, New York.
Kraus, J. K., G. Reed, and J. C. Villettaz. 1983/1984. Active dry wine yeasts (French). *Conn. Vigne Vin* 17(2):93-103; 18(1):1-26.
Kraus, J. K., R. Scopp, and S. L. Chen. 1981. Effect of rehydration on dry wine yeast activity. *Am. J. Enol. Vitic.* 32(2):132-134.
Kreger van Rij, N. J. W. 1984. *The Yeasts: A Taxonomic Study,* 3rd. ed. Elsevier, Amsterdam.
Kung, S., G. F. Russell, B. Stackler, and A. D. Webb. 1980. Concentration changes in some volatiles through six stages of a Spanish solera. *Am. J. Enol. Vitic.* 31(2):187-191.
Kunkee, R. E. 1975. A second enzymatic activity for decomposition of malic acid by malolactic bacteria. In *Lactic Acid Bacteria in Beverages and Food,* J. G. Carr, C. V. Cutting, and G. C. Whiting (eds.). Academic Press, New York.
Kunkee, R. E., and M. A. Amerine. 1970. Yeasts in wine making. In *The Yeasts,* vol. 3, A. H. Rose and J. S. Harrison (eds.). Academic Press, New York.
Kunkee, R. E., and R. W. Goswell. 1977. Table wines. In *Economic Microbiology,* vol. 1, A. H. Rose (ed.). Academic Press, New York.
Kunkee, R. E., and C. S. Ough. 1966. Multiplication and fermentation of *Saccharomyces cerevisiae* under carbon dioxide pressure in wine. *Appl. Microbiology* 14:643-648.
Kunkee, R. E., E. Bordeu, S. E. Kerns, and M. R. Vilas. 1987. Brandy production with decreased fusel oil (higher alcohol) content by use of a leucine-less yeast mutant for fermentation. *Abstr. Am. Soc. Enol. Vitic., Ann. Meeting,* June, Anaheim, Calif.
Lafon-Lafourcade, S. 1983. Wine and brandy. In *Biotechnology,* vol. 5., H. J. Rehm and G. Reed (eds.). VCH Publishing Company, New York.
Lafon-Lafourcade, S., C. Geneix, and P. Ribéreau-Gayon. 1984. Inhibition of the alcoholic fermentation of grape must by fatty acids produced by yeasts and their elimination by yeast ghosts. *Appl. Environ. Microbiol.* 47(6):1246-1249.
Lafon-Lafourcade, S., and P. Ribéreau-Gayon. 1984. Wine spoilage by acetic acid and lactic acid bacteria (French) *Conn. Vigne Vin* 18(1):67-82.
Lafon-Lafourcade, S., F. Larue, P. Bréchot, and P. Ribéreau-Gayon. 1977. Steroids as survival factors for yeasts during the alcoholic fermentation of grape musts (French). *C. R. Acad. Sci.* 284:1938-1942.
Laranidis, P., and S. Lafon-Lafourcade. 1982/1983. Alcohol yield based on sugar during the fermentation of grape must (French). *Rapport des Activités et Recherches,* Inst. d'Oenologie, Univ. de Bordeaux.
Larue, F., C. Geneix, S. Lafon-Lafourcade, and P. Ribéreau-Gayon. 1985. First observations on the mode of action of yeast cell wall material (French). *Conn. Vigne Vin* 18:155-163.
Larue, F., J. N. Murakami, Boidron, and L. Fohr. 1986. First observations on the use of

octanoic and decanoic acids as substitutes for sulfur dioxide in the stabilization of sweet wines (French). *Conn. Vigne Vin* **20**(2):87-95.

Liu, J. R., and J. F. Gallander. 1983. Effect of pH and sulfur dioxide on the rate of fermentation in red table wines. *Am. J. Enol. Vitic.* **34**(1):44-46.

Long, Z. R. 1981. White table wine production in California's North Coast region. In *Wine Production Technology in the United States*, M. A. Amerine (ed.). American Chemical Society, Washington, D.C.

Lonvaud-Funel, A., C. Desens, and A. Joyeux. 1985. Stimulation of the malolactic fermentation by addition to the wine of yeast cell wall material or other adjuvants such as polysaccharides or nitrogenous materials (French). *Conn. Vigne Vin* **19**(4):229-240.

Lonvaud-Funel, A., and A. M. Strasser de Saad. 1982. Purification and properties of a malolactic enzyme from a strain of *Leuconostoc mesenteroides* isolated from grapes. *Appl. Environ. Microbiol.* **43**:357-361.

Luthi, J., and Ch. Schlater. 1983. Biogenic amines in foods. The effect of histamine, tryamine and phenylethylamine on humans (German) *Z. Lebensm. Unters. u. Forsch.* **177**:439-443.

Malik, F., E. Minarik, and K. Kutlik. 1984. Fermentation activity of dry, pure-culture yeasts (German). *Wein-Wissensch.* **39**(3):178-183.

Martinez de la Ossa, E., L. Perez, and I. Caro. 1987. Variations of the major volatiles through aging of sherry. *Am. J. Enol. Vitic.* **38**(4):293-297.

Martini, L. P. 1981. Red wine production in the coastal counties of California 1960-1980. In *Wine Production in the United States*, M. A. Amerine (ed.). American Chemical Society, Washington, D.C.

Mascarenhas, M. A. 1984. The occurrence of malolactic fermentation and diacetyl contents of dry table wines from northeastern Portugal. *Am. J. Enol. Vitic.* **35**(1):49-51.

Masyczerk, R., and C. S. Ough. 1983. The red wine syndrome. *Am. J. Enol. Vitic.* **34**(4):260-264.

Mattick, L. R., and W. B. Robinson. 1960. Changes in the volatile acids during the baking of sherry wine by the Tressler process. *Am. J. Enol. Vitic.* **11**:113-116.

Mayer, K. 1974. Important microbiological and technological findings regarding the microbial reduction of acids (German). *Schweiz. Z. Obst. u. Weinb.* **110**:385-391.

Minarik, E. 1983. Activation of the alcoholic fermentation of musts rich in sugar (German). *Die Wein-Wissensch.* **38**(3):202-209.

Minarik, E., V. Kubalova, and Z. Silharova. 1986. Further knowledge on the influence of yeast starter amount and the activator of *B. cineria* on the course of fermentation under unfavorable conditions (Czech.). *Kvasny Prumysl* **32**(3):58-61.

Minarik, E., and A. Navara. 1986. *Chemistry and Microbiology of Wine* (Slovak). Priroda, Bratislava.

Mooser, J. 1958. The occurrence of yeasts in bees, bumble bees and wasps (German). *Zentralbl. Bakteriol. Parasitenk. u. Infektionskr. Hyg. Abt. 2* **111**:101-115.

Mrak, E. M., and L. S. McClung. 1940. Yeasts occurring on grapes and in grape products in California. *J. Bacteriol.* **40**:395-407.

Nagodawithana, T. W., and J. M. Cuzner. 1979. Method of fermenting brewer's wort. U.S. Patent 4,140,799, Feb. 20.

Nagodawithana, T. W., and K. Steinkraus. 1976. Influence of the rate of ethanol production and accumulation on the viability of *Saccharomyces cerevisiae* in rapid fermentation. *Appl. Environ. Microbiol.* **31**:158-162.

Nagodawithana, T. W., J. T. Whitt, and A. J. Cutaia. 1977. Study of the feedback effect of ethanol on selected enzymes of the glycolytic pathway. *J. Am. Soc. Brewing Chemists* **35**:179-183.

Nelson, K. E., and M. A. Amerine. 1957. The use of *Botrytis cineria* Pers. in the production of sweet table wines. *Hilgardia* **26**:521-563.

Nelson, K. E., and M. S. Nightingale. 1959. Studies in the commercial production of natural sweet wines from botrytized grapes. *Am. J. Enol. Vitic.* **10**:135-141.

Nishino, H., S. Miyazaki, and K. Tohio. 1985. Effect of osmotic pressure on the growth rate and fermentation activity of wine yeasts. *Am. J. Enol. Vitic.* **36**(2):170-174.

Novak, M., P. Strehaiano, M. Morena, and G. Goma. 1981. Alcoholic fermentation: On the inhibitory effect of ethanol. *Biotechnol. Bioeng.* **23**:201-211.

Nykanen, L. 1986. Formation and occurrence of flavor compounds in wine and distilled alcoholic beverages. *Am. J. Enol. Vitic.* **37**(1):84-96.

Ough, C. S. 1964. Fermentation rates of grape juice. I. Effects of temperature and composition on white juice fermentation rates. *Am. J. Enol. Vitic.* **15**:167-177.

Ough, C. S. 1966a. Fermentation rates of grape juice. II. Effect of initial °Brix, pH and fermentation temperature. *Am. J. Enol. Vitic.* **17**:20-26.

Ough, C. S. 1966b. Fermentation rates of grape juice. III. Effects of initial alcohol, pH, and fermentation temperature. *Am. J. Enol. Vitic.* **17**:74-81.

Ough, C. S., and H. W. Berg. 1969. Pressure fermentation of red wines. *Am. J. Enol. Vitic.* **20**:118-119.

Ough, C. S., E. A. Crowell, and B. R. Gutlove. 1988. Carbamyl compound reactions with ethanol. *Am. J. Enol. Vitic.* **39**(3):239-243.

Ough, C. S., E. A. Crowell, and L. A. Mooney. 1988. Formation of ethyl carbamate precursors during grape juice (Chardonnay) fermentation. *Am. J. Enol. Vitic.* **39**(3):243-249.

Parrish, M. E., and D. E. Carroll. 1985. Indigenous yeasts associated with Muscadine (*Vitis rotundifolia*) grapes and musts. *Am. J. Enol. Vitic.* **36**(2):165-169.

Pascual, C., A. Alonso, I. Garia, C. Romay, and A. Kotyk. 1988. Effect of ethanol on glucose transport, key glycolytic enzymes, and proton extrusion in *Saccharomyces cerevisiae*. *Biotechnol. Bioeng.* **32**:374-378.

Peynaud, E., and S. Domercq. 1953. The yeasts of the Gironde (French). *Ann. Inst. Natl. Rech. Agron.* **4**:265-300.

Pichova, A., K. Beran, B. Behalova, and J. Zajicek. 1985. Ergosterol synthesis and population analysis of a fed-batch fermentation of *Saccharomyces cerevisiae*. *Folia Microbiol.* **30**:134-140.

Piton, F., M. Charpentier, and D. Troton. 1988. Cell wall and lipid changes in *Saccharomyces cerevisiae* during aging of champagne wine. *Am. J. Enol. Vitic.* **39**(3):221-226.

Popper, K., F. S. Nury, W. M. Camirand, and W. N. Stanley. 1964. *Development of Botrytis Character in Must by Aerated Submerged Culture*. Wine Institute Technological Advisory Committee, Dec. 11.

Posson, P. 1981. Production of baked and submerged culture sherry-type wines in California 1960-1980. In *Wine Production Technology in the United States*, M. A. Amerine (ed.). American Chemical Society, Washington, D.C.

Postel, W., and U. Guvenc. 1976. Gas chromatographic determination of diacetyl, acetoin and 2,3 pentadione in wine (German). *Z. Lebensm. Unters. u. Forsch.* **161**:35-44.

Postel, W., F. Drawert, and L. Adam. 1972. Gas chromatographic determination of beverage components. III (German). *Chem. Mikrobiol. Technol. Lebensm.* **1**:224-235.

Prasad, R., and A. H. Rose. 1986. Involvement of lipids in solute transport in yeasts. *Yeast* **2**:205-220.

Radler, F. 1975. The metabolism of organic acids by lactic acid bacteria. In *Lactic Acid Bacteria in Beverages and Food*, J. G. Carr, C. V. Cutting, and G. C. Whiting (eds.). Academic Press, New York.

Radler, F., K. Dietrich, and I. Schonig. 1985. Microbiological testing of active dry wine yeasts (German). *Dt. Lebensm. Rundschau* 81:73-77.
Radler, F., and M. Schmitt. 1987. Killer toxins of yeasts: Inhibitors of fermentation and their adsorption. *J. Food Protection* 50:234-238.
Radler, F., and H. Schütz. 1982. Glycerol production of various strains of *Saccharomyces*. *Am. J. Enol. Vitic.* 33(1):36-40.
Rankine, B. C. 1967. Formation of higher alcohols by wine yeasts and relation to taste thresholds. *J. Sci. Food Agric.* 18:583-589.
Rankine, B. C., and B. Lloyd. 1963. Quantitative assessment of dominance of added yeast in wine fermentations. *J. Sci. Food Agric.* 14:793-798.
Rapp, A., M. Gunther, and J. Almy. 1985. Identification and significance of several sulfur-containing compounds in wine. *Am. J. Enol. Vitic.* 36(3):219-221.
Reed, G. 1974. Comparison of the use of commercial yeasts. *Proc. Biochem.* 9(9):11-12, 32.
Reed, G., and S. L. Chen. 1978. Evaluating commercial wine yeasts by fermentation activity. *Am. J. Enol. Vitic.* 29(3):165-168.
Reed, G., and T. W. Nagodawithana. 1988. Technology of yeast usage in wine making. *Am. J. Enol. Vitic.* 39(1):83-90.
Ribéreau-Gayon, P. 1985. New developments in wine microbiology. *Am. J. Enol. Vitic.* 36(1):1-10.
Ribéreau-Gayon, J., E. Peynaud, P. Ribereau-Gayon, and P. Sudraud. 1976. *Science and Technology of Wine* (French). Dunod Edition, Paris.
Ribéreau-Gayon, P., S. Lafon-Lafourcade, D. Dubourdieu, V. Lucmaret, and I. Larue. 1979. Metabolism of *Saccharomyces cerevisiae* in the must of *Botrytis cineria* infected grapes. Inhibition of the fermentation: Formation of acetic acid and glycerol (French). *C. R. Acad. Sci.* 289:441-444.
Riddel, J. L., and M. S. Nury. 1958. Continuous fermentation of wine at Vie-Del. *Wines and Vines* 39(5):35.
Rodriguez, S. B. 1987. A system for identifying spoilage yeast in packaged wine. *Am. J. Enol. Vitic.* 38(4):273-276.
Romano, P., M. G. Soli, G. Suzzi, L. Grazia, and C. Zambonelli. 1985. Improvement of a wine *Saccharomyces cerevisiae* strain by a breeding program. *Appl. Environ. Microbiol.* 50(4):1064-1067.
Rossi, J., and F. Clementi. 1984. L-malic acid catabolism by polyacrylamide entrapped *Leuconostoc oenos*. *Am. J. Enol. Vitic.* 35(2):100-102.
Rossini, G., M. Bertoluccioli, and E. R. Pasquale. 1981. Vinification with commercial active dry wine yeast: Vintage 1979 (Italian). *Vini d'Italia* 130:21-26.
Rous, C. V., and R. Snow. 1983. Reduction of higher alcohols by fermentation with a leucine auxotrophic mutant of wine yeast. *J. Inst. Brew.* 89:274-278.
Rupela, O. P., and P. Tauro. 1984. Isolation and characterization of low hydrogen sulfide producing wine yeast. *Enzyme Microbiol. Technol.* 6:419-421.
Sa-Correia, I., and N. Van Uden. 1983. Temperature profiles of ethanol tolerance: Effects of ethanol on the minimum and maximum temperature for growth of *Saccharomyces cerevisiae* and *Kluyveromyces fragilis*. *Biotech. Bioeng.* 25:1665-1667.
Salo, P. 1970. Determining the odor threshold for some compounds in alcoholic beverages. *J. Food Sci.* 35:95-98.
Schmitt, A., K. Curshmann, and H. Koehler. 1979. On fermentation and the flavor defects of wines (German). *Rebe Wine* 32(9):364-367.

Schmitt, A., K. Curshmann, A. Miltenberger, and A. Koehler. 1984. Active dry wine yeasts compared over several years (German). *Der Deutsche Weinbau* **25/26:**1126-1138.
Schreier, P. 1979. Flavor composition of wines. *Crit. Rev. Food Sci. Nutr.* **12:**59-111.
Shimazu, Y., and M. Watanabe. 1981. Effects of yeast strains and environmental conditions on forming of organic acids in must during fermentation (Japanese). *J. Ferm. Technol.* **59:**27-32.
Sikovec, S. 1966. Effect of some polyphenols on the physiology of wine yeasts. II. Effect of polyphenols on propagation and respiration of yeasts (German). *Mitt. Rebe Wein (Klosterneuburg)* **16:**227-281.
Silver, J., and T. Leighton. 1981. Control of malolactic fermentations in wine. 2. Isolation and characterization of a new malolactic organism. *Am. J. Enol. Vitic.* **32**(1):64-72.
Skofis, E. 1981. Production of table wines in the interior valley (of California). In *Wine Production Technology in the United States*, M. A. Amerine (ed.). American Chemical Society, Washington, D.C.
Snow, R. 1979. Toward genetic improvement of wine yeast. *Am. J. Enol. Vitic.* **30**(1):33-37.
Snow, R. 1983. Genetic improvement of wine yeast. In *Yeast Genetics*, J. F. T. Spencer et al. (eds.). Springer, New York.
Snow, P. G., and J. F. Gallander. 1979. Deacidification of white table wines through partial fermentation with *Schizosaccharomyces pombe*. *Am. J. Enol. Vitic.* **30**(1):45-48.
Soles, R. M., C. S. Ough, and R. E. Kunkee. 1982. Ester concentration differences in wine fermented by various species and strains of yeast. *Am. J. Enol. Vitic.* **33**(2):94-98.
Sols, A. 1956. Selective fermentation and phosphorylation of sugars by Sauternes yeast. *Biochim. Biophys. Acta* **20:**62-68.
Spencer, J. T. F., and D. M. Spencer. 1983. Genetic improvement of industrial yeasts. *Ann. Rev. Microbiol.* **37:**121-142.
Subden, R. E., R. Cornell, and A. C. Noble. 1980. Evaluation of API20C clinical yeast identification system for must and wine yeast identification. *Am. J. Enol. Vitic.* **31**(4):364-366.
Suzzi, G., P. Romano, and C. Zambonelli. 1985. *Saccharomyces* strain selection in minimizing $SO_2$ requirement during vinification. *Am. J. Enol. Vitic.* **36**(3):199-202.
Thomas, A. S., and A. H. Rose. 1979. Inhibitory effect of ethanol on growth and solute accumulation by *Saccharomyces cerevisiae* as affected by plasma membrane lipid composition. *Arch. Microbiol.* **122:**49-55.
Thornton, R. J. 1982. Selective hybridization of pure culture wine yeasts. II. Improvement of fermentation efficiency and $SO_2$ tolerance. *Eur. J. Appl. Microbiol. Biotechnol.* **14:**150-164.
Thornton, R. J. 1985. The introduction of flocculation into a homothallic wine yeast. *Am. J. Enol. Vitic.* **36**(1):47-49.
Thoukis, G. 1958. The mechanism of isoamyl formation using tracer techniques. *Am. J. Enol. Vitic.* **9:**161-167.
Thoukis, G., G. Reed, and R. J. Bouthilet. 1963. Production and use of compressed yeast for winery fermentation. *Am. J. Enol. Vitic.* **14:**148-154.
Thoukis, G., M. Ueda, and D. Wright. 1965. The formation of succinic acid during alcoholic fermentation. *Am. J. Enol. Vitic.* **16**(1):1-8.
Totsuka, A., and S. Hara. 1983. Decomposition of malic acid in red wine by immobilized yeast cells (Japanese). *Hakkokogaku* **56:**231-237.
Traverso-Rueda, S., and R. E. Kunkee. 1982. The role of sterols on growth and fermentation of wine yeasts under vinification conditions. *Dev. Ind. Microbiol.* **23:**131-143.
Tredoux, H. G., J. L. F. Kock, P. M. Lategan, and H. B. Muller. 1987. A rapid identification

technique to differentiate between *S. cerevisiae* strains and other yeast species in the wine industry. *Am. J. Enol. Vitic.* **38**(23):161-164.

Troost, G. 1980. *Technology of Wine* (German). Eugen Ulmer, Stuttgart.

Ulbrich, M., and W. Saller. 1951. Investigations of the practicality of aeration of pure cultures of commercial wine yeasts (German). *Mitt. Rebe Wein, Ser. A. (Klosterneuburg)* **1**:94-104.

Unterholzner, O., M. Aurich, and K. Platter. 1980. Taste and flavor flaws of red wines caused by *Schizosaccharomyces pombe* (German). *Mitt. Klosterneuburg* **38**:66-70.

Ussegli-Tomasset, L., and R. di Stefano. 1981. Variabilities in the production of volatile components with the same yeast strain (Italian). *Vini d'Italia* **23**:249-264.

Van der Merwe, C. A., and C. J. van Wyck. 1981. The contribution of some fermentation products to the odor of dry white wine. *Am. J. Enol. Vitic.* **32**(1):41-46.

Van Vuuren, H. J. J., and L. Van der Meer. 1987. Fingerprinting of yeasts by protein electrophoresis. *Am. J. Enol. Vitic.* **38**(1):49-53.

Vezinet, F. 1981. Application of yeast genetics to wine making. Methodology and objectives (French). *Bull. de l'OIV* **54**(608):830-832.

Viegas, C. A., I. Sa-Correia, and J. M. Novais. 1985. Synergistic inhibition of the growth of *Saccharomyces bayanus* by ethanol and octanoic and decanoic acids. *Biotech. Lett.* **7**(8):611-614.

Vojketova, G., and E. Minarik. 1985. Changes in the composition of the yeast flora of grapes, musts and wines in the wine region of the small Carpathian mountains in the course of 20 years (German). *Mitt. Klosterneuburg* **35**:82-88.

Vos, P. J. A., and R. S. Gray. 1979. The origin and control of $H_2S$ during fermentation of grape must. *Am. J. Enol. Vitic.* **30**(3):187-196.

Wagner, P. 1981. Grapes and wine production in the East. In *Wine Production Technology in the United States*, M. A. Amerine (ed.). American Chemical Society, Washington, D.C.

Wagner, K., P. Kreutzer, and K. Mahlmeister. 1986. Malic acid reduction produced by different pure culture wine yeasts (German). *Die Weinwirtsch.-Technik* **5**:197-198, 201, May 13.

Wahlstrom, V. L., and K. C. Fugelsang. 1988. Utilization of yeast hulls in wine making observed. *Res. Bull. Calif. State Univ.*, Fresno, 6 p.

Webb, A. D., and R. E. Kepner. 1961. Fusel oil analysis by means of gas liquid partition chromatography. *Am. J. Enol. Vitic.* **12**:51-59.

Watanabe, M., and Y. Shimazu. 1981. Quality of wine made from cAMP-added botrytized must. *Am. J. Enol. Vitic.* **32**(1):73-75.

Webb, A. D., and A. C. Noble. 1976. Aroma of sherry wines. *Biotech. Bioeng.* **18**:939-952.

Wibowo, D., R. Eschenbruch, C. R. Davis, G. H. Fleet, and T. H. Lee. 1985. Occurrence and growth of lactic acid bacteria in wine: A review. *Am. J. Enol. Vitic.* **36**(4):302-313.

Wick, E. 1968. Penetration of yeast by sugar during grape juice fermentations. *Am. J. Enol. Vitic.* **19**:273-281.

Wolf, E., and I. Benda. 1967. Differentiation of yeast strains by *Drosophila melanogaster* with regard to representatives of the genus *Schizosaccharomyces* (German). *Weinberg Keller* **14**:163-166.

Zee, J. A., R. E. Simard, L. L'Heureux, and J. Tremblay. 1983. Biogenic amines in wines. *Am. J. Enol. Vitic.* **34**(1):6-9.

CHAPTER
5

# DISTILLER'S YEASTS

Distilled alcoholic spirits may be produced from any alcoholic materials such as fermented mashes of cereal grains, fermented fruit juice, sugar cane juice, molasses, honey, and cactus juice. Although the art of making wine and beer can be traced back to 5000 to 6000 B.C., the history of distillation is much shorter. It is, however, uncertain where and when it originated. In Europe, distillation was known to the alchemists, and it began to spread in the thirteenth and fourteenth centuries. The best-known prototypes of distilled beverages are whisky, based on the distillation of beer, and brandy, based on the distillation of wine. Rose (1977) has provided a short introduction to the history of distilled beverages with some appropriate references.

Figure 5-1a shows a schematic drawing of a pot still used today for the production of some whiskies, rum, and tequila. It resembles in principle the alembics of the alchemists. Figure 5-1b shows the simplest form of a continuous still consisting of a beer still and a rectifying column. It provides for removal or partial removal of fusel oil (tails) and heads (aldehydes and esters). Distillation affects the flavor of the beverage, but does not directly concern the art and science of yeast fermentation.

The fermentation of the various raw materials is best understood by dividing them into the fermentation of cereal mashes, which contain all the constituents of the cereal (whisky and sake), and the fermentation of juices and liquid extracts, which contain all carbohydrates in the form of fermentable sugars (fruit juice, molasses, fully hydrolyzed starches). Cereal

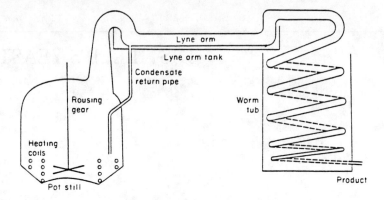

**Figure 5-1a.** Irish distillery pot still. (*From Lyons and Rose 1977*)

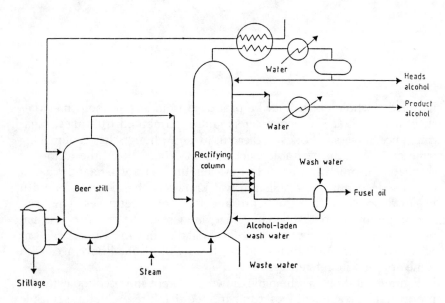

**Figure 5-1b.** Two-column distillation system. (*After Horak et al. 1974*)

mashes are characterized by the concurrent enzymatic formation of sugars and their fermentation, at least in the latter stages of the process.

The earliest distilled beverages must have been produced from thin beer and wine of relatively low alcohol content. Today the producer of distilled alcoholic beverages tries to obtain the highest possible concentration of ethanol consistent with reasonable productivity in order to reduce the costs of distillation and capital investment in tanks. The production of distilled beverages is a year-round operation, similar to the production of baked goods and in contrast to the highly seasonal fermentation of wines and the production of beer, which peaks during the summer months.

Distilled alcoholic beverages may be stored for long periods without any danger of microbial spoilage, and some must be stored (matured) for several years. Therefore, microbial spoilage of the final product need not be considered.

Distiller's yeasts have to carry out the alcoholic fermentation under conditions that vary greatly for the many diverse substrates and processes. Therefore, it is difficult to generalize regarding their efficiency under differing conditions of substrate concentration, osmotic pressure, temperature, and pH. These characteristics will first be treated in some detail as they apply to the traditional whisky fermentation. Deviations from these characteristics or requirements will then be mentioned as they apply to the production of other distilled beverages.

The production of fuel ethanol from agricultural raw materials is not the subject of this chapter. But some of the newer techniques developed in this field in the past decade may be useful in the production of potable alcoholic beverages, and to that extent they will be mentioned.

## WHISKY

### Bourbon

Table 5-1 (Lyons and Rose 1977) is a very useful schematic presentation of the differences in raw materials and processes for the different distilled alcoholic beverages. Bourbon whisky, named after its origin in Bourbon County, Kentucky, is made with a grain bill containing at least 51% corn (maize) and lesser percentages of "small grains": barley, rye, and wheat. The percentage of barley malt is between 10% and 15%. Brewer's malt generally contains 50 SKB alpha amylase units per gram. Distillers prefer to hydrolyze all available starch to fermentable sugar. They often use a gibberellin-treated malt with about 75-100 SKB units per gram. Gibberellic acid is required for the de novo synthesis of alpha amylase during malting, and very small concentrations occur naturally in barley (Sfat and Doncheck 1981). They permit a decrease in the percentage of malt to a total of about 5% of the grain bill, but require addition of a fungal alpha-1,4-glucosidase during the conversion step (Brandt 1975).

The grains are ground to a meal prior to mashing. The mashing ratio, that is, the number of gallons of liquid per bushel of grain (56 lb), varies somewhat. A higher mashing ratio facilitates heating and cooling of the slurry, but it also requires more energy and results in a lower concentra-

Table 5-1. Raw Materials and Unit Processes for the Production of Different Types of Whisky

| | Scotch Malt | Scotch Grain | Irish | Bourbon | Am. Grain |
|---|---|---|---|---|---|
| Raw Materials | Peated, malted barley | Corn and a small proportion of malted barley | Barley and unpeated malted barley | Corn, small grains, and unpeated malted barley | Corn, rye, and unpeated unmalted barley |
| Conversion | Infusion mash | Mash cook, conversion stand | Infusion mash | Mash cook, conversion stand | Mash cook, conversion stand |
| Fermentation | Distiller's yeast and brewer's yeast | Distiller's yeast | Distiller's yeast | Distiller's yeast | Distiller's yeast |
| Distillation | Two-pot stills | Patent still | Three-pot stills | Patent still | Patent still |
| Maturation | At 11° overproof in charred oak casks at least three years | Up to 20° overproof in used cherry casks at least three years | At 25° overproof in sherry casks or uncharred oak casks at least three years | At 125° (U.S.) proof for at least three years | At or above 190° proof in oak containers for at least one year |

*Source:* Slightly modified from Lyons and Rose 1977.

tion of alcohol. For bourbon the mashing ratio is about 25 gal/bu, or 100 L/26.8kg. The liquid used consists of water and a variable amount of backset stillage. Stillage is the liquid-fermented beer from which the alcohol has been distilled. Its pH is 4.0. Use of backset stillage as part of the makeup water lowers the pH of the mash from about 5.8 to about 4.8-5.2. It serves to minimize bacterial contamination and provides buffer capacity to the mash. It also saves some water.

The slurry of grains must be cooked to gelatinize the starch in order to make it available for hydrolysis by amylase. About 0.5% to 1.0% of barley malt (based on total grain bill) is added prior to cooking in order to reduce the viscosity at the time of starch gelatinization and during subsequent cooling. The enzymes of this so called "premalt" do not survive the cooking process. The slurry is cooked batchwise at atmospheric pressure or in pressurized vessels at 120°-150°C. The cooked mash is then cooled to a temperature of 60°-65°C. At this point the major portion of the malt, the "conversion malt," which has been slurried separately, is added to the mash. Conversion of the dextrinized starch by the action of $\alpha$-amylase, $\beta$-amylase, and limit dextrinase proceeds rapidly while the mash is cooled further to the fermentation set temperature of 20°-24°C.

The conversion of distiller's mashes with malt requires about 30 minutes at 55°-57°C. From 70-80% of the starch will be converted to fermentable sugars—glucose, maltose, and maltotriose—by the time the mash has been cooled to the set fermentation temperature. The remaining 20-30% consists of dextrins that are slowly hydrolyzed during the fermentation period. Figure 5-2 is an instructive example of the fermentation of carbohydrates in two distiller's mashes converted with either malt or fungal amylases. Yeasts ferment the preformed sugars rapidly during the first 10-30 hours of the fermentation period. After 40 hours of the malt fermentation all of the preformed sugars have been fermented, and from this point on the rate of fermentation is limited by the enzymatic hydrolysis of residual dextrins, and the extent of alcohol formation is also limited by the ability of the malt enzymes to hydrolyze these dextrins (Pan, Andreasen, and Kolachov 1950). Figure 5-2 also shows the decreasing concentration of sugar during a fermentation of a glucose solution. It is completed within about 30 hours. In this case, yeast activity is the only factor limiting the rate of fermentation.

Yields of ethanol are expressed in the United States as *proof gallons* per bushel of grain. A proof gallon contains 50% by volume of ethanol. Such yields may range from 5-6 proof gallons per bushel of grain depending on the type of grain bill and the efficiency of the process. The so-called fermentation efficiency of a distiller's process is expressed as a percentage of the available starch and depends on the efficiency of both conversion and fermentation. In Britain, the concentration of sugars in a distiller's

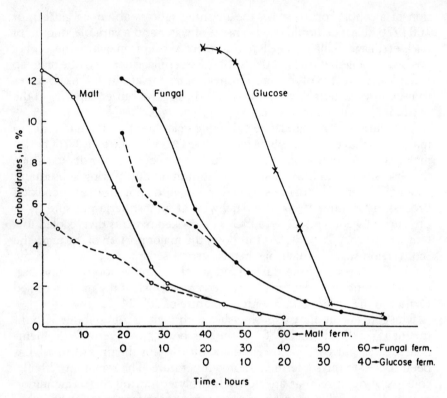

**Figure 5-2.** Change of carbohydrate concentrations during fermentation of malt- and fungal-enzyme-converted corn mashes and during a glucose fermentation.

x———x———x  glucose fermentation
o———o———o  malt-converted mashes, total carbohydrates
o----o----o  malt-converted mashes, dextrins
●———●———●  fungal-enzyme-converted mashes, total carbohydrates
●----●----●  fungal-enzyme-converted mashes, dextrins

(From Pan, Andreasen, and Kolachov 1950)

wort (see Scotch Whiskey) is expressed as its gravity. A wort gravity of 1.06-1.07 normally results in an ethanol concentration of 9-10% by volume (Simpson 1977).

In continental Europe yields are expressed as liters of alcohol per 100 kilograms raw material. Typical values are 21-23.1 L for wheat, 20-23.1 L for maize, 22-25 L for rye, and 10-12 L for potatoes. Based on the actual starch content of these raw materials, the efficiency of ethanol production is high—generally above 90%. It has been mentioned that corn, wheat, and rye mashes must be cooked to make their starch available for enzyme hydrolysis, whereas barley and barley malt may be hydrolyzed by their starches' intrinsic amylases at about 63°C. This difference is a reflection of the different gelatinization temperatures of the starches and an expression of the different organization of the starch molecules in the granules. Even within a given sample of starch, individual granules may gelatinize at different temperatures that account for the ranges in gelatinization temperatures shown in the literature. Pomeranz (1984) reported the following ranges: barley, 59°-64°C; corn, 62°-72°C; potato, 56°-69°C; wheat, 62°-75°C; rice, 61°-78°C; rye, 57°-70°C.

Yeast for the fermentation may be grown on the same mash as is used for the final fermentation except for a somewhat higher proportion of barley malt. Distillers keep proprietary cultures in their own culture collection for inoculation of various mashes. A sour mash culture may be prepared by inoculating with a lactic-acid-producing culture, usually *Lactobacillus delbrueckii*.

The bacterial fermentation proceeds at about 50°C until a pH of 3.9-4.0 is reached. At this point, the soured mash is pasteurized and the temperature is lowered to 22°-24°C for inoculation by the pure yeast culture. Yeast fermentation and growth proceed in this mash for several hours, at least until the Balling has dropped to one-half of its original value. The inoculating culture may contain from $90\text{-}150 \times 10^6$ live cells per milliliter (Lyons and Rose 1977). It is used at a rate of 2-3% of the liquid fermenter volume, and hence provides about $2\text{-}5 \times 10^6$ cells for the start of the commercial fermentation.

Distiller's yeasts, strains of *S. cerevisiae*, are also available commercially in the form of active dry yeasts. Such active dried distiller's yeasts (DADY) contain approximately $25\text{-}30 \times 10^9$ cells per gram. Inoculation with 10 g/hl (1.5 lb/1000 gal) results in a cell density of about $3 \times 10 \times 10^3$ per milliliter. Requirements for the rehydration of DADY are identical to those practiced in the baking and wine industries. In general, whisky fermentations last from 48-72 hours. The set temperature should be sufficiently low (e.g., 20°C) to keep the final temperature of the fermentation below 32°C) and preferably below 30°C. High temperatures accelerate the rate of yeast fermentation but favor the growth of lactic-acid-producing organisms, which are contaminants and may lead to serious flavor defects.

Some species of *Lactobacillus* convert glycerol to $\beta$-hydroxypropionaldehyde, which breaks down to acrolein during distillation and gives the distillate an acrid (peppery) odor. Contaminating microorganisms may enter the fermenter in at least three ways: in the cooked mash as surviving bacterial spores, in the barley malt slurry that is not heated beyond 65°C, and in the yeast inoculum. The yeast population can readily attain a live cell count of 100-150 × $10^6$ cells per milliliter. The bacterial population as shown in Figure 5-3 can easily reach similar counts (Bluhm 1983).

Figure 5-4 shows the growth of the yeast population to a total of about 100 × $10^6$ cells per milliliter during the first day of the bourbon fermentation followed by a decline in the number of live cells until the end of the three-day fermentation. This decline is characteristic of other fermentation for the production of distilled beverages and bears a considerable

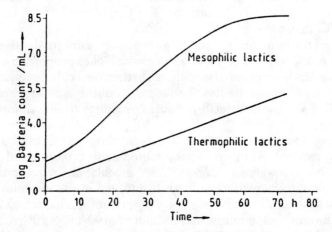

**Figure 5-3.** Lactic acid bacteria during a bourbon fermentation. (*From Bluhm 1983*)

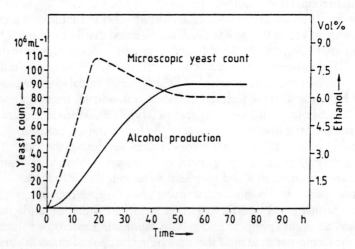

**Figure 5-4.** Yeast count and alcohol concentration during a bourbon fermentation. (*From Bluhm 1983*)

resemblance to beer and wine fermentations. The increase in ethanol also follows in general curves that have previously been shown for wine and beer fermentations.

At the end of the fermentation period, alcohol and volatile flavor substances are removed from the mash and concentrated by distillation. This procedure is usually done in a beer still that may be divided into a stripping section, a section for the retention of particulate matter, and a rectifying column. This operation may be followed by additional columns to increase the concentration of ethanol and/or to separate volatile compounds with lower boiling points (heads) and those of higher boiling points (tails).

The Coffey stills and other early Patent stills used in the production of whiskies have not changed greatly, but stills for the production of grain neutral spirits for the production of gin and vodka have been greatly improved. The composition of the vapor at 30 different plates of the rectifying column of a Coffey still has been reported by Pyke (1965). Variations in the composition of the vapors in ethanol, butanol, isobutanol, propanol, amyl alcohol, esters, aldehydes, furfural, and acids at the different plates makes it obvious that the final composition in congeners (flavor compounds) depends largely on the complexity of the stills and on their operating conditions. Figure 5-5 shows the composition of the vapor at different plates.

Distilled alcoholic beverages are matured for varying time periods of one to several years. Storage in oak barrels results in the extraction of additional flavor compounds from the wood such as furfural, 5-methyl furfural, furfuryl alcohol, and $\beta$-methyl-$\gamma$-octolactone (also known as oak lactone or whisky lactone) (Marsal et al. 1988). The processes of distillation and maturation will not be described in detail since they are postfermentation operations. They have been treated by Kreipe (1963), Pyke (1965), Lyons and Rose (1977), Bluhm (1983), and others.

Distillation carried out in the beer still permits separation of the insoluble material of the fermented mash. This material with a protein content of 20-22% (based on dry weight) is dried and sold as a valuable feed as distiller's spent grains. The stillage of the final distillation may be concentrated and dried. It is also used as a feed known as distiller's dried solubles.

## Scotch Whisky

Two distinct types of whisky are produced in the Scottish Highlands: malt whisky and grain whisky. The two types are usually blended in proportions that vary from distillery to distillery, but they are also available as scotch malt whisky and scotch grain whisky. Scotch malt whisky is made with the use of *peated* malt, fermentation of an all-malt mash, and distilla-

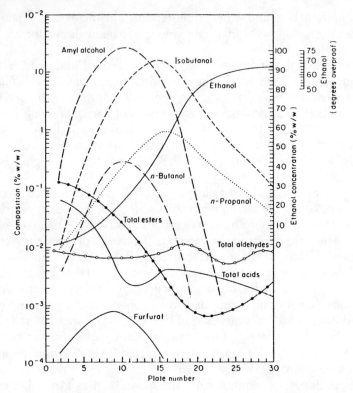

**Figure 5-5.** Changes in the composition of vapor at different plates in a Coffey still rectifier used in the production of Scotch grain whisky. (From Pyke 1965)

tion in a pot still. The malt is dried in kilns in which peat is burned. The smoke is adsorbed on the malt and ultimately imparts to the whisky a slightly smoky flavor. These flavor compounds are phenolic in nature.

Peated malt may contain 2-20 ppm of such substances (determined as phenol). The mashing ratio is about 31-32. The all-malt mash is held at 64°C for about 1 hour. It is never boiled. Conversion is nearly complete because of the high activity of the malt amylases. The mash is drained by passage of the liquid wort through the settled, insoluble material similar to the procedure for a beer wort. However, the wort for the production of scotch malt whisky is never boiled. The spent grains are washed several times with hot water to recover all of the soluble solids, and the washings are used for preparation of subsequent mashes. The spent grains (*draff* in Britain) are dried for use as a feed supplement.

The wort is almost free from suspended insoluble materials. It is fermented with distiller's yeast or baker's yeast in a manner similar to that of the bourbon fermentation. There is ample opportunity for the development of a bacterial flora since the worts are not sterilized. Rapid bacterial growth occurs at the end of the fermentation based on the metabolism of the products of yeast autolysis. The bacterial flora often has a desirable effect on flavor by forming lactic acid and lactic acid esters. It may also have an undesirable effect through the conversion of glycerol to $\beta$-hydroxypropionaldehyde, which has already been mentioned. A normal flora of $10^4$ cells/ml is desirable, and cell counts exceeding $10^8$ cells/ml are undesirable. Unfortunately, desirable or harmful strains of *Lactobacilli* could not be attributed to a particular species since they frequently do not conform to established species characteristics (Barbour and Priest 1988).

Fermentations of scotch whisky yield approximately 8-9 % ethanol by volume. The whisky is distilled twice in pot stills and matured for at least three years—often much longer—in charred oak casks, usually at 110° proof.

The grain bill for the production of scotch grain whisky consists of unpeated barley malt, usually about 15%, plus other grains such as barley, corn, rye, and more recently, wheat. The grains are cooked, and the malt is slurried at 60°-65°C. The conversion, fermentation, and distilling procedures resemble those described above for bourbon and more specifically those for American light whisky. Distillation is carried out in a continuous Patent still, and maturing requires at least three years.

### Irish Whisky

Irish malt whisky is made with a grain bill consisting largely (80%) or entirely of barley malt. The grain is not cooked and preparation of the mash and conversion are carried out by the infusion process. The malt may be unpeated or only lightly peated. The drawn, fermented wort, now called the wash, is distilled three times in pot stills. At the end of the distillation process the proof is about 144°. It is matured for at least three years in casks at a proof of about 125°. Ireland also produces a product similar to scotch grain whisky.

### Rye Whisky, Corn Whisky, and Light Whisky

These whiskies are produced by procedures similar to those described for bourbon. Rye whisky must contain at least 51% rye in the grain bill and corn whisky must contain more than 80% corn. Light whisky may be made from any desired grain bill. It usually contains a very large percentage of corn. It is distilled to 160°-190° proof. As with some other whiskies, the flavor may vary considerably depending on the extent to which congeners are removed in the distillation.

## Canadian Whisky

The grain bills and procedures are similar to those used in the United States. However, there are no specific requirements for the grain bill, distillation procedures, or cooperage. Canadian whiskies are usually blends of grain-neutral spirits (see below) and smaller proportions of more heavily flavored whiskies such as bourbon or rye whisky. Production of these whiskies follows the trend of consumption of lighter, that is, less heavily flavored, whiskies.

## Grain-Neutral Spirits

Grain-neutral spirits are used for blending with flavorful whiskies and for the production of so-called "white goods": gin and vodka. They may be made from any combination of grains. In the United States, the most cost-effective cereals are corn and milo. There is no particular requirement for the formation of desirable flavors since congeners are removed almost completely during distillation. Thus, the grain-neutral spirits are almost odorless and tasteless.

Cooking of the grain bill with water and backset is generally carried out with jet heaters at atmospheric pressure or at higher pressure in a continuous process. Conversion is accomplished with microbial enzymes and/or malt. The emphasis is on efficiency, yield, and productivity of the operation, and commercial compressed baker's yeast or active dry distiller's yeast are widely used. Distillation results in a high-proof spirit of 190° proof. A schematic drawing of a four-column still suitable for the removal of congeners and a high-proof final product is shown in Figure 5-6 (Bluhm 1983).

Neutral spirits of the same quality may also be produced from whey permeate (Tzeng et al. 1979). They may, however, not be designated as grain-neutral spirits, and they are not widely used. The yeast used must be a lactose-fermenting yeast such as *Kluyveromyces marxianus*, or the lactose must be prehydrolyzed by a microbial $\beta$-galactosidase.

All of the grain-derived potable spirits mentioned so far have been produced from dry-milled grains. With corn as raw material, there is an alternative method, wet milling, that results in the isolation of corn starch and produces the following by-products: corn germ, corn oil, corn steep water, and corn gluten feed. Wet milling is now used widely for the production of corn syrups consisting almost entirely of glucose (96 DE [dextrose equivalent]). Such enzyme-hydrolyzed syrups may also be used effectively for the production of neutral spirits.

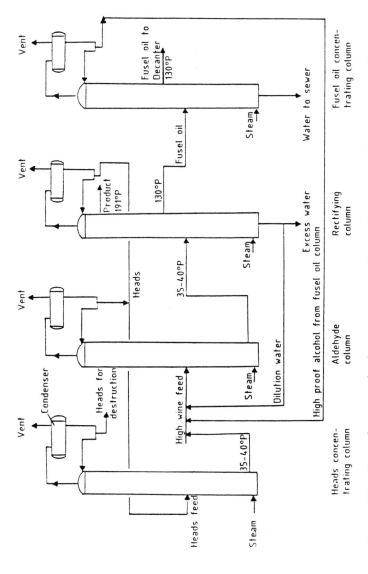

**Figure 5-6.** Four-column unit for the continuous production of grain-neutral spirits. (*From Bluhm 1983*)

## Vodka and Gin

These white goods are produced with grain-neutral spirits. *Vodka,* derived from the slavic word for water, is produced from potatoes or any other fermentable product, but in the United States grain-neutral spirits are used throughout. Residual congeners may be removed from the spirits by filtration through active charcoal. With modern methods of distillation and efficient removal of congeners the charcoal filtration is not required.

Gin is produced from grain-neutral spirits by the addition of flavor extracts of various botanicals. The principal flavor note is supplied by extracts of juniper berries. The designation *distilled gin* may be used if the botanicals are added to the spirits and distilled with them in a gin still. The botanicals are juniper berries as well as coriander seeds, dried orange peel, angelica, cardamom, and others (Brandt 1982).

## DISTILLATES FROM SUGAR-CONTAINING RAW MATERIALS

In principle, alcoholic drinks can be produced by fermentation from any sugar-containing raw material, and distillates with high ethanol concentrations can then be obtained by appropriate distilling methods. The sugar-containing raw materials are fruits, such as grapes, berries, pomes, and dates, as well as cane juice, cactus juice, and honey. Wine and brandy have even been produced from milk sugar containing whey permeate. In practice, the important distilled beverages are brandy (from grape wine), fruit brandy, rum, and tequila.

For all of these fermentations, preformed fermentable sugar (sucrose, glucose, and fructose) are the substrates for the alcoholic fermentation. Sometimes sugar from other sources may be added, such as the use of beet or cane sugar to ameliorate grape juice in some U.S. states, or the addition of brown sugar to agave juice. For all of these fermentations graphs showing the decrease of fermentable sugars, the increase in ethanol concentration, and the development of the yeast population as a function of time have a striking similarity. Sucrose is almost immediately hydrolyzed to glucose and fructose at the start of the fermentation by yeast. Glucose and fructose are fermented at a similar rate, although most industrial yeasts are glucophilic and ferment glucose at a slightly faster rate. This fermentation is in contrast to the fermentation of grain-based mashes for which the disappearance of glucose, maltose, and maltotriose is almost, but not quite, sequential and in the order shown (see Fig. 1 in Chapter 3).

### Brandy

Brandies are preferably produced from white grapes. French cognac is produced in the Charente region. The fermentations of the wine are carried out without $SO_2$ and without clarification of the must. In natural fermentations several species of yeast other than S. cerevisiae have been identified by Park (1974). They include S. rosei, S. capensis, as well as some other species now merged into S. cerevisiae, and Saccharomycodes ludwigii. The wines frequently undergo the malolactic fermentation. At the start of the fermentation the population consists of Leuconostoc mesenteroides, L. oenos, and Lactobacillus plantarum. At the end of the fermentation L. oenos is the surviving species. As in other wine fermentations this species is largely responsible for carrying out the malolactic fermentation.

Wines for the production of cognac are distilled in traditional copper pot stills. Some of the lees that contain yeast cells are kept in the wine during distillation. The first distillation concentrates the material to 28% ethanol, and a portion of the heads and tails are removed. A second distillation concentrates the cognac to 70% ethanol by volume. It is matured first in new oak casks and later in old ones. During maturation new flavor compounds are formed, in part by extraction from the wood, and in part by reaction of the extracted compounds with congeners. The traditional distillation of cognac has been described by Lafon, Couillaud, and Gaybellile (1973).

Armagnac is produced at a lesser volume than cognac, but warrants mentioning because of important differences in the method of production. It is produced in the Gers district among the upper tributaries of the Garonne. The wine containing 8-9% ethanol by volume is produced without the addition of $SO_2$. It is stored on the lees, but in contrast to cognac, the lees are removed before distillation. The distillation is carried out in continuous copper stills to an ethanol concentration of between 50% and 70% by volume. Maturation is in oak casks with a high concentration of tannins.

In the United States, brandy is produced in several states, but by far the largest volume is produced in California. White grape varieties are generally used, such as Thompson Seedless, French Colombard, Emperor, and Tokay. The grape variety is not as critical for the production of a goodquality brandy as the following factors: sound, immature grapes; clean fermentation at a low temperature; minimum use of $SO_2$; and care in distillation and maturing (Onishi, Crowell, and Guymon 1978). In California, wines are generally distilled in continuous stills. Partial removal of the fusel oil fraction during distillation is desirable, particularly for the production of wine spirits. Beverage brandy is distilled at 170° proof or below, matured in oak casks two years or longer at 102°-130° proof, and bottled at 80°-100° proof as beverage brandy.

Wine spirits used in the fortification of dessert wines may be distilled at higher proof as neutral spirits (above 190° proof), neutral brandy (171°-189° proof), and brandy (170° proof or below). Wine spirits designated as neutral spirits contain the lowest concentration of congeners. Brandy may also be produced from the lees of table wine fermentations (lees brandy) or grape pomace (pomace brandy).

Fruit brandy may be produced from any sugar-containing fruit, but must be designated by the name of that fruit, for example, plum brandy. Apple brandy is produced in larger volumes in France under the name calvados. In the United States, it may be designated as applejack. It should be produced from apple juice, not from the pulp. Apple pulp contains a relatively high concentration of pectin whose hydrolysis by pectin methyl esterase leads to higher concentrations of methanol in the distillate. To a lesser extent, the same is true of grape pomace brandy. Fruit brandies from stone fruit (plums, cherries, apricots) contain small concentrations of benzaldehyde (2-10 ppm) that contribute to the characteristic flavor of the brandy (Bandion, Valenta, and Kain 1976). The production of fruit brandies has been described in detail by Pieper, Bruchmann, and Egorov (1977). An excellent and comprehensive description of brandy production has been published by Amerine et al. (1980).

The fermentation of wines used for brandy production will not be described since the principal features have been treated in Chapter 4. The formation of flavor will be discussed below in conjunction with the flavors of other distilled beverages.

### Tequila

Tequila production has a long tradition of production in Mexico. Within the last decade it has been widely distributed in the North American market. It is rarely mentioned in the European literature.

The substrate is made from a cactus, *Agave tequilana*. The core of the cactus tissue is isolated, chopped, and cooked for a day in ovens at 93°C. The sugary syrup exuded by the pulp is collected. The pulp is further extracted with water to collect residual sugar. The fermentable sugars consist largely of D-fructose derived from the enzymatic conversion of inulin. Additional fermentable sugar in the form of brown (cane) sugar may be added. The fermentation is carried out at a low Balling (9.5°) that produces about 4-5% ethanol by volume. The alcohol is distilled twice in pot stills and is drawn between 76° and 106° proof. It is bottled immediately for sale as white tequila or aged in barrels for a year and sold as tequila anejo.

Figure 5-7 shows the flow diagram for tequila (Bluhm 1983). The figure has been included because tequila production is rarely described in the literature (see also Brandt 1982). An alcoholic drink (not distilled), pulque, produced by simple fermentation of agave sap, is widely sold in Mexico.

### Rum

Rum is produced either from sugar cane juice or molasses. The typical raw material is blackstrap molasses, a by-product of cane sugar production.

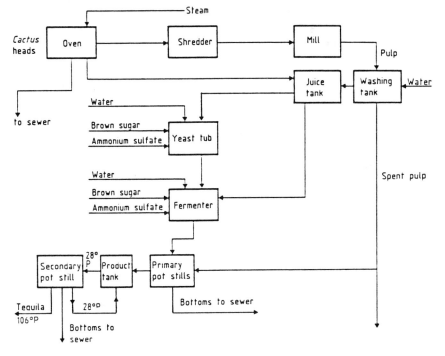

Figure 5-7. Tequila production. (*From Bluhm 1983*)

The composition of molasses at about 80° Brix with about 50-55% fermentable sugars has been shown in Table 6-2. The molasses are usually clarified by addition of sulfuric acid followed by sterilization in a heat exchanger and by centrifuging. The molasses are diluted to a fermentable sugar concentration of 12-20%. The pH is adjusted to about 4, and a nitrogen source, usually ammonium sulfate, is added.

The yeast strains used could be isolated from molasses. Lehtonen and Suomalainen (1977) listed 34 species reported by earlier authors and refer to the use of *Schizosaccharomyces pombe* for the production of heavy flavored rum. However, for almost all rum fermentations *S. cerevisiae* is now used, either by growing the yeast in stages from a pure culture slant or by using dry yeasts. If the molasses wort is well clarified the yeast may be recycled for use in subsequent fermentations.

The percentage of ethanol in the fermented wort may vary from 7-12% by volume. The fermentation is usually carried out at temperatures not exceeding 30°-32°C. Rum is produced in tropical or subtropical areas in which sugar cane is grown — Mexico, the Caribbean, Louisiana — and cooling of the fermentation is required. Rum fermentations are relatively

Table 5-2. Analysis of Congeners of Pot Still Rums, Medium Light Continuous Still Rums, and Very Light Continuous Still Rums (in mg/100 ml absolute ethanol)

|  | Pot Still Rums | Medium Continuous Still Rums | Very Light Continuous Still Rums |
|---|---|---|---|
| Demerara |  |  |  |
| Esters | 24.3 | 9.5 | 1.1 |
| Aldehydes | 18.1 | 4.0 | 0.4 |
| Higher alcohols | 363.0 | 84.5 | 3.7 |
| Jamaica |  |  |  |
| Esters | 120.0 | 49.0 | 4.1 |
| Aldehydes | 16.0 | 32.1 | 0.4 |
| Higher alcohols | 290.0 | 117.0 | 1.1 |

Source: From L'Anson 1971.

short—about 36 hours—since all of the fermentable sugar is preformed in the wort.

Rum is distilled in pot stills or continuous stills at less than 190° proof. As with other distilled beverages concentration of congeners can be manipulated by the type and operation of the still. Table 5-2 shows the highly significant changes in flavorants that can be achieved in this manner. With a tendency toward the consumption of lighter-flavored beverages the use of continuous stills is now common.

## CHARACTERISTICS AND STRAINS OF DISTILLER'S YEASTS

The requirements for specific properties of yeasts vary considerably for the widely varying substrates used in the production of distilled beverages. The fermentations vary greatly in temperature and pH. They also vary greatly in osmotic pressure, which is low in fruit juices and relatively high in molasses media. Finally, there are specific requirements for the production of a desired flavor, although the contribution of the yeast to the flavor of the distillate may not be well understood. However, some general features of yeast characteristics can be discussed.

There are two questions that apply to all yeast fermentations and that require good judgment on the part of the producer. The first question concerns the balance between productivity and yield of ethanol. Toward the end of the fermentation productivity (grams of ethanol produced per liter per hour) declines greatly, but yield continues to increase slightly.

Therefore, one has to decide at which time the fermentation should be terminated and the wash sent to the still. This decision is particularly important in fermentation of grain mashes from the dry milling process where the rate of fermentation toward the end is likely to be limited by the rate of enzymatic hydrolysis of dextrin to fermentable sugar.

A second rather difficult question concerns the concentration of fermentable sugars in the substrate. In grain mashes the concentration can be controlled to some extent by the mashing ratio; in grape juice it is determined at least in part by the maturity of the fruit; and in molasses media it can be precisely determined by the dilution of the molasses. Traditionally most fermentations resulted in the production of 8-9% ethanol by volume, which was about as high a concentration as could be achieved in grain mashes since the high viscosity of the mashes limited the substrate concentration that could be achieved.

However, today viscosities can be lowered fairly well by use of bacterial heat-stable $\alpha$-amylases. Yet alcohol concentrations are usually not higher, certainly not much higher, than in traditional fermentations. Achieving a high alcohol concentration saves energy for the distillation, but results in low productivities toward the end of the fermentation. Figure 5-8 shows

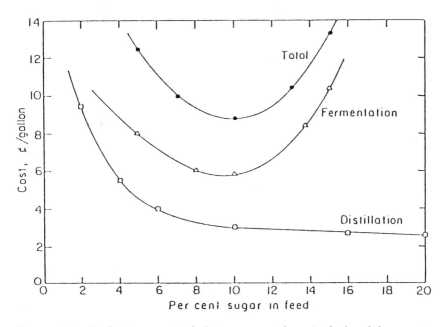

**Figure 5-8.** Production costs (excluding raw material costs) of ethanol fermentation ($2.37 \times 10^4$ gal/day of 95% ethanol) and of distillation as a function of feed sugar concentration. (*From Cysewski and Wilke 1976*)

the cost of distillation as a function of the percent sugar in the feed. This cost declines with increasing ethanol concentrations, but the decline is quite small at ethanol concentrations higher than 6-7% by volume. Although the costs shown in the graph are not current, and the costs may vary with different systems of distillation, the principle is applicable to all processes of fermentation followed by distillation.

## Strains

Distiller's yeasts are strains of *S. cerevisiae* except for natural fermentations in which any number of species may be involved. For natural wine fermentations a succession of species during the course of fermentation has been described. For natural fermentations (for instance, rum fermentations) such successions may exist but have not been described. Yeasts suitable for grain fermentations can be baker's compressed yeast, distiller's active dry yeast (often proprietary strains of individual distilleries), or strains grown in the distillery from selected pure culture slants. When pitching yeasts are grown in the plants of distillers, they are generally grown in a high-malt medium, and they are transferred to the next larger vessel when the fermentation is vigorous and when about half of the fermentable sugars have been consumed.

A common description of the characteristics required for a distiller's yeast is as follows: a yeast that is very alcohol-tolerant and gives a good yield; a yeast that ferments rapidly and hence minimizes the risk of contamination; and finally a yeast that produces the best concentration and balance of congeners desired in the distilled beverage. Such a description may be correct, but it is not very meaningful. Good tolerance for ethanol and good fermentation activity are required of all industrial yeasts; and the requirement for good flavor production or for a characteristic flavor production cannot be assessed except by trial and error and within rather wide limits.

Some attempts have been made to improve yeast, especially for use in distilled beverages. A leucineless mutant that decreases fusel oil production has already been described in Chapter 4. The hydrolysis of starches and dextrin are of critical importance in both brewing and the production of distilled beverages. Attempts have been made to use a yeast with native amylolytic capability: *Saccharomyces diastaticus*.

Table 5-3 shows the effect of starch concentration on the production of ethanol by *S. diastaticus* in the absence of any added amylases. The optimum temperature for a 72-hour fermentation was 37°C, which is highly advantageous, but lack of alcohol tolerance limited alcohol concentrations (Bannerjee, Debnath, and Majumdar 1988). An alternative route is the introduction of amyloglucosidase genes into industrial distiller's strains of

Table 5-3. Effect of Starch Concentration on the Production of Alcohol by S. diastaticus

| Starch (%) | Alcohol (% w/v) | Conversion (%) |
|---|---|---|
| 5.0 | 2.3 | 81.3 |
| 10.0 | 4.48 | 79.2 |
| 12.5 | 5.68 | 80.3 |
| 15.0 | 4.7 | 55.4 |
| 17.5 | 4.56 | 46.0 |

Source: From Bannerjee, Debnath, and Majumdar 1988.

S. cerevisiae by genetic engineering. The principle has already been described in Chapter 3. Inlow, McRae, and Ben-Bassat (1988) introduced the glucoamylase gene from Aspergillus awamori into a distiller's strain of S. cerevisiae. It is too early to predict the usefulness of this approach for commercial starch fermentations.

Traditional methods of selection of strains continue to be used. D'Amore et al. (1989) tested 15 species of yeasts for their ability to produce ethanol at elevated temperatures. A strain of S. diastaticus was the only one that fermented 15% glucose solution completely at 40°C within 24 hours. It formed 6.4% (w/v) ethanol. At higher glucose concentrations additional medium supplementation was required to obtain complete utilization of glucose. Inhibition of the alcoholic fermentation by furfural and hydroxymethyl furfural is a problem in some natural media. S. cerevisiae could be acclimatized to the presence of the inhibitor by frequent (20 times) transfer in media containing 2 mg/ml each of the two inhibitors (Azhar et al. 1982). Introduction of the killer factor into industrial brewer's and wine yeasts has already been treated in preceding chapters (see also Young 1981). Little is known about its potential usefulness in distiller's fermentations.

Han and Steinberg (1989) compared a sake strain and a commercial active dry distiller's yeast in starch fermentations in which starch hydrolysis was carried out by koji extracts using raw (uncooked), ground corn as the substrate. At 20% starch concentrations the yeasts produced 11% w/w ethanol (DADY) and 11.5% w/w ethanol (sake yeast). The process is interesting, but the economics are in doubt.

The use of bacterial strains will not be discussed further. Their frequent participation in yeast fermentations is well recognized (Wood and Hodge 1985).

Table 5-4. Ethanol Formation in Glucose Syrups as a Function of Glucose Concentration and Yeast Concentration

| °Brix | Yeast DADY (g/L) | % ethanol (w/v; final) | Fermentation time (hr) | EtOH productivity (g/L/hr) |
|---|---|---|---|---|
| 18 | 24 | 8.80 | 9.1 | 9.19 |
| 18 | 64 | 9.20 | 5.8 | 15.07 |
| 30 | 24 | 15.33 | 36.5 | 3.99 |
| 30 | 64 | 16.33 | 24.9 | 6.23 |
| 24 | 4 | 12.13 | 29.9 | 3.85 |

Conditions: 25°C; pH 4.5; glucose syrup supplemented with minerals and 16 g yeast extract per L (except 24° Brix syrup with 28 g yeast extract). Fermentation time and ethanol productivity are calculated for the period at which the fermentation was 95% complete.

*Source:* Modified from Chen 1981.

## Fermentation Parameters

The principal parameters that govern the performance of distiller's yeast, like that of other industrial yeasts, include temperature, pH, concentration of substrate, ethanol tolerance, availability of nutrients, and osmotic pressure. They can rarely be evaluated in individual sets of experiments. Chen (1981) determined the effect of substrate concentration of a glucose syrup, yeast concentration (DADY), and nutrient supplementation on final ethanol concentration, fermentation time, and ethanol productivity. Table 5-4 shows only a small fraction of his data.

The yeast obviously has good ethanol tolerance because it produced 16% ethanol (w/v) from the 30° Brix glucose solution in near theoretical yield. The few data shown in the table demonstrate the decrease in fermentation activity at substrate levels exceeding 18° Brix. A drastic decrease in productivity is, of course, merely a reflection of the rate of fermentation. Table 5-4 expresses productivity as grams ethanol produced per liter per hour. If it is calculated on a per gram of DADY basis, the productivity is greatest at the lowest yeast concentration.

Chen (1981) also determined the effect of nutrient supplementation with a fixed amount of added minerals and variable amounts of yeast extracts. Figure 5-9 shows the effect of the addition of 0.8% yeast extract on shortening the fermentation time. Such drastic effects are observed in fermentations in which considerable yeast growth takes place as in all

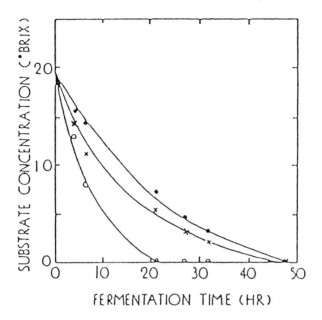

**Figure 5-9.** Effect of nutrient supplementation on the fermentation rate of a 95 DE corn syrup.

●—●—●   control, unsupplemented
×—×—×   supplemented with inorganic salts
○—○—○   supplemented with inorganic salts and yeast extract (8 g/L)

(*From Chen 1981*)

distiller's fermentations. With a nominal inoculum of $5 \times 10^6$ cells per milliliter, the final yeast concentration may reach $150 \times 10^6$ cells per milliliter (De Becze 1967). In practice, inoculation of fermentations with 15 grams of distiller's active dry yeast per hectoliter produces a cell count of approximately $5 \times 10^6$ cells per milliliter.* Modeling of the kinetics of the alcoholic fermentation is generally carried out with continuous fermentations under steady-state conditions (Jayanata and Reuss 1984). The usefulness of such data for the alcoholic batch fermentations is limited.

The effect of temperature on the rate of ethanol formation does not differ materially from the effects in wine and brewer's fermentations. Verduyn et al. (1984) observed the rate of ethanol fermentation by a baker's yeast for low glucose concentrations with an enzyme electrode.

---

*Yeast levels indicated in Figure 5-9 and Table 5-4 are unrealistically high, but support the principles expressed by the author.

Table 5-5. Carbohydrate Content and Ethanol Yields for Raw Materials Used in the Production of Distilled Beverages

|  | Fermentable Carbohydrate as Glucose | | Weight of Raw Material (kg) per kg Absolute EtOH Produced | |
|---|---|---|---|---|
|  | % of Whole Weight | % of Dry Weight | Whole Weight (kg) | Weight, Solids (kg) |
| Beet molasses | 50 | 63 | 4.2 | 3.4 |
| Cane molasses | 35 | 44 | 6.0 | 4.8 |
| Potato | 19 | 76 | 11.0 | 2.8 |
| Barley | 57 | 66 | 3.7 | 3.2 |
| Rye | 65 | 76 | 3.2 | 2.8 |
| Wheat | 66 | 77 | 3.2 | 2.8 |
| Corn | 67 | 77 | 3.1 | 2.7 |
| Millet | 67 | 76 | 3.1 | 2.7 |

Source: Abbreviated and modified from Suomalainen et al. 1968.

Maximal ethanol formation expressed as mMole per gram of wet yeast cells per hour was as follows: 20°C, 1.9; 25°C, 2.8; 30°C, 4.3; 35°C, 5.4. It must be remembered that such values are only valid for low sugar and alcohol concentrations. At high ethanol concentrations fermentations are increasingly inhibited at higher temperatures. The same caution has to be applied to a determination of maximum temperatures for yeast growth (Walsh and Martin 1977).

Yields of ethanol based on glucose as substrate are always less than the 51% as predicted by the Gay-Lussac equation. These amounts are the result of the formation of fermentation by-products and of the requirements of carbon sources for yeast growth and maintenance. Practical yields for substrates used in the production of distilled beverages are shown in Table 5-5 (Suomalainen et al. 1968). The values given are approximations. American beet and cane molasses contain higher percentages of fermentable sugars than those shown in Table 5-5.

The effect of higher osmotic pressure on yeast fermentations is particularly important in doughs. With few exceptions it is of lesser importance in the production of beer and distilled beverages.

## FLAVOR COMPOUNDS

Flavor and odor compounds produced during anaerobic or microaerobic yeast fermentations are quite similar in bread doughs, beer, and wine. Identical compounds appear in the numerous compilations, and the differences are usually quantitative rather than qualitative. These flavor compounds are usually grouped into the following classes: alcohols other than the flavorless and odorless ethanol (fusel oil), organic acids, esters, and carbonyl compounds. The number of these compounds is very large and it keeps growing with refinements of analytical techniques (Lehtonen and Suomalainen 1979). In 1977, Lehtonen and Suomalainen listed 20 alcohols, 33 organic acids, 73 esters, 82 carbonyl compounds and acetals, 24 phenolic compounds, 15 nitrogenous compounds (mostly pyrazines), 6 sulfur compounds, 7 lactones, and 7 miscellaneous compounds in rum.

In distilled beverages the flavor compounds derived from the raw material and the fermentation are greatly modified by distillation and additionally by maturation. Neutral spirits are produced for the production of vodka or gin, or for blending, and they are distilled in such a manner that flavor compounds are completely or almost completely removed. Of course, they do not have to be matured, and they will not be further discussed in this section.

The phrase *flavor compounds* is used loosely since a high percentage of these compounds are present in such small concentrations that it is questionable whether they can be perceived when the beverage is tasted. Salo, Nykanen, and Suomalainen (1972) determined the relationship of the threshold of odor perception of individual compounds with their actual concentration in whisky. The data are shown in Table 5-6 only for those compounds whose concentration exceeded the odor threshold value by a factor of 1 to 36. For the alcohols these factors were 2.3 for isobutyl alcohol, 2.8 for optically active amyl alcohol, and 36 for isoamyl alcohol.

Ten other alcohols were determined, but are not included in Table 5-6 since their concentration individually was below the odor threshold. This omission does not mean that collectively they did not contribute to overall odor. Indeed the authors show that they did, but it is difficult if not impossible to assign to them any quantitative estimate of their contribution to the flavor of whisky. A determination of groups of flavor compounds or of individual flavor compounds does not permit a sound judgment on the quality of the flavor. Such analyses with gas chromatographic methods have furthered the understanding of flavor formation but are not entirely satisfactory in characterizing the perceived odor or flavor of the beverage. Piggott and Jardine (1979) developed a descriptive sensory analysis of whisky flavor based on the use of descriptor terms by a trained panel.

Numerous compilations of flavor and odor compounds for various distilled beverages including some quantitative data are available from previously cited publications. These data are not presented in this chapter. Instead some specific examples will be used to illustrate the effect on flavor of fermentation, distillation, and maturation in oak barrels.

Table 5-6. Concentrations and Odor Threshold Values of Individual Model Whisky Compounds Whose Concentration Matches or Exceeds the Odor Threshold

| Acids | Acetic | Isovaleric | Caprylic | Capric | Lauric | | |
|---|---|---|---|---|---|---|---|
| Concentration, mg/L | 36.0 | 3.2 | 14.0 | 17.0 | 8.8 | | |
| Odor threshold, mg/L | 26.0 | 0.75 | 15.0 | 9.4 | 7.2 | | |
| Esters | Ethyl Acetate | Ethyl Caproate | Ethyl Caprylate | Ethyl Caprate | Ethyl Laurate | Isoamyl Acetate | |
| Concentration, mg/L | 100 | 0.85 | 2.80 | 7.70 | 5.50 | 2.60 | |
| Odor threshold, mg/L | 17.0 | 0.076 | 0.24 | 1.10 | 0.64 | 0.23 | |
| Carbonyl Compounds | Acetaldehyde | Pentanedione | Isobutyraldehyde | n-Butyraldehyde | Isovaleraldehyde | n-Valeraldehyde | Diacetyl |
| Concentration, mg/L | 2.80 | 0.12 | 8.00 | 0.56 | 2.50 | 1.20 | 0.60 |
| Odor threshold, mg/L | 1.20 | 0.078 | 1.30 | 0.028 | 0.12 | 0.11 | 0.020 |
| Alcohols | Isobutyl | Optically Active Amyl | Isoamyl | | | | |
| Concentration, mg/L | 174 | 91 | 232 | | | | |
| Odor threshold, mg/L | 75.0 | 32.0 | 6.5 | | | | |

Source: Modified from Salo, Nykanen, and Suomalainen 1972.

A large number of flavor compounds are produced during the fermentation and released into the fermenter liquid (wash). Table 5-7 shows the formation of organic acids and fusel oil as a function of the nitrogen source. The presence of amino acids in the malt wort stimulates formation of the higher alcohols, presumably through the Ehrlich reaction. In media containing only ammonium salts or no assimilable nitrogen the formation of fusel oils is slight, but the production of organic acids and glycerol is higher. Harrison and Graham (1970) reported these results and have also dealt with the pathways responsible for the differentiation.

The effect of distillation on flavor and odor can be dramatic. It is possible to remove perceptible odor almost completely by quantitative removal of the heads and tails fraction during distillation. Table 5-2 shows a comparison of the congener fractions of rums distilled in pot stills and in continuous stills. The ability of continuous stills to remove both higher boiling and lower boiling congeners is well demonstrated. Bluhm (1983) also reports the presence of 10-fold higher concentrations of fusel oils in pot-distilled dark rum than in continuously distilled light rum.

The effect of maturation in wood is yet more complex. It consists of the reaction of flavor compounds with each other during the long storage period, the extraction of flavor compounds (and color) from the wood, and the reaction of the extracted compounds with the other flavor compounds. The extracted materials are principally tannins, color, furfural, esters, and total solids. The rate of congener formation or loss is greatest during the first year of storage in wood. Thereafter, the rate of change decreases, except for esters, which are formed at an approximately equal rate throughout the entire storage period.

Table 5-7. Effect of Nitrogen Source in Fermentation Media on the Concentrations of Ethanol and Fermentation By-products

| Product | Amino Acids as N Source (1) | Ammonium Salts as N Source (2) | Absence of Assimilable N (3) |
|---|---|---|---|
| Ethanol | 5.130 mg | 4.563 mg | 4.723 mg |
| Glycerol | 216 mg | 518 mg | 425 mg |
| Organic acids (as succinic) | 54 mg | 140 mg | 150 mg |
| Fusel oils (as butanol) | 20 mg | 5 mg | 9 mg |
| Yeast dry matter (final minus initial) | 208 mg | 190 mg | 38 mg |

*Source:* Modified from Harrison and Graham 1970.
(1) 100 ml malt wort containing 15 mg of assimilable N and 10.1 g of assimilable carbohydrate (as glucose).
(2) 100 ml synthetic medium, with adequate minerals and growth factors, containing 10.1 g glucose and 15 mg N from ammonium salts.
(3) 100 ml synthetic medium; same as (2) but without ammonium salts.

Table 5-8. Components of Aged and Unaged Whisky (in g per 100 L of 100° proof)

| | Bourbon | | Rye | | Light | | Scotch | |
|---|---|---|---|---|---|---|---|---|
| | Unaged | Aged | Unaged | Aged | Unaged | Aged | Unaged | Aged |
| Total acids as acetic | | 44.7 | | 71.2 | | 27.9 | 5.3 | 37.5 |
| Volatile acids as acetic | | 39.4 | | 61.4 | | 19.6 | 5.3 | 27.8 |
| Tannins | | 53.4 | | 54.9 | | 67.0 | 0.0 | 27.3 |
| pH | | 3.9 | | 3.9 | | 3.9 | 5.5 | 4.0 |
| Ethyl acetate | 4.9 | 27.8 | 5.0 | 63.7 | 0.3 | 11.2 | 25.1 | 20.3 |
| Total fusel oils | 173.9 | 177.6 | 276.5 | 269.9 | 11.0 | 51.5 | 143.5 | 198.6 |
| Furfural | 0.0 | 0.83 | 0.0 | 1.5 | | 0.4 | trace | 0.8 |
| 2-phenyl ethanol | 3.13 | 2.5 | 0.9 | 2.1 | | 0.6 | 2.0 | 2.2 |

*Source:* Modified from Brandt 1982.

The following variables affect the extraction of stored whisky: type of wood (oak), new or reused barrels, charred or uncharred barrels, proof of stored whisky, temperature, and, of course, length of storage. In general, extraction is more effective at higher temperature, at lower proof, in charred barrels (versus not charred), in new barrels (versus reused), and with increased storage time. The origin of the extracted compounds and their reactions with whisky components is not always known. Ethanol lignin is derived from the reaction of the lignin of the wood with ethanol. Pentoses and hexoses are probably derived from hemicelluloses.

Some of the aromatic aldehydes such as synapaldehyde, syringaldehyde, coniferaldehyde, and vanillin are obtained by oxidation of wood compounds. Acetic acid is formed by oxidation of acetaldehyde, which in turn leads to a great increase in ethyl acetate during maturation. These and other reactions have been reviewed in an excellent paper by Reazin (1981). In-house data of a distilled beverage manufacturer comparing the congener concentrations of unaged and aged whiskies are shown in Table 5-8.

## FERMENTATION ALCOHOL AS FUEL

Fuel ethanol per se is not the subject of this section. But some newer methods of production used in the fuel ethanol industry are likely to have an impact on beverage ethanol production, and such methods will be briefly mentioned.

Fuel ethanol for blending with gasoline (gasohol) or for direct use in motor cars requires a 200 proof (100%) product. It is used to extend crude oil supplies, to replace lead in gasoline, and to increase the value of low-octane gasoline. It can be produced from 96% ethanol by extractive distillation, which need not be discussed here.

The major emphasis in fuel ethanol production is on a low-cost product, that is, the most efficient methods of fermentation and distillation and the lowest raw material cost. In the United States, this raw material is starch from corn obtained by wet milling or dry milling. Other raw materials are waste materials, particularly potato waste. Cheese whey permeate is also used, but the most important material is corn starch. The use of other raw materials such as sweet sorghum (Day and Sarkar 1982) and fodder beets (Gibbons, Westby, and Arnold 1988) have been investigated. In Brazil, a major producer of fuel ethanol, the raw material is cane juice or cane molasses.

The following three approaches to the fermentation that will be briefly considered are directed toward an improvement of ethanol productivity, that is, the production of ethanol per fermenter volume per hour. They are the continuous fermentation, yeast cell immobilization, and extraction of ethanol during the fermentation. In some investigations these methods have been combined.

## Immobilization

Immobilization of yeast cells sometimes reduces their fermentation activity, but an increase in metabolic activity has been reported by several investigators. Galazzo and Bailey (1989) report that glucose uptake and ethanol and glycerin production are twice as fast in immobilized cells of *Saccharomyces cerevisiae*. They conclude that the increase in glucose utilization reflects an increase in glucose uptake and/or glucose phosphorylation. They ascribe the effect to the lower internal pH of immobilized cells. Vieira et al. (1989) speculate that the improvement of the fermentation rate of immobilized cells is due to media supplementation with compounds contained in the support materials. Although this explanation may not appear to be very likely, it should be recalled that the presence of particulate matter in grape juice accelerates ethanol production.

Immobilized cells are usually used in packed beds in columns that are fed with the substrate solution at the bottom. The fermented liquid exits at the top. This arrangement approaches the conditions of a heteroge-

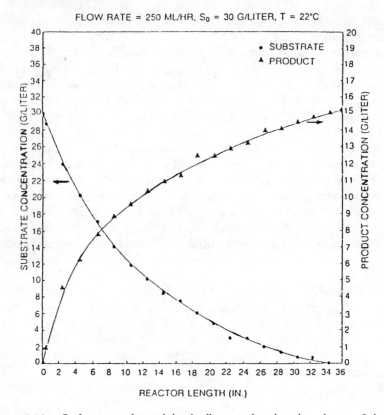

**Figure 5-10.** Performance of immobilized cell reactor for ethanol production. Substrate (●—●—●) and ethanol (▲—▲—▲) were determined at 2-inch intervals of a 36-column after 1 day of continuous operation. (*From Sitton, Foutch, and Gaddy 1979*)

neous (plug flow) fermentation. It invariably increases ethanol productivity because ethanol inhibition is only encountered in the upper part of the column (Fig. 5-10) (Sitton, Foutch, and Gaddy 1979). A major problem with packed bed columns is the disruptive effect of the large volumes of $CO_2$ generated in the column. Some relief can probably be obtained by using special configurations, for instance, by the tapered column used by Hamamci and Ryu (1987). However, the ultimate practicability of packed columns of immobilized cells is doubtful.

### Extraction of Ethanol During Fermentation

Inhibition of the fermentation rate at higher alcohol concentrations limits the productivity of all systems used for the alcoholic fermentation. Several means for removing this inhibition by continuous extraction of ethanol during the fermentation have been suggested. Cysewski and Wilke (1977) used vacuum distillation of the ethanol during a fermentation at nominal temperatures. Ethanol may also be removed by extraction with water-immiscible solvents. Aires Barros, Cabral, and Novais (1987) used oleic acid to extract ethanol during an alcoholic fermentation by immobilized yeast cells. L'Italien, Thibault, and LeDuy (1989) used extraction of ethanol with $CO_2$ at hyperbaric pressure. The required pressure is by itself inhibiting. But their report contains valuable information on the inhibiting effect of high pressure per se and on the specific effect of the gas used to pressurize the fermenter.

Finally, use of membrane fermenters for continuous removal of ethanol and retention of substrate and yeast cells on the inside of the membrane has been described by Lee and Chang (1987). None of the mentioned methods have found industrial use, although increased productivity could be demonstrated. The reason must be sought in the high cost of additional equipment and process and material costs. Biofilm reactors may provide the most promising solution (Vega, Clausen, and Gaddy 1988).

## Continuous Fermentation

Continuous fermentation of glucose to ethanol in single-stirred fermenters, that is, homogeneous fermentations, has been frequently described. Figure 5-11 (Cysewski and Wilke 1976) shows the concentration of ethanol, cell mass productivity, and glucose concentration as functions of the dilution rate. At a dilution rate of 0.1, the ethanol concentration was about 4% (w/v), productivity was 3.5 g/L/hr, and the glucose concentration was negligible. At higher dilution rates productivity can be greatly increased, but ethanol concentration decreases, and since glucose concentrations increase, the yield must decline because of loss of glucose in the effluent stream.

Jayanata and Reuss (1984) have modeled the kinetics of a continuous (homogeneous) molasses fermentation and verified their model. Their results do not differ greatly from the earlier work. A better solution, and one that is used industrially, is the cascade continuous fermentation. It is a multistage continuous fermentation with cell recycle. Such multistage operations approach the conditions found in heterogeneous (plug flow) fermenters in which ethanol inhibition occurs only in the last fermenters of the cascade.

Recycling of the yeast is carried out by centrifugal separation of the cells and their return to the first fermenter. Good sanitation, cooling of the yeast cream if it is to be held for some time, and possibly washing of the cells are required (Nagodawithana 1986; Lyons 1981).

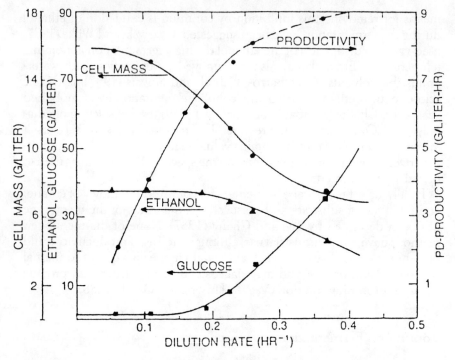

**Figure 5-11.** Effect of dilution rate on ethanol productivity (♦—♦—♦) and on the concentrations of ethanol (▲—▲—▲), glucose (■—■—■), and cell mass (●—●—●). Continuous fermentation with adapted yeast. Glucose concentration = 8.9%; oxygen tension = 0.07 mmHg. (*From Cysewski and Wilke 1976*)

## FURTHER READINGS

The following sources deal with the microbiology of distilled beverages (including brandy): Harrison and Graham (1970); Rose (1977); Humphrey and Stewart (1978); Packowski (1978); Maisch, Sobolov, and Petricola (1979); Amerine et al. (1980); Brandt (1982); Reed (1983); Bluhm (1983); Cosaric et al. (1983); Lafon-Lafourcade (1983); Nagodawithana (1986); and Hepner and Male (1987).

## REFERENCES

Aires Barros, M. R., J. M. S. Cabral, and J. M. Novais. 1987. Production of ethanol by immobilized *S. bayanus* in an extractive fermentation system. *Biotechnol. Bioeng.* **29:**1097-1104.

Amerine, M. A., H. W. Berg, R. E. Kunkee, C. S. Ough, V. L. Singleton, and A. D. Webb. 1980. *The Technology of Wine Making*, 4th ed. AVI Publishing Company, Westport, Conn.

Azhar, A. P., M. K. Bery, A. R. Colcord, and R. S. Roberts. 1982. Development of a yeast strain for the efficient ethanol fermentation of wood hydrolysate. *Dev. Industrial. Microbiol.* **23**:351-360.
Bandion, F., M. Valenta, and W. Kain. 1976. Study of benzaldehyde content in brandies and liqueurs of stone fruit (in German). *Mitt. Rebe Wein, Obstbau, Fruchteverw., Klosterneuburg* **26**:43-50.
Bannerjee, M., S. Debnath, and S. K. Majumdar. 1988. Production of ethanol from starch by direct fermentation. *Biotechnol. Bioeng.* **32**:831-834.
Barbour, E. A., and F. G. Priest. 1988. Some effects of *Lactobacillus* contamination of scotch whisky fermentations. *J. Inst. Brewing* **94**(2):89-92.
Bluhm, L. 1983. Distilled beverages. In *Biotechnology*, vol. 5, H. J. Rehm and G. Reed (eds.). VCH Publishing Company, Weinheim, West Gemany.
Brandt, D. A. 1975. Distilled alcoholic beverages. In *Enzymes in Food Processing*, 2nd ed., G. Reed (ed.). Academic Press, New York.
Brandt, D. A. 1982. Distilled beverage alcohol. In *Prescott and Dunn's Industrial Microbiology*, 4th ed., G. Reed (ed.). AVI Publishing Company, Westport, Conn.
Chen, S. L. 1981. Optimization of batch alcoholic fermentation of glucose syrup substrate. *Biotechnol. Bioeng.* **23**:1827-1836.
Cosaric, N., A. Wieczorek, G. P. Consentino, R. J. Magee, and J. E. Prenosil. 1983. Ethanol fermentation. In *Biotechnology*, vol. 3, H. Dellweg (ed.). VCH Publishing Company, Weinheim, West Germany.
Cysewski, G. R., and C. R. Wilke. 1976. Utilization of cellulosic materials through enzymatic hydrolysis. I. Fermentation to ethanol and single cell protein. *Biotechnol. Bioeng.* **18**:1297-1313.
Cysewski, G. R., and C. R. Wilke. 1977. Rapid ethanol fermentations using vacuum and cell recycle. *Biotechnol. Bioeng.* **19**:1125-1143.
D'Amore, T., G. Celotto, I. Russell, and G. G. Stewart. 1989. Selection and optimization of yeast suitable for ethanol production at 40°C. *Enzyme Microbiol. Technol.* **11**(7):411-416.
Day, D. F., and D. Sarkar. 1982. Fuel alcohol from sweet sorghum. *Dev. Industrial. Microbiol.* **23**:361-366.
De Becze, G. I. 1967. Alcoholic beverages, distilled. In *Encyclopedia of Industrial Chemical Analysis*, vol. 4. Wiley, New York.
Galazzo, J. L., and J. E. Bailey. 1989. In vivo nuclear magnetic resonance analysis of immobilization effects on glucose metabolism of the yeast *S. cerevisiae*. *Biotechnol. Bioeng.* **33**:1283-1289.
Gibbons, W. R., C. A. Westby, and E. Arnold. 1988. Semicontinuous diffusion fermentation of fodder beets for fueled ethanol and cubed protein feed product. *Biotechnol. Bioeng.* **31**:696-704.
Hamamci, H., and D. D. Y. Ryu. 1987. Performance of tapered column packed-bed bioreactor for ethanol production. *Biotechnol. Bioeng.* **29**:994-1002.
Han, I. Y., and M. P. Steinberg. 1989. Simultaneous hydrolysis and fermentation of raw dent and high lysine corn and their starches. *Biotechnol. Bioeng.* **33**:906-911.
Harrison, J. S., and C. J. Graham. 1970. Yeasts in distillery practice. In *The Yeasts*, A. H. Rose and J. S. Harrison (eds.). Academic Press, London.
Hepner, L., and C. Male. 1987. Economic aspects of fermentation processes. In *Fundamentals of Biotechnology*, P. Praeve et al. (eds.). VCH Publishing Company, New York.
Horak, W., F. Drawert, P. Schreier, W. Heitmann, and H. Lang. 1974. In *Ullmanns Encyclopadie der Technischen Chemie* (in German). VCH Publishing Company, Weinheim, West Germany.
Humphrey, T. W., and G. G. Stewart. 1978. Alcoholic beverages. In *Food and Beverage Mycology*, L. R. Beuchat (ed.). AVI Publishing Company, Westport, Conn.

Inlow, D., J. McRae, and A. Ben-Bassat. 1988. Fermentation of corn starch to ethanol with genetically engineered yeast. *Biotechnol. Bioeng.* **32**:227-234.

Jayanata, V., and M. Reuss. 1984. Kinetics of ethanol fermentation of molasses. *American Institute of Chemical Engineers Symposium*, April 18, St. Louis, MO.

Kreipe, H. 1963. Grain and potato distilleries. In *Die Hefen*, 2 vols., F. Reiff et al. (eds.). Verlag Hans Carl, Nuremberg, West Germany.

Lafon, J., P. Couillaud, and F. Gaybellile. 1973. *Le Cognac. Sa Distillation*, 5th ed. (in French). J.-B. Bailliere et Fils, Paris.

Lafon-Lafourcade, S. 1983. Wine and brandy. In *Biotechnology*, vol. 5, G. Reed (ed.). VCH Publishing Company, Weinheim, West Germany.

L'Anson, J. A. P. 1971. Rum manufacture. *Proc. Biochem.* **6**(7):35-39.

Lee, C. W., and H. N. Chang. 1987. Kinetics of ethanol fermentations in membrane recycle fermentors. *Biotechnol. Bioeng.* **29**:1105-1112.

Lehtonen, M., and H. Suomalainen. 1977. Rum. In *Economic Microbiology*, vol. 1, A. H. Rose (ed.). Academic Press, New York.

Lehtonen, M., and H. Suomalainen. 1979. The analytical profile of some whisky brands. *Proc. Biochem.* **14**(2):5-6, 8-9.

L'Italien, Y., J. Thibault, and A. LeDuy. 1989. Improvement of ethanol fermentation under hyperbaric conditions. *Biotechnol. Bioeng.* **33**:471-476.

Lyons, T. P. 1981. Batch and continuous fermentation. *Gasohol USA* **3**(8):10-11.

Lyons, T. P., and A. H. Rose. 1977. Whisky. In *Economic Microbiology*, vol. 1, A. H. Rose (ed.). Academic Press, New York.

Maisch, W. F., M. Sobolov, and A. J. Petricola. 1979. Distilled alcoholic beverages. In *Microbial Technology*, 2nd. ed., D. Perlman and H. J. Peppler (eds.). Academic Press, New York.

Marsal, F., C. Sarre, D. Dubourdieu, and J. N. Boldron. 1988. Role of yeasts in the transformation of some volatile constituents of the wood during production of white wines in oak casks (in French). *Conn. Vigne Vin* **22**(1):33-38.

Nagodawithana, T. W. 1986. Yeasts. Their role in modified cereal fermentations. *Adv. Cereal Sci. Technol.* **8**:15-104.

Onishi, M., E. A. Crowell, and J. F. Guymon. 1978. Comparative composition of brandies from Thompson Seedless and three white wine grape varieties. *Am. J. Enol. Vitic.* **29**:54-58.

Packowski, G. W. 1978. Distilled beverage spirits. In *Kirk-Othmer Encyclopedia of Chemical Technology*, 3rd ed., vol. 3, pp. 830-863. Wiley, New York.

Pan, S. C., A. A. Andreasen, and P. Kolachov. 1950. Rate of secondary fermentation of corn mashes converted with *Aspergillus niger*. *Ind. Eng. Chem.* **42**:1783-1789.

Park, Y. H. 1974. Doctoral thesis (in French). Université de Bordeaux II, France. As cited by Lafon-Lafourcade (1983).

Pieper, H.-J., E. E. Bruchmann, and I. A. Egorov. 1977. *Technologie der Obstbrennerei* (Technology of Fruit Brandies, in German). Eugen Ulmer, Stuttgart.

Piggott, J. R., and S. P. Jardine. 1979. Descriptive sensory analysis of whisky flavour. *J. Inst. Brewing* **85**:82-85.

Pomeranz, Y. 1984. *Functional Properties of Food Components*. Academic Press, Orlando, Fla.

Pyke, M. 1965. The manufacture of scotch grain whisky. *J. Inst. Brewing* **71**:209-218.

Reazin, G. H. 1981. Chemical mechanism of whiskey maturation. *Am. J. Enol. Vitic.* **32**(4):283-289.

Reed, G. 1983. Production of fermentation alcohol as fuel. In *Prescott and Dunn's Industrial Microbiology*, 4th ed., G. Reed (ed.). AVI Publishing Company, Westport, Conn.

Rose, A. H. (ed.). 1977. *Economic Microbiology*, vol. 1. Academic Press, New York.
Salo, P., L. Nykanen, and H. Suomalainen. 1972. Odor thresholds and relative intensities of volatile aroma compounds in an artificial beverage imitating whiskey. *J. Food Sci.* **37**:394-398.
Sfat, M. R., and J. A. Doncheck. 1981. Malts and malting. In *Kirk-Othmer Encyclopedia Chemical Technology.* Wiley, New York.
Simpson, A. C. 1977. Gin and vodka. In *Economic Microbiology*, vol. 1, A. H. Rose (ed.). Academic Press, New York.
Sitton, O. C., G. I. Foutch, and J. L. Gaddy. 1979. Ethanol from agricultural residues. *Proc. Biochem.* **14**(9):7-10.
Suomalainen, H., O. Kauppila, L. Nykanen, and R. L. Peltonen. 1968. Alcoholic beverages (in German). In *Handbuch der Lebensmittelchemie*, vol. 7, J. Schormuller (ed.). Springer Verlag, Berlin.
Tzeng, C. H., B. A. Bernstein, S. W. Chu, and S. Bernstein. 1979. Commercial production of ethanol by the continuous two-stage fermentation of cheese whey. Paper presented at the U.S.-Japan Intersoc. Microbiol. Congress, Honolulu, May 1979.
Vega, J. L., E. C. Clausen, and J. L. Gaddy. 1988. Biofilm reactors for ethanol production. *Enzyme Microbiol. Technol.* **10**(7):390-402.
Verduyn, C., T. P. L. Zomerdijk, J. P. van Dijken, and W. A. Scheffers. 1984. Continuous measurement of alcohol production by aerobic yeast suspensions with an enzyme electrode. *Appl. Microbiol. Biotechnol.* **19**:181-185.
Vieira, A. M., I. S. Correia, J. M. Novais, and J. M. S. Cabral. 1989. Could the improvement in the alcoholic fermentation of high glucose concentrations by yeast immobilization be explained by media supplementations? *Biotechnol. Lett.* **11**(2):137-140.
Walsh, R. M., and P. A. Martin. 1977. Growth of *S. cerevisiae* and *S. uvarum* in a temperature gradient incubator. *J. Inst. Brewing* **83**:169-172.
Wood, B. J. B., and M. M. Hodge. 1985. Yeast-lactic bacteria interactions and their contribution to fermented foodstuffs. In *Microbiology of Fermented Foods*, vol. 1, B. J. B Wood (ed.). Elsevier, London.
Young, T. W. 1981. The genetic manipulation of killer character into brewing yeast. *J. Inst. Brewing* **87**:292-295.

CHAPTER

# 6

# BAKER'S YEAST PRODUCTION

Bread doughs are fermented for very short periods of time with a range of 30 minutes to 4 hours. They are inoculated with $300 \times 10^6$ cells per gram and there is little or no yeast growth during the fermentation. In contrast, wine, beer, and distiller's mashes are fermented for periods ranging from several days to several weeks. Inoculation levels are in the range of $2\text{-}10 \times 10^6$ cells per milliliter and there is a 5- to 10-fold multiplication of yeast cells during the fermentation. In addition, yeast cells may be recycled for use in succeeding batches of beer or wine fermentations. Baker's yeast cannot be recycled because the yeast is killed during the baking process. Consequently, the production of baker's yeast can be carried out on a very large industrial scale, and since the latter part of the nineteenth century, baker's yeast has been produced by companies that specialize in its production.

It is not quite correct to say that baker's yeast cannot be recycled. For several millenia, baker's yeast has been recycled in part by retaining a portion of the yeasted dough (generally from a third to a tenth) and blending it with fresh water and flour for formation of the next dough. This method is still practiced in some countries and is used in the United States for production of San Francisco sour dough bread. It will be discussed in Chapter 7.

In countries that produce beer it was soon found that ale yeast could be used in bread production, and prior to 1850, it was widely used for that purpose. It required only separation from the beer foam and pressing. The level of inoculation of bread doughs with this ale yeast was carried out at low levels and the fermentation times were long, generally overnight.

Since about 1848, brewers in the United States started to produce lager beer instead of ale. The lager beer yeast, *Saccharomyces carlsbergensis* (later *S. uvarum*), was not suitable for leavening doughs because it does not tolerate the high osmotic pressure. The production of baker's yeast exclusively for the production of bread doughs dates from the latter part of the nineteenth century. It started with production of distiller's yeast on grain mashes, which were later replaced with the least expensive source of assimilable sugar, molasses.

During the last century the major advances in the production of baker's yeast have been the following:

1. *Aeration.* The stimulating effect of aeration on yeast growth was well known toward the end of the nineteenth century, and continuous aeration of mashes was used in Britain in 1886.
2. *Fed-batch process.* The use of incremental feeding (called Zulauf process in German) was introduced between 1910 and 1920 by Danish and German scientists. This process is used today because it is the only practical method that permits production of yeast biomass without simultaneous production of sizable quantities of ethanol.
3. *Molasses.* At the turn of the century the mash bill consisted principally of corn, malt, and malt sprouts. During the 1920s and 1930s, the grains were slowly replaced with molasses as carbon and energy source for yeast growth.
4. *Active dry yeast.* Progress was made in the late 1930s in the drying of compressed yeast (CY) to the more stable active dry yeast (ADY). In the 1940s, ADY began to penetrate the consumer yeast market. Slow but continued progress in drying methods has led to the point where ADY has replaced CY in several bakery applications, particularly in institutional baking and in pizzerias.
5. *Automation.* During the past decade there has been a rapid shift to automatic control of the fermentation.

The history of baker's yeast production has been reviewed by Kiby (1912) and Butschek and Kautzmann (1962). The development of ADY has been reviewed by Frey (1957).

## MANUFACTURING PROCESS OUTLINE

Cane or beet molasses supply not only sugars as a carbon and energy source but also some organic nitrogen, minerals, sulfur, vitamins, and trace elements. The liquid growth medium must be supplemented with additional nitrogen (usually ammonia or ammonium salts) and phosphate, as well as with additional minerals (Ca, Mg), vitamins, and trace elements.

Yeast is grown in large fermenters by the fed batch process. The fermenters are equipped with cooling coils and with means for vigorous aeration to maintain highly aerobic growth. The growth rate is restricted to a growth rate coefficient of less than 0.25. In the fermenter liquid 4-6% of yeast solids can be produced.

Postfermentation processing begins with centrifuging to produce a concentrate (yeast cream) with 18-20% solids. The yeast cream is washed and either pressed or filtered to a semisolid yeast mass of 30% solids. This press or filter cake is packaged as a crumbly mass in bags or extruded in blocks that are wax wrapped. It is cooled and shipped refrigerated to bakeries. Some bakeries also accept shipment of the pumpable, refrigerated yeast cream.

ADY is produced from the yeast cake by extrusion in the form of thin strands. These are air dried on a continuous belt or in air-lift driers. ADY with a solids content of 92-96% is shipped as such or in vacuum-packed or nitrogen-flushed pouches. It does not require refrigerated shipment or storage.

## STRAINS

Until the early 1970s, two strains of *Saccharomyces cerevisiae* were used in the United States. The strain used for the production of baker's compressed yeast was grown to a nitrogen level of about 8.2-8.8% (dry weight basis). The ADY strain, typified by strain no. 7752 in the American Type Culture Collection, was grown to a nitrogen level rarely exceeding 7%. The lower nitrogen level favored the stability of the yeast, but resulted in a somewhat lower fermentation activity in comparison with compressed yeast on an equivalent solids basis.

Since about 1970, there has been a need for additional strains based on the following requirements of the baking industry: a dry yeast strain with improved fermentation activity, a yeast strain with exceptional fermentation activity in high-sugar doughs, and a yeast with improved performance in yeast-leavened frozen doughs. Some new strains have now become available, but acceptance by the baking industry has been slow, partly because bakers do not like to deal with several strains in their bakeries, and partly because the production and distribution of several strains by yeast companies are more costly. Nevertheless, a trend toward more yeast strains for specialty applications can be expected. The development of yeasts for special applications will be briefly discussed later. However, some additional comments on other strain developments are warranted.

Hybridization methods for strain development have generally been traditional. In many instances, protoplast fusion is now used. Newer methods of genetic engineering (DNA modification) are tempting research workers, but have not yet led to demonstrable practical results. For instance, it was thought that insertion of genes forming one or more enzymes of the glycolytic pathway might result in increased fermentation activity. But it is now apparent that the "activities of most of the glycolytic enzymes are not rate limiting, not even in the case of the last two reactions from pyruvate to ethanol" (Schaaf 1988).

Some yeast strains can ferment starch (McCann and Barnett 1986), and the insertion of genes that form amylases in brewer's yeasts has been achieved. There is some doubt whether this procedure would be particularly useful to the baking industry, but it deserves consideration. Killer yeasts that kill sensitive strains of wild yeasts may have application in the brewing and wine industries, but are not likely to be useful in baking. In Japan, the species *Torulaspora delbrueckii* (syn. *Saccharomyces rosei*) has been used, and hybrids of *S. cerevisiae* and *T. delbrueckii* have been developed in Germany. Some of these strains have been claimed to be tolerant to osmotic pressures prevalent in "heavy" doughs such as cookie doughs and cake doughs normally leavened with chemical leavening (Kowalski, Zander, and Windisch 1981).

Strains of baker's yeast for industrial production are preserved as pure cultures, usually on agar slants in test tubes. They are frequently transferred by well-known pure culture methods. Commercial fermentations are started from such slants. Industrial strains of baker's yeast are quite stable.

Baker's yeast strains are available from public and private culture collections as listed by Kirsop and Kurtzman (1988). But they can also be isolated from commercial baker's yeast since this yeast enters commerce as a viable product. There are several patents protecting specific strains. Enforcement of such patent rights depends, of course, on the ability to identify the patented strain with certainty, which is often difficult.

## PRINCIPLES OF AEROBIC GROWTH

### Growth Rate

The growth rate is defined as the increase in the number of yeast cells or the increase of the weight of the cells for a given time period. It is usually expressed as the increase in dry weight of the biomass in a 1-hour period. If cell weight for a given volume is plotted against time, the growth rate is represented as the slope of the curve. The specific growth rate constant $\mu$ is defined by the following equation:

$$dP/dt = \mu \times P$$

in which $d$ is differential, $P$ is the mass of yeast, and $t$ is time. Upon integration one obtains $\ln (P_t/P_0) = \mu \times t$; and for $t = 1$ hour one obtains $\ln \mu = (P_t/P_0)$ or $2.31 \times \log_{10} (P_t/P_0)$, which is the specific growth rate

Table 6-1. Relation of the Generation Time to the Specific Growth Rate Constant

| Generation Time (hr) | $P_t/P_0$ | Specific Growth Rate Constant, $\mu$ |
|---|---|---|
| 1 | 2.00 | 0.693 |
| 2 | 1.41 | 0.345 |
| 3 | 1.26 | 0.230 |
| 4 | 1.19 | 0.173 |
| 5 | 1.15 | 0.139 |

Notes: $P_t$ = mass of product at time $t$. $P_0$ = mass of product at time 0.

constant for exponential growth. For continuous fermentation in steady state its value is identical with that of the dilution rate.

Commercial yeast fermentations are always carried out as fed-batch fermentations. They are neither batch fermentations in which the fermenter is filled with substrate medium followed by inoculation with a starter culture of yeast nor are they continuous fermentations in which the medium is fed continuously and in which an equal volume of the fermenter content is continuously removed. In fed-batch fermentations the growth medium is fed incrementally to a starter culture of yeast over a given time period (usually 12-20 hours) until the fermenter is filled. The fermentation is ended, and all of the yeast in the fermenter is harvested at this time.

It is very difficult to determine the specific growth rate constant with precision at any time during the fed-batch fermentation because it decreases with time. Thus, there is no steady state. The rate of progress of yeast growth is often expressed as the generation time, that is, the time required for a doubling of the cell mass. The relation of $\mu$ to generation time is expressed as follows:

$$\text{Generation time (hr)} = 2.303 \times \log_{10} 2/\mu$$

Table 6-1 shows some numbers relating $\mu$ to generation time over a practical range. Normal generation times for baker's yeast production are often 3 hours at the beginning, decreasing to 5 hours, and finally to 7 hours toward the end. There is approximately an 8-fold multiplication of yeast mass during a 15-hour fermentation.

## Oxygen Requirement and Aeration

Under anaerobic conditions the yield of yeast biomass (based on the weight of fermented sugar) is low. $Y_s$, the yield coefficient, is generally no more than 0.075, or 7.5 kilograms of yeast solids per 100 kilograms of glucose. Under strictly aerobic conditions the best yield is 0.54.

The purpose of a baker's yeast plant is to maximize growth of yeast and to minimize alcoholic fermentation. Several conditions have to be met to maximize yeast yield. The fermentation must be carbohydrate-limiting, that is, sugar levels in the fermenter must be kept very low to avoid catabolite repression. Sufficient oxygen must be supplied by aeration, and the growth rate coefficient must be kept below $0.2\ hr^{-1}$. If these conditions are not met ethanol will be formed during the fermentation regardless of the aerobicity of the process.

One gram of oxygen is required for the formation of 1 gram of dry yeast mass (Mateles 1971). Wang, Cooney, and Wang (1977) show the elemental composition of yeast as $C = 45\%$; $H = 6.8\%$; $N = 9.0\%$; $O = 30.6\%$. These percentages differ slightly from the ones that can be calculated from the molecular composition reported by Harrison (1967) of $C_6H_{10}NO_3$. About one-half of the required oxygen is incorporated in yeast cell mass, and the other half escapes as $CO_2$. Oxygen is supplied to the fermenter by aeration. It may be supplied by using air enriched with oxygen, and it may even be supplied by hydrogen peroxide that is converted to nascent oxygen by yeast catalase.

Oxygen is an essential nutrient in aerobic fermentations. In contrast to other nutrients it is most difficult to supply because of its low solubility in water. In small shake flasks it may be supplied by mechanical motion of the flask, but in large commercial fermenters it must be supplied by spargers under pressure, often supplemented by mechanical stirrers.

The efficiency of a system of transfer of $O_2$ from air into the fermenter is expressed as the volumetric oxygen transfer coefficient $K_L a$ $(hr^{-1})$, where $K_L$ is the oxygen transfer coefficient (m/hr) and $a$ is the interfacial area between the air bubbles and the liquid per unit volume of liquid ($m^2/m^3$ or l/m). In practice, the rate of oxygen transfer in a given fermentation system and with a given medium is expressed as the millimoles of $O_2$ transferred per liter per hour. Based on the knowledge that it takes 2 g $O_2$ to produce 1 g yeast solids, one can readily calculate that an oxygen transfer rate of 140 mM/L-hr is required to produce 4.5 g yeast solids/L-hr. This particular figure has been used as an example because it represents the approximate required oxygen transferred during peak growth in a commercial yeast fermentation.

For fermenters without agitation, that is, for simple sparging with air, the rate of oxygen utilization may not exceed 20% of the oxygen. Thus, for incoming air with an oxygen concentration of 22%, about 17% of oxygen can be found in the effluent stream (the amount of $CO_2$ produced during yeast growth must be considered). Also, about five times the amount of air must be blown as that calculated from the oxygen requirement for growing the yeast. The use of stirrers causes better dispersion of the bubbles of air and hence a larger surface area for oxygen transfer.

Many such systems have been described in the literature including air lift loop reactors, bubble column reactors, stirred tank reactors, and jet loop reactors (Blenke 1987). These systems are more efficient than simple spargers because of the much smaller bubble size and the larger interfacial area available for oxygen transfer. Smaller bubbles also rise more slowly in the fermenter, resulting in an increased holdup that reduces the liquid volume fill of the fermenter.

The concentration of dissolved oxygen in the fermenter can be determined quite well with oxygen electrodes. At 30°C one can assume a concentration of 7 ppm and use this figure for calibration of the electrode. Figure 6-1 shows the concentration of oxygen during a typical baker's fed-batch fermentation expressed as percent of saturation. It drops as the rate of biomass formation increases and may reach levels below 0.4 ppm or less than 5% of saturation.

The concentration of oxygen increases again as the rate of molasses feed to the fermenter is reduced from the eleventh hour on. There is a particular oxygen concentration below which yeast growth is limited by the oxygen supply: the critical oxygen concentration. It varies for various microorganisms. For yeasts it has been determined by Finn (1967) to be 4.6 $\mu$M at 35°C (0.073 ppm) and 3.7 $\mu$M at 20°C (0.059 ppm).

It is very difficult to predict the performance of an aeration system. One

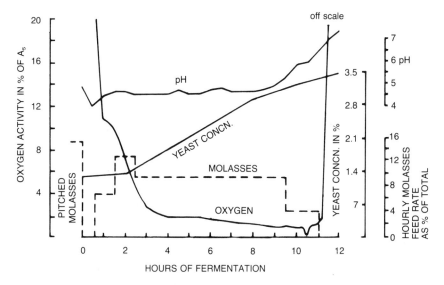

**Figure 6-1.** The fed-batch production of baker's yeast in a pilot plant showing concentration of yeast solids, rate of molasses feed, oxygen concentration (as % of saturation), and pH. (*After Strohm and Dale 1961*)

can use the sulfite oxidation method of Cooper, Fernstrom, and Miller (1944). It measures the rate of oxidation of a sulfite solution in the fermenter in the presence of a metal catalyst (Cu or Co) and permits a comparison between different fermenter configurations, aeration and agitation devices, stirrer speeds, and so forth. However, it does not permit a precise calculation of the rate of oxygen transfer during an actual fermentation. This rate varies with the viscosity, interfacial tension (gas/liquid), bubble size, and so on. Limitations of the sulfite oxidation method have been discussed by Linek and Benes (1978). The measurement and regulation of the dissolved oxygen concentration have been lucidly described by Sikyta (1983).

The specific growth rate coefficient, $\mu$, must not exceed approximately 0.2 ($hr^{-1}$) if biomass yield is to be optimized. This effect can best be demonstrated under steady-state conditions of a continuous fermentation as shown in Figure 6-2 (Dellweg, Bronn, and Hartmeier 1977). Above a growth rate of 0.18, yield drops sharply, and the respiratory quotient, $RQ$, increases to levels normal for anaerobic fermentations. Meyenburg (1969)

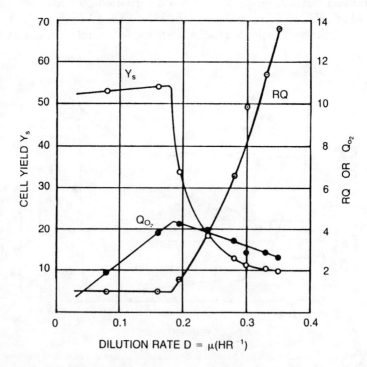

**Figure 6-2.** Effect of dilution rate on yeast metabolism in continuous culture: $Y_s$, cell yield in g solids per 100 g of substrate; $Q_{O2}$, respiration rate in M/g yeast solids per hr; $RQ$, respiratory quotient. (*From Dellweg, Bronn, and Hartmeier 1977*)

reported similar results except that his yields did not decrease until a value of $\mu = 0.23$ had been reached. At higher growth rates ethanol was formed, a phenomenon called aerobic fermentation.

The driving force of oxygen from gas to liquid can be increased by increasing the pressure of the gas. To some extent this increase is realized in commercial fermentations by increasing the liquid height of the fermenter, that is, by building a taller fermenter. The driving force can also be increased by raising the concentration of oxygen in the sparged gas, although this practice is not economical at present. Figure 6-3 illustrates

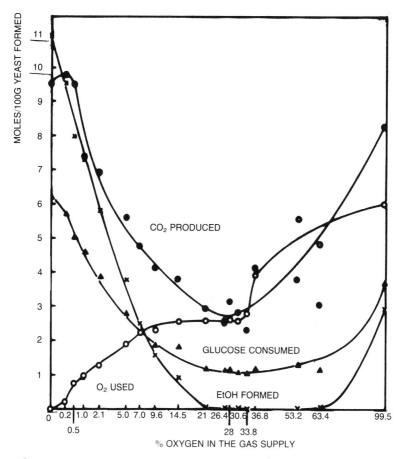

**Figure 6-3.** Effect of the concentration of oxygen in the influent sparged gas during the continuous culture of baker's yeast: ▲ = consumption of glucose per 100 g of yeast solids formed; ○ = oxygen uptake per 100 g of yeast solids formed; ● = carbon dioxide formed per 100 g of yeast solids produced; X = ethanol formed per 100 g of yeast solids produced. (*After Oura 1974*)

the effect. The critical curve indicating the effect on yield is the one labeled "glucose consumed." It shows the amount of glucose in moles consumed for the production of 100 grams of yeast solids.

The optimum yield between oxygen levels of 28% and 34% in the sparged gas corresponds to a 55% substrate yield. The curve is fairly flat between 21% and 63% oxygen. At lower and higher levels of oxygen in the sparged gas, ethanol is formed with a corresponding loss in the yield of yeast solids. The region between 21% and 30% oxygen in the gas is also the region where oxygen uptake rate and carbon dioxide production rate are equal, that is, a $RQ$ of 1.

## Catabolite Repression

The presence of carbon sources above a critical level switches the respiratory growth metabolism of sensitive cells to a fermentative metabolism. Baker's yeast is sensitive to glucose concentrations above 0.2%. This repres-

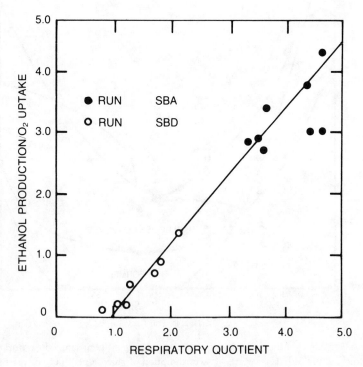

**Figure 6-4.** Experimental correlation between the ethanol production rate and the $RQ$. The slope of 1 is consistent with the theoretical analysis. (*From Wang, Cooney, and Wang 1977*)

sion of the enzymes of the respiratory pathways is known as catabolite repression or the Crabtree effect. It results in the excretion of intermediary products of metabolism such as ethanol. The repression also takes place in the presence of dissolved oxygen.

Various authors have found critical glucose concentrations from 0.02-0.85 mM of glucose (3.6-210 mg/L). The variation may be due to particular strains or to physiological conditions of the cells (Woehrer and Roehr 1981). It will now be recognized that the catabolite repression (Crabtree effect*) is the reason for the fed-batch fermentation. Sugar must be fed incrementally and continuously at a slow rate so that it can be taken up by the cells continuously, and so that it will not surpass the critical concentration.

The respiratory quotient, $RQ$, is a quantitative indicator of ethanol formation as shown in Figure 6-4. Above an $RQ$ value of 1 there is some ethanol formation. Between an $RQ$ of 0.9 and 1 there is oxidative growth. $RQ$ values between 0.7 and 0.8 correspond to endogenous metabolism, and values below 0.6 correspond to the utilization of ethanol (Wang, Cooney, and Wang 1977).

## RAW MATERIALS

### Carbon and Energy Sources

Historically the major source of carbon for yeast growth was maltose derived from malt-converted grain mashes. Today glucose, fructose, sucrose, and a portion of raffinose from molasses serve universally as the carbon source. The utilization of various sugars for yeast growth has been extensively reviewed in the literature, for example, by Barnett (1976). The discussion here is restricted to a discussion of the growth of *Saccharomyces cerevisiae* on molasses and to a brief indication of potential sugar sources from starches and from whey.

Table 6-2 shows the composition of beet and cane molasses typical of the raw material used by yeast producers. Cane molasses is generally used in subtropical areas where cane is grown or in places to which it can be readily shipped by sea. In northern climates where sugar beets are grown extensively, and in areas distant from ocean shipping lanes, beet molasses is the major constituent of the mash bill. The price of molasses is largely determined by its value as a feed. In the fermentation industry its major

---

*The Crabtree effect has sometimes been called the reverse Pasteur effect. The term is misleading (Lagunas 1986) and will not be used.

Table 6-2. Molasses Composition (as percent of total solids)

| Composition | Cane Molasses | Beet Molasses |
|---|---|---|
| Sugars | 73.1 | 66.5 |
| sucrose | 45.5 | 63.5 |
| raffinose | 0 | 1.5 |
| invert sugar | 22.1 | 0 |
| other | 5.5 | 1.5 |
| Organic | 15.5 | 23.0 |
| GA & PY* | 2.4 | 4.0 |
| other N | 3.1 | 0 |
| other amino acids | 0 | 3.0 |
| betaine | 0 | 5.5 |
| organic acids | 7.0 | 5.5 |
| pectin, etc. | 2.7 | 5.0 |
| Inorganic | 11.7 | 10.5 |
| $K_2O$ | 5.3 | 6.0 |
| $Na_2O$ | 0.1 | 1.0 |
| $CaO$ | 0.2† | 0.2 |
| $MgO$ | 1.0 | 0.2 |
| $Al_2O_3 \cdot Fe_2O_3$ | 0 | 0.1 |
| $SiO_2$ | 0 | 0.1 |
| Cl | 1.1 | 1.7 |
| $SO_2 + SO_3$ | 2.3 | 0.5 |
| $P_2O_5$ | 0.8 | 0.1 |
| $N_2O_5$ | 0 | 0.4 |
| others | 0.9 | 0.2 |

Source: Data from Hongisto snd Laakso 1978.
*Glutamic acid + pyrrolidin carboxylic acid.
†After Ca precipitation.

use is for the production of baker's yeast and glutamate. At present, molasses sugar is the cheapest source of fermentable sugars.

In the United States, mash bills contain at least 20% of cane molasses to supply enough biotin for yeast growth. Apart from price of the raw material there are some additional considerations. Beet molasses is relatively easy to clarify whereas cane molasses contains colloidal substances that make clarification more difficult. The cost of waste disposal will be discussed in more detail below. It is mentioned here only because it is an important factor in any consideration of the cost of molasses.

Table 6-3. Composition of Baker's Compressed Yeast (all values expressed on a solids basis)

| | |
|---|---|
| Protein (N × 5.7) | 47% |
| Carbohydrates | 33% |
| Minerals | 8% |
| Nucleic acids | 8% |
| Lipids | 4% |

Approximate composition of minerals (total solids basis)

| | | | | |
|---|---|---|---|---|
| Na | 0.12 mg/g | Cu | 8.0 | $\mu$/g |
| Ca | 0.75 | Se | 0.1 | |
| Fe | 0.02 | Mn | 0.02 | |
| Mg | 1.65 | Cr | 2.2 | |
| K | 21.0 | Ni | 3.0 | |
| P | 13.5 | Va | 0.04 | |
| S | 3.9 | Mo | 0.4 | |
| Zn | 0.17 | Sn | 3.0 | |
| Si | 0.03 | Li | 0.17 | |

Approximate composition of vitamins (total solids basis)

| | |
|---|---|
| Thiamine | 60-100 $\mu$/g |
| Riboflavin | 35-50 |
| Niacin | 300-500 |
| Pyridoxine HCl | 28 |
| Pantothenate | 70 |
| Biotin | 1.3 |
| Choline | 4000 |
| Folic acid | 5-13 |
| Vit. $B_{12}$ | 0.001 |

Notes: Approximately 1% of lipids extractable by solvents as is; 4% after acid hydrolysis.
Crude protein is generally expressed as N × 6.25.

Many authors have attempted to show the relationship between the composition of yeast biomass and the amount of various nutrients required to achieve that composition. The composition of a typical baker's yeast widely used for the production of compressed yeast is shown in Table 6-3. The gross composition shown as protein, nucleic acids, and major mineral constituents is quite representative; the presence of vitamins is less so.

The concentrations of trace minerals as shown in Table 6-3 are certainly no guide to a desirable composition of nutrients. Some of them may not be essential for yeast growth, and those that are essential for yeast growth may be present in concentrations higher than required to sustain that growth. Elements in growth media may not always be available for yeast growth or they may react with one another, which is an additional complicating factor.

Nevertheless, the comparison of elements in yeast and in an amount of molasses required to grow that yeast have some merit (Table 6-4). The table indicates a yield of 54% of yeast solids based on molasses sugar, which is probably a little too high. It also indicates quite correctly that molasses is deficient in nitrogen, phosphorus, and magnesium. These and some trace minerals and vitamins have to be supplied by supplementation of the molasses or by separate addition to the fed-batch fermentation.

Beet and cane molasses are shipped at 80-85 Brix. They have a sugar content that varies between 50% and 55% by weight. As a rule, one can assume that 100 pounds of molasses will produce 25 pounds of yeast solids. Molasses can be stored well in large tanks often holding several million pounds. It can be pumped at temperatures above 20°C. Molasses contains some insoluble materials and some precipitate forms on heating. Therefore, molasses is clarified. Cane molasses cannot be filtered well. It is clarified by addition of phosphates and by sterilization followed by centrifugation. Beet molasses may be clarified in the same manner or by filtration instead of centrifugation. Prior to clarification, molasses is diluted with water to 30-40 Brix, which facilitates the process.

Molasses as agricultural raw material is somewhat variable in composition. The composition of molasses of various origins and derived from different sugar production methods has been compiled by Ceijka (1983). In practice, some molasses is not satisfactory for the production of yeast, and the reason cannot always be established. It may be due to the absence or insufficient concentration of a nutrient or to the presence of a yeast growth inhibitor. The latter category may include fatty acids, nitrites, various insecticides or herbicides used in the field, or bactericides used in the sugar factory (Bergander 1969; Notkina, Balyberdina, and Lavrenchuk 1975).

Table 6-4. Elemental Composition of Yeast and Beet Molasses

|  | C (g) | N (g) | P (g) | K (g) | Mg (g) | Ca (g) |
|---|---|---|---|---|---|---|
| 100 g beet molasses | 33 | 1.5-2 | 0.03 | 6 | 0.025 | 0.3 |
| 100 g compressed yeast (27% solids) | 12.3 | 2.3 | 0.28 | 0.54 | 0.03 | 0.01 |

Source: Data from Oura 1983.
Note: The 100 g of beet molasses are a sufficient carbon and energy source for the growth of 27 g of yeast solids.

Beet molasses contains approximately 1-1.5% raffinose. Baker's yeast strains assimilate only the fructose moiety. Use of an α-galactosidase (meli-

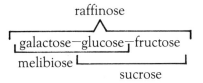

biase) hydrolyzes melibiose and permits complete assimilation of raffinose (Liljestrom-Suominen, Joutsjoki, and Korhola 1988). The economics of using this enzymatic process for the treatment of beet molasses have not yet been established.

### Nitrogen

Beet or cane molasses contains small concentrations of nitrogenous materials. Betaine nitrogen in beet molasses is not assimilated. A portion of the small amino acid fraction is assimilated (Kautzman 1969), but yeast producers cannot rely on this minor fraction. In the production of beer and wine the concentration of assimilable amino nitrogen is of critical importance. In the production of baker's yeast the element is supplied readily and in assimilable form as ammonia, ammonium salts (sulfate or phosphate), or urea.

Baker's yeast contains from 6-9% nitrogen. The preferred concentration depends on an estimate of the desired qualities. With higher nitrogen levels the yeast is usually more active but less stable, which applies to both compressed and active dry yeasts. The amount of nitrogen to be fed during yeast growth can be calculated from the estimated total yield. However, if the actual yield obtained by a given fermentation is lower than assumed, the yeast will contain higher concentrations of nitrogen than desired, and vice versa.

Feeding of nitrogen in the form of ammonium sulfate or phosphate leads to the liberation of sulfuric or phosphoric acids, which must be neutralized to stay in the desired pH range. Urea may be used as a nitrogen source, but requires higher levels of biotin, and it decomposes more readily upon thermal sterilization. For the metabolism of other organic nitrogen compounds by yeasts see Large (1986).

### Minerals

Molasses is grossly deficient in phosphorus whose concentration is usually expressed as $P_2O_5$. As a rule, the concentration of $P_2O_5$ should be one-third that of the nitrogen concentration, that is, between 2% and 3%. Phosphorus may be supplied in the form of phosphoric acid or its salts, for example, ammonium phosphate. Molasses contains sufficient potassium, calcium, and sulfur, but some magnesium must be added to the growth medium.

Yeast growth requires the presence of various trace elements. These have been quite difficult to establish. The presence of a trace metal in yeast does not prove that the element is essential. Most of the media reported in the literature indicate the presence of trace minerals on a percentage basis without reference to the amount of yeast that is grown on the medium; hence, such data do not permit a quantitative estimate of the actual amounts required. Oura (1974) has listed the composition of nine media widely used for growing yeasts, but these are not media used by industry.

Table 6-5 shows the composition of media used by three authors who have grown baker's yeast by the fed-batch process in synthetic media (glucose-based). The values have been recalculated to a common basis, 500 grams of glucose, to facilitate comparison. Even these concentrations of minerals show considerable differences, and some trace minerals have been found necessary by one author but not by others.

Woehrer and Roehr (1981) and Oura (1974) obtained optimal growth and yields with these media, but did not try to quantify the requirement of trace minerals or vitamins. In contrast, Bronn (1985) deliberately established the minimum requirements of trace minerals for use in commercial fed-batch fermentations of baker's yeast on corn syrup. These criteria do not exclude the possibility that some trace minerals that were not added to the medium were present in minimal concentrations in water or were leached from the equipment.

### Vitamins

Baker's yeast strains require biotin for growth. Cane molasses supplies ample amounts of this vitamin, usually 0.5-0.8 ppm. Beet molasses contains only 0.01-0.02 ppm of biotin. Therefore, at least 20% cane molasses is used in beet/cane blends for yeast production. For fermentations of carbohydrate sources that contain little or no biotin (corn syrup, date concentrate) the vitamin may be added as the synthetic compound at a rate of 60-100 $\mu$g for each 100 g of yeast solids grown. Biotin may be replaced by biocytin, D-desthiobiotin, or D-biotin methyl ester.

Other vitamins shown in Table 6-5 are usually present in sufficient quantity in molasses. There is no absolute requirement by baker's yeast for thiamin that can synthesize the vitamin from thiazole and pyrimidine compounds. However, concentrations above 100$\mu$g/g of yeast solids in American commercial baker's yeast indicate that some thiamin has been added to the growth medium. Such high levels of thiamin do not improve

Table 6-5. Mineral and Vitamin Requirements for Aerobic Growth of Yeast

|  | I | II | III |
|---|---|---|---|
| Element (g) | | | |
| N | 19.0 | 25.4 | |
| S | 28.5 | 36 | |
| P | 4.9 | 5 | |
| K | 4.7 | 0.6 | 9.2 |
| Mg | 0.9 | 0.62 | 0.92 |
| Ca | 0.24 | 0.24 | 9.2 |
| Na | | 0.23 | 0.46 |
| | | | |
| Trace element (mg) | | | |
| Zn | 20.5 | 23 | 92.5 |
| Fe | 25.8 | 44.9 | 46.2 |
| Cu | 1.4 | 1.3 | 3.2 |
| Mn | | 10.4 | 9.25 |
| Co | | 4.8 | 1.85 |
| Mo | | 12.9 | |
| B | | 12.8 | |
| I | | 4 | |
| Ni | | 5.6 | |
| | | | |
| Vitamins (mg) | | | |
| Biotin | 1.4 | 1.25 | 0.3 |
| Pantothenate | 20 | 62.5 | 40 |
| m-inositol | 900 | 1250 | 600 |
| Thiamin | 181 | 50 | 11 |
| Pyridoxine | 36 | 62.5 | |
| Nicotinic acid | | 50 | |
| | | | |
| Yeast extract (g) | 1.8 | | |

Note: All values calculated for growth on 500 g glucose in synthetic media.
(I) Woehrer and Roehr 1981.
(II) Oura 1974.
(III) Bronn 1985. The author did not list the well-known requirements for ammonium, phosphate, and sulfate ions.

growth rate or yield, but have a small but demonstrable effect on yeast fermentation activity.

## Alternate Carbon Substrates

At present, molasses is the least expensive source of sugar for growing baker's yeast. But prices for molasses fluctuate greatly and on occasion they have approached those for corn-derived sugars. Therefore, other carbohydrate sources have been investigated. Ultimately, the most abundant source of fermentable sugar will be glucose derived from cellulose. However, processing costs are still prohibitive. Corn syrups consisting almost entirely of glucose (96 DE) are potentially a good source. Bronn (1985) provided convincing evidence that corn syrup is a suitable raw material providing growth rates, yields, and bake performance equal to molasses-grown yeast. He estimated that replacement of molasses with high DE corn syrup will lower the chemical oxygen demand (COD) of waste streams (fermenter beer without yeast, water used in washing yeast) by about 80%.

Cheese whey is an abundantly available by-product of the cheese industry. At present, it must be disposed of as a high biological oxygen demand (BOD) waste or dried and sold as feed yeast, often at a loss. Lactose-fermenting yeasts such as *Kluyveromyces marxianus* (syn. *Saccharomyces fragilis, K. fragilis*) can readily be grown on lactose as a carbon and energy source. Baker's yeast does not assimilate lactose, but grows on glucose and galactose resulting from lactose hydrolysis (Moulin and Galzy 1984; Sanderson and Reed 1985).

Attempts can be made to introduce the LAC gene or genes into baker's yeast (Sreekrishna and Dickson 1985), but they have not yet led to practical results; or lactose can be hydrolyzed to the monosaccharides with commercially available immobilized $\beta$-galactosidase. A plant based on growth of baker's yeast on hydrolyzed lactose has operated in the United States (Anon. 1982; Stineman, Edwards, and Grosskopf 1980), but production has now been discontinued. Whey and whey-ultrafiltered permeate are costly to transport because they contain only about 4.8% lactose. The materials are difficult to store in concentrated form because of the poor solubility of lactose.

A process for the fermentation of cheese whey permeate to lactic acid and subsequent growth of baker's yeast on lactic acid and galactose has been reported (Champagne and Goulet 1988). Growth of baker's yeast on this substrate is shown in Figure 6-5. Galactose was assimilated before lactic acid, but while galactose was still present it did not repress lactic acid assimilation completely.

Ethanol is also assimilated by baker's yeast strains. For instance, ethanol produced anaerobically or aerobically during a fed-batch fermentation

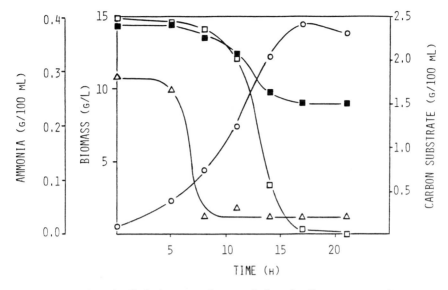

**Figure 6-5.** Growth of baker's yeast on fermented whey ultrafiltrate: ○ = yeast biomass; △ = galactose utilization; □ = lactic acid utilization; ■ = ammonia utilization. (*Courtesy of Prof. J. Goulet, Laval University, Quebec*)

may later be assimilated. Yields of yeast solids on ethanol are the same as yields on sugar if the two substrates are compared on an equivalent carbon basis. The continuous cultivation of S. cerevisiae on ethanol under steady-state conditions was described by Mor and Fiechter (1968). The substrate is too expensive for industrial use.

A fairly large number of yeast strains that can hydrolize starch have been reported (McCann and Barnett 1986), but S. cerevisiae cannot hydrolyze starch and hence cannot grow on it. The insertion of genes for glucoamylase formation by S. cerevisiae has been reported for brewer's strains and may have utility in that industry. For growth of baker's yeast the hydrolysis of starch by commercial amylases would be more cost effective.

## Fermentation Activators and Inhibitors

Occasionally activators of yeast growth are reported in the literature. For instance, Jakubowska and Wlodarczyk (1969) have suggested the stimulating effect of indolyl acetic acid. In many instances, it is not clear whether a substance reported as an activator does not simply function as a required nutrient. This case is particularly true of trace elements. Table 6-6 shows molar concentrations of elements that activate and inhibit yeast growth

Table 6-6. Yeast Ionic Nutrition: Activation and Inhibition

| Activation: concentration required for optimum growth | | | | | |
|---|---|---|---|---|---|
| $B^+$ | 0.4 μM | $Ca^{2+}$ | 4.5 M | $Co^{2+}$ | 0.1 μM |
| $Cu^{2+}$ | 1.5 μM | $Fe^{2+}$ | 1-3 μM | $H^+$ | 1 μM (pH 6) |
| $K^+$ | 2-4 mM | $Mg^{2+}$ | 2-4 mM | $Mn^{2+}$ | 2-4 μM |
| $Mo^{2+}$ | 1.5 μM | $Ni^{2+}$ | 10-90 μM | $Zn^{2+}$ | 4-8 μM |
| Inhibition: observed at concentrations exceeding those listed below | | | | | |
| Au | 10 μM | Au | 10 μM | $UO_2^{2+}$ | 10 μM |
| $Cd^{2+}$ | 10 μM | $Pd^{2+}$ | 10 μM | $Os^{2+}$ | 10 μM |
| $Th^{2+}$ | 2 mM | $Al^{3+}$ | 0.3 mM | $Cr^{3+}$ | 10 μM |
| $Hg^{2+}$ | 0.1 mM | $Va^{2+}$ | 0.4 mM | $Pb^{2+}$ | 1 mM |
| $Sn^{2+}$ | 1 mM | $Li^+$ | 0.2 M | $Na^+$ | 0.1 M |

Source: Data from Rodney and Greenfield 1984.

(Rodney and Greenfield 1984). Most of the "activators" have previously been discussed as nutrients.

The table also shows the minimum concentrations of elements inhibiting yeast growth. Some of the heavy metals are inhibiting at very low concentrations. $SO_2$ inhibits yeast growth, but there is good adaptation even at concentrations of 80-100 ppm. Molasses contains variable amounts of nitrate that can be reduced to nitrite by bacterial action during yeast growth. Yield losses due to nitrite inhibition have been reported for nitrite concentrations of 0.001-0.004% (Notkina, Balyberdina, and Lavrenchuk 1975).

The high pressure developed by carbon dioxide formation during the secondary champagne fermentation inhibits yeast growth and yeast fermentation. Generally the literature does not distinguish between the effect of pressure per se and that of dissolved $CO_2$. Chen and Gutmanis (1976) investigated the effect of $CO_2$ in the influent air during the fed-batch production of baker's yeast. Slight inhibition was noted at 40% $CO_2$ in the influent air and strong inhibition was noted at above 50% $CO_2$.

## ENVIRONMENTAL PARAMETERS

### pH and Temperature

The protoplasm of the yeast cell does not show the same pH at various locations in the cell. Yeast can maintain such internal pH values quite well in solutions of a different pH, and it can be grown quite well at pH levels between 3.6 and 6. The optimum pH for growth is between pH 4.5 and 5. Lower pH levels minimize growth of contaminating bacteria, but enhance adsorption of coloring material from molasses. Therefore, commercial fermentations are usually conducted at pH levels that are shown in Figure 6-1, that is, from a starting pH of 4.5 to a final pH above 6.

Baker's yeasts are generally grown at a controlled temperature of 30°C. In contrast to some anaerobic fermentations in bread doughs, and in smaller tanks of grape must or beer wort, the aerobic growth of baker's yeast generates so much heat that temperature control is absolutely required. Growth rates as expressed in generation times were determined by White (1954) to be as follows: 20°C, 5 hr; 24.5°C, 3 hr; 30°C, 2.2 hr; 36°C, 2.1 hr; 40°C, 4 hr. Such values cannot be literally translated into practice because they reflect generation times at the start of the fermentation. In practice, generation times increase drastically during the fermentation.

**Yield and Development of Heat**

Yields of baker's yeast are best expressed as the yield coefficient based on the carbon substrate (glucose). A coefficient of 0.54 has been indicated above. Chen and Gutmanis (1976) obtained a yield coefficient of 0.5; Oura (1974) and Dellweg, Bronn, and Hartmeier (1977) reported values of 0.52 and 0.54, respectively. These values include the energy required by the cell for maintenance of all of its metabolic functions, the so-called maintenance energy. This energy is required whether the culture is growing or static.

Cooney (1981) estimated that the carbon used for maintenance in a fast baker's yeast fermentation is less than 5% of the total carbon used. In practice, yields of compressed yeast (by weight) are expressed as a percentage of the weight of molasses used. This measurement is useful for accounting purposes since molasses costs are based on pounds purchased, not on the amount of fermentable sugar purchased. But it makes it difficult to compare the efficiency of plant operations since the sugar content of molasses may vary somewhat. Also, in the United States, baker's compressed yeast has a solids content of 30% versus 27-28% in Europe.

The ability to remove heat from rapid baker's yeast fermentations is—next to the problem of supplying enough oxygen—a major problem for large fermenters. Data have been reviewed by Reed and Peppler (1973). The rate of heat evolution is a function of growth rate and concentration of cell mass in the fermenter.

Figure 6-6 shows heat evolution in kcal/L/hr as a function of growth in g/L/hr. The slope of the curve shows that 4.4 kcal are evolved per gram of cell solids produced (Cooney 1981). This value is somewhat higher than 3.9 kcal $g^{-1}$ calculated by Harrison (1967), or 15.6 million BTU per ton of yeast solids. The requirement for heat removal increases during the conduct of a fed-batch fermentation in proportion with the rate of molasses feed. It is reduced toward the end of the fermentation when the feed rate is lowered again; that is, it is proportional to productivity, not to yeast concentration.

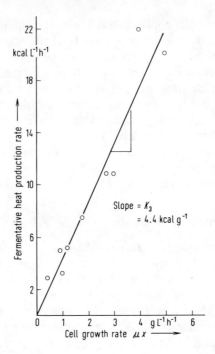

**Figure 6-6.** Heat of fermentation versus cell growth rate for *Saccharomyces cerevisiae* grown on molasses. (*From Cooney 1981*)

### Yeast Concentration in the Fermenter

In the 1970s, yeast solid concentrations in the final fermentation stage reached 3.5-4.5%, or 10-13% by cell volume. At present, final concentrations of 5-6% are not uncommon. This amount has a twofold impact on costs. It reduces the gallonage of fermenter liquid that has to be centrifuged, and it reduces the gallonage of spent beer (though not the total BOD), which requires disposal as waste.

It is well known that exponential feeding of the fermentation cannot be continued beyond the first few hours. Generation times of an 18-hour commercial fermentation increased from 4 hours to 6 hours to 8 hours with an eight-fold increase in cells (Fries 1962). Various reasons have been advanced for this lowering of the growth rate. The basic problem is merely the inability to supply sufficient oxygen. In small, efficiently aerated laboratory fermenters the concentration of baker's yeast solids could readily be increased to 10%. This increase corresponds to a cell volume of 25% of the total liquid, and results in a sizable increase in viscosity, which reduces oxygen diffusion.

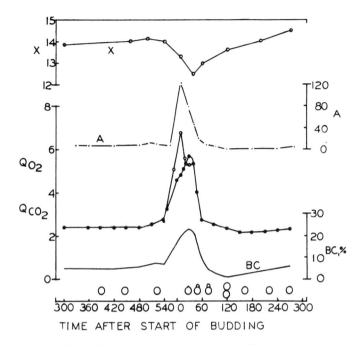

**Figure 6-7.** Budding cycle during synchronous growth of *Saccharomyces cerevisiae*: X, cell dry weight in mg/ml; A, ethanol concentration in $\mu$/ml; $Q_{CO_2}$, $CO_2$ formation in mM/g dry weight/hr; $Q_{O_2}$, $O_2$ uptake in mM/g dry weight/hr; BC, % of initial budding cells; mean generation time 9.5 hr ($D = 0.073$). (*From Meyenburg 1969*)

## Periodicity and Budding

During the propagation of baker's yeast the nutrient feed is reduced toward the end in order to "mature" the yeast, that is, to reduce the number of actively budding yeast cells to no more than 5-10%. The withholding of nutrients from the rapidly growing yeast population within the last hour of the fed-batch fermentation inhibits budding and induces a certain degree of synchrony into the cell cycle of the yeasts. In continuous culture systems synchrony can indeed be induced by pulsed rises in temperature or by pulsed nutrient additions.

The final baker's yeast trade fermentation is usually started with an inoculum in which the yeast cells are in the same phase of their budding cycle. Fries (1962) showed that the growth rate coefficient in such commercial fermentations decreases to a minimum value during budding, and increases again as the separate mother and daughter cells resume growth.

Figure 6-7 shows some of the changes that take place at the onset of budding in more detail. At the start of the budding cycle there is a sharp rise in respiration indicated by a steep rise in oxygen consumption ($Q_{O_2}$) and an even steeper rise of $CO_2$ evolution ($Q_{co_2}$) (Meyenburg 1969). This time coincides with a short period of ethanol formation. It may be at least a partial explanation of why mature baker's yeast, that is, a culture of nonbudding cells, is more stable but has a slightly lesser fermentation activity than yeast harvested during a period of rapid growth.

## PRACTICE OF AEROBIC GROWTH

### Preparation of Nutrient Feed

Cane molasses or beet molasses is received in freight cars or trucks, or in tankers if the factory is located on a waterway. It is usually stored in large vats with a capacity of several million pounds. With a usual Brix of 80, molasses cannot be pumped if the temperature is too low. For the process of clarification the molasses is diluted with water to 30-40 Brix. Dilution permits easier pumping and facilitates clarification by providing a greater gradient between the liquid and suspended solids. Molasses is fed to the fermenter at this concentration. The pH of the diluted wort is adjusted to about 5 with acids such as sulfuric acid, and the insoluble solids are removed in a desludger centrifuge, that is, a solid-bowl centrifuge with intermittent discharge of the accumulated sludge.

Beet molasses may also be clarified by filtration, but cane molasses is very hard to filter even with the addition of filter aids. The clarified molasses wort is then sterilized in a heat exchanger. From a temperature of 95°C, the wort is heated to 140°-145°C in 1 second. It is held at this temperature for about 4 seconds and then cooled back to 95°C by flashing into a wort storage tank (Rosen 1977). The mineral supplements described in the previous section are either added to the molasses wort or they are fed separately. Ammonia as a nitrogen source is almost always fed separately into the fermenter. It is important that the molasses wort not be heated excessively or stored hot for prolonged periods, since heat leads to caramelization and loss of fermentable sugars.

### Fermentation

Baker's yeast is produced commercially by a sequence of fermentations with transfer of the grown yeast from the smallest vessel to successively larger vessels. There are often as many as six stages. The last stage that produces the yeast for sale to bakeries is called the trade fermentation. Production in earlier stages has some bearing on the quality of the yeast, but the final trade fermentation is decisive for overall productivity, yield, and quality. This fermentation will be discussed first, and the preceding sequential fermentations will be discussed later.

## Fermentation Tanks

Fermenters are constructed of stainless steel. Their size may vary; a frequently used size is 150 m$^3$ (40,000 gal). The usable liquid volume is usually only 75% of the total volume because of the expansion of volume by dispersed air bubbles and by foam. Internal cooling coils also reduce the usable volume. In contrast to fermentations that require complete sterility (such as fermentations for the production of amino acids, antibiotics, or enzymes), baker's yeast is produced under conditions that minimize but do not prevent growth of contaminants. Of course, this condition is characteristic of all food fermentations. Therefore, fermenters for the production of baker's yeast need not be pressurized. However, it is important to use construction that permits easy cleaning and access to all parts of the fermenter, and permits easy sanitizing of all pipes, valves, and instrument sensors that may be in contact with the fermenter content.

Baker's yeast fermentations generate a great deal of heat, about 3.5-4.4 kcal/g of yeast solids grown. Extensive cooling is required that is generally accomplished with internal cooling coils wound in spiral fashion along the fermenter wall. The stainless steel pipes have a diameter of 2-2.5 inches, and cold water is used as a coolant. It is important to prevent leaks of cooling water in order to avoid contamination of the fermenter content.

A great number of fermenter designs for aerobic fermentations are available for biomass production (Brauer 1985; Blenke 1987; Schuegerl and Sittig 1987). The commonly used types for baker's yeast production are quite simple. They are air-sparged with or without agitation. The 150 m$^3$ air-sparged fermenter described by Rosen (1977) has a horizontal center pipe with 24 side tubes provided with about 30,000 holes of 1.5-mm diameter. Air is supplied by mechanical blowers and aeration and stirring is due entirely to the air bubbles that pass the fermenter liquid. The volume of air blown through the fermenter per minute is often the same as or 1.5 times the volume of fermenter liquid, and stirring is vigorous.

Such systems are quite simple because they have no moving parts. A great deal of attention has been paid to the size of the holes in the aerating device in the belief that smaller bubbles will transmit more oxygen than an equal volume of larger bubbles. This belief is indeed true, but it neglects the fact that bubble size depends largely on turbulence of the liquid about 10 cm from the orifice. Figure 6-8 shows the arrangement of the aeration tubes.

Mechanical agitation by top- or bottom-mounted motors greatly increases the efficiency of air utilization. The simplest system of such agitated fermenters uses a circular perforated pipe to supply air and an agitator whose blades are mounted above the air outlets. Figure 6-9 shows the general layout of a conventional stirred vessel reactor. Such fermenters normally require baffles to prevent the liquid from rotating.

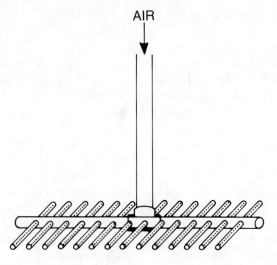

**Figure 6-8.** Network of perforated aeration tubes. (*After DeBeczy and Liebmann 1944*)

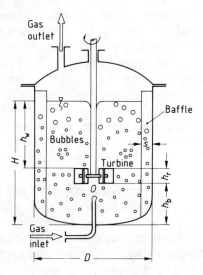

**Figure 6-9.** General layout of a conventional stirred vessel reactor. (*From Brauer 1985*)

The efficiency of air utilization in such fermenters reaches 25-30%, which refers to the percentage of oxygen removed from the air during its passage through the fermenter liquid (Stros, Caslavsky, and Tomisek 1968). Table 6-7 shows a compilation by Chen and Chiger (1985) of literature data on the aeration capacity and economy of common fermenter systems. Fermenters with agitators afford better oxygen utilization,

Table 6-7. Aeration Capacity and Economy of Fermenter Systems

| | Normal Operating $O_2$ Transfer Capacity $(mM\ O_2\ L^{-1}\ hr^{-1})$ | Aeration Economy $(lb\ O_2\ HP^{-1}\ h^{-1})$ | Air Utilization (%) |
|---|---|---|---|
| Mechanically agitated | | | |
| submerged turbine | 300-350 | 2.4 | 14-45 |
| Waldhof aeration wheel | | 2.4-3.2 | |
| Phrix | | 1.7-2.5 | 14-22 |
| self-priming (Frings) | | 2.4 | 14 |
| aeration wing | 140 | 2.3-2.7 | 14-19 |
| deep jet aerator | 315-385 | 3.3 | 28 |
| Diffused air systems | | | |
| gas-sparged fermenter | 100-150 | 2.7 | 10-15 |
| air-lift fermenter | 120 | 4.3-7.0 | 7-8 |

Source: Abstracted by Chen and Chiger 1985 from Cooney et al. 1977; Hatch 1975; Hospodka et al. 1962; Schreier 1974; Reed and Peppler 1973; Reed 1982.

but at a higher energy cost and with a somewhat higher heat removal cost.

A self-priming aerator that is suitable for smaller fermentation vessels has been described by Ebner, Pohl, and Enenkel (1967). The device operates with a turbine that sucks air through a hollow, vertical shaft into the fermenter liquid. It eliminates the need for blowers or air compressors.

It should still be mentioned that aeration efficiency is often expressed as the volume of air pumped through a fermenter per volume of fermenter liquid or VVM. This expression can be grossly misleading. Since the percentage of oxygen in the air bubbles is usually not greatly reduced during passage, it is obvious that a taller fermenter will permit better oxygen transfer than a shorter fermenter of equal volume for the same VVM. An expression such as volume of air per cross-sectional area of the fermenter would make more sense, but it is also deficient because it neglects many variables that have previously been mentioned. One of the best ways of determining the true oxygen transfer rate of a fermenter design is to relate it to the maximum yeast productivity of the system.

Many other fermenter designs for aerobic fermentations permit better oxygen utilization and a vastly increased oxygen transfer rate. To the extent that they increase bubble holdup time in the liquid, they increase the volume of air and decrease the usable liquid volume of the system.

*Feed Rates*

The rate at which the wort is fed into the fermenter has been developed in the early part of the century by trial and error. Such feed curves are shown

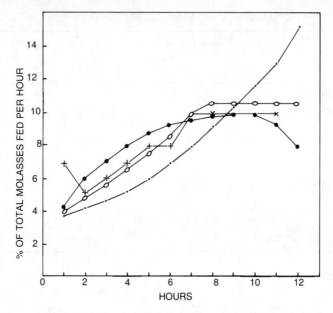

**Figure 6-10.** Molasses feed curves: experimental and commercial:

x—x—x   commercial (1)
●—●—●   commercial (2)
•—•—•   exponential, experimental (2)
o—o—o   exponential and constant, experimental (2)

(1) Butschek and Kautzmann 1962; (2) Drews, Specht, and Herbst 1962.

in Figures 6-1 and 6-10. They are designed to optimize yield and productivity, but they must also produce a yeast that is mature, that is, reasonably stable upon refrigerated storage. The last-mentioned requirement is reflected in the deviation from the exponential feed curve shown in Figure 6-10, and particularly in the reduction of feed toward the end of the commercial fermentation. It has always been desired to regulate feed rates automatically, at least until the final hour of the fermentation. The means for such regulation will be described below.

*Sequence of Fermentations*

All baker's yeast fermentations are started from pure culture slants and carried on through a series of stages of increasing volume. The yeast produced in any stage is used for "pitching" (that is, inoculating) the next stage. The number of stages as well as their designations vary for different manufacturing plants, but the scheme shown in Figure 6-11 is fairly representative of commercial usage.

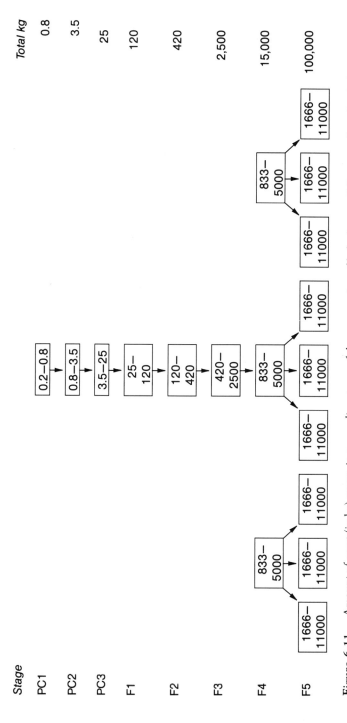

Figure 6-11. Amount of yeast (in kg) grown in succeeding stages of the propagation of baker's yeast. PC, pure culture batch fermentation; F1 and F2, batch fermentations; F3, F4, F5, fed-batch fermentations. (*After Suomalainen 1963*)

The first two or three stages are always pure culture stages carried out batchwise with a set medium sterilized in the small fermenters. The following stages are carried out in larger, "open" fermenters, that is, in fermenters in which complete sterility cannot be maintained and in which the level of contaminants (generally lactic acid bacteria) increases. The pitching of yeast from one stage to the next may be carried out by merely pumping the contents into the next larger vessel, or by centrifuging the fermenter contents to obtain a concentrated yeast slurry (called yeast cream) and using this cream as pitch. The centrifuging operation permits washing the yeast cells with water or acidified water that reduces the level of contaminants appreciably.

Yields of yeast in the first stages of the process are usually quite low, but the last three stages are always carried out as fed-batch fermentations. The yield achieved in these fermentations is decisive for the overall yield of the plant. Centrifuging of yeast from any stage also permits refrigerated storage of the cream and facilitates staggered schedules for the largest fermentations.

*Defoaming*

The large volumes of air used in aerobic fermentations require use of defoaming compounds. These are usually silicones or fatty acid derivatives. Addition of defoamers is done automatically through a probe inserted above the liquid level of the fermenter. As foam reaches the probe, release of a given amount of defoaming compound is triggered. Antifoam agents affect oxygen transfer in a complex manner. They tend to depress oxygen transfer coefficients, but increase the total interfacial area between air bubbles and fermenter liquid (Finn 1967).

*Utilization of Ethanol*

Overfeeding or underaeration of a fermentation leads to the formation of ethanol. Alcohol levels up to 0.05% indicate satisfactory performance, but levels above 0.1% result in some loss of yield. Ethanol formed during the early stages of the process may be metabolized later. In some plants early formation of ethanol to about 2.5% has been induced deliberately in order to use the sanitizing effect of the alcohol.

*Automatic Process Control*

Instruments for control of the temperature, impeller speed, and liquid and gas flow are well known. pH control by addition of ammonia or alkali bicarbonates can also be readily achieved. The interesting problem is control of the wort feed rate based on the ability of the yeast to grow on a somewhat variable substrate. Installation of such a control would not merely replace manual operation with an automatic one. It would permit regulation of the feed rate at any time by the actual nutrient requirements of the yeast in the fermenter, and it would compensate automatically for variations in wort composition, temperature, yeast concentration, and any other variables that affect the growth rate of the yeast.

The most logical parameter for such control is the glucose concentration in the fermenter liquid. Glucose sensors are indeed available but are not sufficiently sensitive. Feed may be controlled by any parameter that is sensitive to the appearance of the alcoholic fermentation. One such parameter is the respiratory quotient, $RQ$. It is the ratio of the carbon dioxide evolution rate, $Q_{CO_2}$, to the oxygen uptake rate, $Q_{O_2}$. An $RQ$ above 1 indicates ethanol formation; 1 to 1.09 is characteristic of oxidative growth; 0.7 to 0.8 indicates endogenous metabolism; and a $RQ$ below 0.6 indicates ethanol utilization. The output from various sensors can be used to control molasses feed rates by computers. Aiba, Nagai, and Nishizawa (1976) have also used $RQ$ for the computer control of baker's yeast fermentations. Whaite et al. (1978) adjusted feed rates to keep the $RQ$ between 1 and 1.2.

The determination of ethanol in the fermenter liquid may also be used to control feed rates. The alcohol diffuses through a semipermeable Teflon tubing inserted into the fermenter. The quantitative determination of ethanol in the tubing could be used as a feedback control system for nutrient addition to the fermenter (Dairaku et al. 1981). The authors achieved a specific growth rate of $\mu$ equals 0.3 per hour and a cell yield of 0.5 g/g of sugar in the feed. This system is used in at least one commercial yeast plant.

Bach, Woehrer, and Roehr (1978) used the continuous determination of ethanol in the effluent gas for automatic control of the feed rate. The sensor calibration curve indicates that ethanol concentrations of up to 3 g/L can be measured. The response time of the sensor (basically a smoke detector) was only 10 seconds, and the sensitivity was 1 ppm ethanol in the fermenter liquid. The sensor is a semiconductor whose electrical conductivity changes when combustible gases are oxidized at its surface. During a baker's yeast fermentation ethanol was the major constituent of such oxidizable gases.

Williams et al. (1986) reported several methods that could be used for on-line adaptive control for fed-batch fermentations of yeast. The parameters considered were the $RQ$ and $CO_2$, $O_2$, and ethanol in the effluent gas. Automatic process control by sensors measuring physical and chemical parameters has been described in detail by Fiechter, Meriners, and Sukatsch (1987).

## Continuous Fermentation

Yeast biomass can be grown in continuous culture with good yield and with higher productivity than can be achieved in fed-batch operations. The principle of continuous culture has also been applied to the production of baker's yeast and a British plant has operated in this manner for some time (Olson 1961; Sher 1961). The system operated as a five-stage, open, homogeneous, and continuous process with each continuous run lasting from 5 to 7 days. The process was later abandoned presumably because of difficulties in preventing buildup of contaminants. Burrows (1970) has described some of the problems with this process, but has not resolved the question of whether future research may lead to a satisfactory continuous process for the production of baker's yeast.

## Harvesting of Yeast Cells

At the end of the fermentation the fermenter liquid consists of about 5% yeast solids and about an equal concentration of soluble solids in the yeast-free liquid. This liquid is frequently called *beer* or *spent beer*, which goes back to a time when the liquid was alcoholic. The water content of the yeast cells is about 58-60%, and the density is approximately 1.13 g/cm$^3$ (Sambuchi et al. 1971). It varies somewhat with the osmotic pressure of the system, but the density of the cells is sufficiently greater than that of the liquid, which permits efficient separation with nozzle-type centrifuges.

Figure 6-12 shows a schematic drawing of such a continuous discharge nozzle centrifuge that develops about 4,000-5,000 g. During a first pass of the fermenter liquid a yeast solids concentration of about 20% may be reached. The yeast slurry contains about 50% yeast cells by volume; the free liquid still contains about 5% soluble solids. Dilution of the cream and a second centrifuging can further reduce the amount of soluble non-yeast solids.

Thorough washing of the yeast cells is particularly important for the production of ADY. The yeast cream containing 18-20% yeast solids is quite viscous, but is still pumpable. At concentrations of 23% solids it cannot be pumped, and at concentrations of 27-30% the yeast assumes a semisolid plastic consistency. Yeast cream may be stored refrigerated at temperatures slightly above 0°C for up to 2 weeks with little or no deterioration.

Additional concentration of the yeast requires filtration in a plate and frame filter. Such filter presses have 24- or 48-inch frames. The applied pressure may be 125-150 psi. The filter cloth is so tightly woven that no filter aid is required, which is an important consideration since the filter

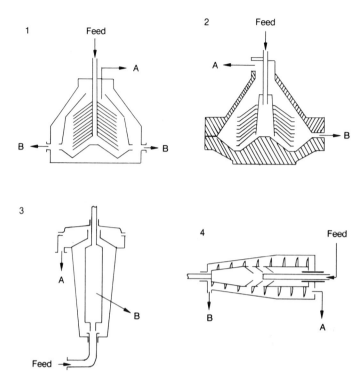

**Figure 6-12.** Schematic representation of solid-bowl centrifuges: 1, nozzle separator; 2, intermittently discharging separator; 3, tubular bowl centrifuge; 4, decanter; A, light phase discharge; B, heavy phase discharge. (*From Schmidt-Kastner and Gölker 1987*)

residue (not the filtrate) is the desired product. The press cake has solid levels of 27-32% solids. Alternatively, a continuously operating, rotary vacuum filter may be used. Figure 6-13 shows a schematic drawing of such a filter. The filter shown has a trough filled with yeast cream. The liquid is sucked by vacuum through the surface of the rotating filter drum while the yeast accumulates on the outside of the filter cloth. The cloth must be coated with a filter aid, in this case an edible material such as potato starch, which has a large granule size. The yeast is removed by a doctor knife.

The filter cake may contain very small amounts of potato starch that will remain in the final product. The moisture of the filter cake can be reduced by adding about 0.5% of salt to the yeast cream just prior to filtration. This process draws additional water out of the cells by the difference in osmotic pressure. The salt is then removed by spraying the filter cake with water while it is still on the drum (Kuestler and Rokitansky 1960). With this procedure the solids level of the yeast can be raised to 33% if desired.

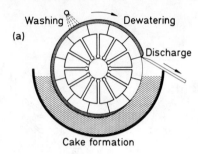

**Figure 6-13.** Principle of the rotary vacuum filter. (*From Kula 1985*)

## Mixing, Extruding, and Packaging of Compressed Yeast

The pressed yeast or the filter cake are dropped into a mixer, generally a ribbon blender. Small amounts of water may be added to adjust the moisture content to 70%. Also, small amounts of edible oil may be added to facilitate extrusion as well as emulsifiers such as diglycerides to prevent water spotting of the yeast cakes. The thoroughly mixed mass is then extruded through Teflon-coated nozzles in the form of a continuous band with a rectangular cross-section. It is then cut automatically into individual 1-pound cakes with a dimension of about 2¼ × 2¼ × 5¼ inches. The yeast cakes are wrapped in wax paper and the ends are heat sealed. Mixing, extruding, and wrapping raise the temperature of the yeast cakes that now require cooling to a temperature below 4°C.

Alternatively, the mixed yeast mass may be merely crumbled and packed into 25- or 50-pound bags. The polyethylene-lined, multiwall bags must be carefully sealed to prevent access of air that would enhance respiration and warm-up. The larger packages of crumbled yeast are more suitable for larger wholesale bakeries. Both yeast cakes and crumbled yeast must be kept refrigerated until delivered to the bakery.

## Yeast Cream

The pressing, mixing, and extruding operations may be bypassed by shipping the centrifuged yeast cream at 18-20% solids to bakeries in liquid, pumpable form. This type of shipment in tank trucks has been practiced occasionally as long as 40 years ago in France, and it has been used for many years in Britain for shipment of yeast to distilleries. The practice seems to be common in the USSR (Volkova, Drobot, and Roiter 1974;

Volkova and Roiter 1973). In the United States, this type of shipment was only introduced a few years ago. Obviously, it is only suitable for very large bakeries. It requires the installation of at least two refrigerated holding tanks at the bakery.

The shipment in liquid form has some advantages. It eliminates several processing steps at the yeast plant, and prevents the unavoidable warming up of the yeast during mixing and extrusion, which gives cream yeast a slight advantage in stability. It also permits wholesale bakers to meter the yeast cream directly into the mixer. Shipping weight of cream yeast is about 50% greater than that of cake or crumbled yeast. The use of cream yeast in Canada has recently been described by Van Horn (1989).

## Stability of Compressed Yeast

Baker's compressed yeast is perishable and must be kept refrigerated until used. Storage temperatures from 3-7°C are satisfactory. Such temperatures can normally be kept in industrial refrigerators. Lower temperatures near 0°C would provide better stability, but cannot be provided under commercial conditions. One can use the straight dough fermentation activity of yeast to measure its stability. With this yardstick the performance of compressed yeast in refrigerated storage may drop 5% during the first week, and show a lesser drop during the next weeks. The shelf life of the yeast may also be limited by mold growth after 3-4 weeks of storage.

Storage at elevated temperatures leads to a rapid disappearance of glycogen in the yeast cell, followed by a drop in trehalose concentration, and ultimately autolysis. Hautera and Lovgren (1975) made extensive storage studies with commercial, molasses-grown baker's yeasts at three temperatures. At 5°C there was little or no loss in activity over a 4-week period. At 23°C the activity started to drop after 4 days and the yeasts were dead after 20 days. At 35°C the yeasts were practically inactive after 5 days.

Unfortunately, these tests were carried out with 7.2% solutions of glucose, fructose, sucrose, and maltose, that is, at osmotic pressures far below those of doughs. It is known that upon storage, fermentation activity drops faster at high osmotic pressures, for instance, in sweet doughs. Therefore, one must qualify the conclusions that can be drawn from this study. The point is merely made to emphasize the fact that performance of different yeasts may vary greatly depending on the conditions encountered in practical usage.

At high temperatures, such as ones that may be encountered in bakeries (35°C), compressed yeast should be used shortly after its removal from the

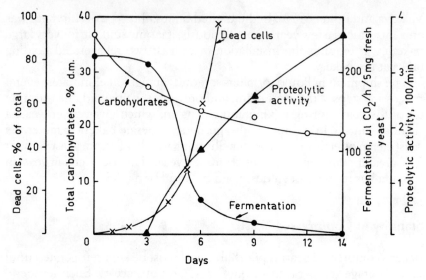

**Figure 6-14.** Decrease of the carbohydrate content and the fermentation activity, and increase of dead cells and of proteolytic activity of baker's compressed yeast during storage at 35°C. (*From Suomalainen 1975*)

refrigerator. Figure 6-14 shows that respiration sets in quickly with a loss of internal carbohydrates followed by a rapid drop in fermentation activity and cell death (Suomalainen 1975).

## Contamination

Baker's yeast always contains some contaminating organisms since it is grown in fermenters that cannot be kept completely sterile. The word *contaminating* does not mean that these organisms are actually or potentially harmful. Specifications for baker's yeast require absence of *Salmonella* and other pathogens, and restrict the number of coliform organisms to less than 1000 per gram and the count of *Escherichia coli* to less than 100 per gram. Baker's yeast plants follow Good Manufacturing Practice (GMP) rules of the Food and Drug Administration, and great efforts are made to keep the number of contaminants to a minimum.

Molasses is generally sterilized in bulk or in continuous heat exchangers. The air used to aerate the fermenter is passed through depth filters, membrane filters, or is heated. Fermentation tanks are thoroughly cleaned by hot alkaline solutions or detergents, usually with clean-in-place (CIP) equipment, and if necessary by scrubbing. It is important to prevent individuals from climbing into fermenters without close supervision because there is potential danger from accumulated $CO_2$.

All pipes, pumps, valves, and other auxiliary equipment must be thoroughly sanitized. It is also important to consider potential contamination from other accessory equipment such as air pipes, leaking cooling coils, probes for pH meters and other instruments, and sampling devices. Particular attention has to be paid to ducts for exit air to prevent condensate from dripping back into the fermenter. Similar attention has to be paid to further processing of the grown yeast, which includes the centrifuges, presses, filters, and all packaging operations. For active dry yeasts it includes the driers. Yeast is an excellent growth medium for bacterial contaminants, and deposits of wet or dry yeasts on equipment become a ready source of infection.

The overwhelming proportion of contaminants in CY or ADY are lactic-acid-producing organisms often called *diplos* (diplococci) in the industry. In the past, these organisms have been thought to be useful as contributors to the flavor of bread. This belief is certainly not true for the very short dough fermentations at very high levels of yeast inocula typical of modern methods of baking. Carlin (1958) reported total bacterial counts of 2-3 × $10^9$ per gram of yeast, but these values are much higher than could be ascertained in continuous sampling of CY and ADY samples. Actual counts are likely to vary between $10^4$ and $10^8$ per gram of yeast.

These organisms always or almost always belong to heterofermentative lactic-acid-producing organisms of the genus *Leuconostoc* or to homofermentative bacteria of the genus *Lactobacillus*. Some coliform organisms and occasionally a few *Escherichia coli* can be found. Active dry yeast made by drying compressed yeast contains the same kinds of microorganisms, but the total count generally decreases during drying and upon subsequent storage of the dried yeast. Specifications for rope spores require an upper limit of 200 per gram of ADY.

Occasionally contamination with *Oidium lactis* or other molds has been reported. Fowell (1965) listed various species of contaminating yeasts. Most of them fall into the genera *Candida* and *Torulopsis* (Podel'ko et al. 1975). Methods for the detection and estimation of bacteria, yeasts, and molds in baker's yeast have been reported by Fowell (1967).

### Active Dry Yeast

Baker's yeast may be dried in its vegetative state to a product having less than 8% moisture without substantial loss of viability, but success of the drying method requires mastery of the various requirements of the methodology. One can indeed dry finely crumbled CY be merely spreading it on adsorbent paper and keeping it uncovered for several days at room temperature, and one can recover a fair number of viable cells. But commercial production requires almost complete recovery of the viability and fermentation activity of the product. The history and development of ADY has been described by Frey (1957). Its first use was in areas in which it was practically impossible to ship CY, for instance, for use by the army in inaccessible areas or for shipment by the Berlin airlift during the blockade in 1948/1949.

The goal of ADY research has been the development of strains and of production methods that produce a dry product with the same fermentation activity as CY on an equivalent solids basis. Thus, 1 gram of ADY should have the same bake activity as 3.1 grams of CY. During the past 40 years this goal has been approached stepwise, but has not quite been reached. Throughout this period ADY has made inroads into markets where its superior stability was more important than its cost per unit activity. This situation exists in the household market for consumer ADY sold in nitrogen-flushed aluminum foil pouches (7 g net weight) with a shelf life exceeding one year. ADY is now also used extensively by retail bakers, institutional users, commissary-type operations (e.g., pizza chains), and, of course, in all areas in which it is impossible or too costly to distribute refrigerated CY. CY has not been replaced in major wholesale bakeries.

All ADYs are grown and harvested by methods described above for CYs, but strains, fermentation schedules, composition, and so forth may differ. Hence, ADY cannot be better than the CY from which it is made by drying. A comparison between the performance of CY and ADY should be made on the basis of equal weights of solids, not on the basis of equal cell numbers, since cell size may vary by a factor of 2 or more.

*Strains for ADY Production*

Until a few years ago almost all of the ADY was produced with a strain typified by strain no. 7752 of the American Type Culture Collection. The strain gives a better yield on a sugar basis. Its cells are larger than those of strains used for the production of compressed yeast. The strain is hardy, but cannot be grown to nitrogen levels exceeding 8.5%. For this reason, its fermentation activity is lower than that of compressed yeast on an equivalent moisture basis. There is some loss of solids during the early stages of drying due to increased respiration. The strain can be dried in continuous tunnel (chamber) driers with minimal loss of fermentation activity (0-5%).

Efforts to improve the properties of dried yeasts have been concerned with questions of yeast composition, strains, and improved methods of drying. With regard to composition, it was found that the level of trehalose should be 12% or higher (solids basis) (Clement 1983). The same author also found that compressed yeast used for the production of ADY must be washed thoroughly in order to reduce the molarity of extracellular water. Retention of residual salt, resulting from an acid treatment or from the use of the salt process during filtration, is particularly harmful during drying.

For a given strain the fermentation activity is considerably higher in yeasts grown to a higher nitrogen level, but such yeasts are also more sensitive to drying. Hence, the search for better methods of drying has assumed particular importance. Methods were sought that would permit drying at a lower temperature, in a shorter period of time, and with much smaller particles of press cake to improve the uniformity of moisture removal. Some success could be achieved with a fluid bed drying method, a drying time of less than 2 hours, and a yeast strain with a nitrogen content of more than 8% (Langejan 1980).

The resulting ADY had a smaller particle size and a more regular shape (cylindrical with rounded ends) than ADY dried in tunnel driers. This type of dry yeast was later called instant active dry yeast (IADY) because it did not require separate rehydration before use. Grylls, Rennie, and Kelly (1980) took the additional step of using mechanical disintegration of the yeast granules in the fluid bed, always with the goal of promoting uniform moisture removal during drying. However, this solution is only a partial one.

The greatest uniformity in the drying of yeast has been accomplished by Johnston (1959), who emulsified liquid yeast cream in warm oil. Drying was done with air blowing through the oil/water/yeast emulsion. The dried yeast was recovered by solvent extraction of the oil. This process permitted the drying of yeast with a high protein concentration and excellent fermentation activity. It has not been used on a commercial scale, possibly because of difficulties in removing the last traces of solvent.

Numerous patents have claimed certain strains of *Saccharomyces cerevisiae* as particularly useful for producing ADY, often in conjunction with definite processing conditions. Such strains may have been produced by classical hybridization, protoplast fusion, or mutation. Yeasts described in these patents will not be discussed in this chapter because evaluation of such yeasts in comparison with commercially available products is either missing or too uncertain.

*Biology of Yeast Drying*

There is no better source on the biology of yeast drying than the extensive review by Beker and Rapoport (1987). It deals with the structural changes that can be observed by electron microscopy of cell cross-sections showing, for instance, the rupture of the cytoplasmic membrane of a dehydrated cell

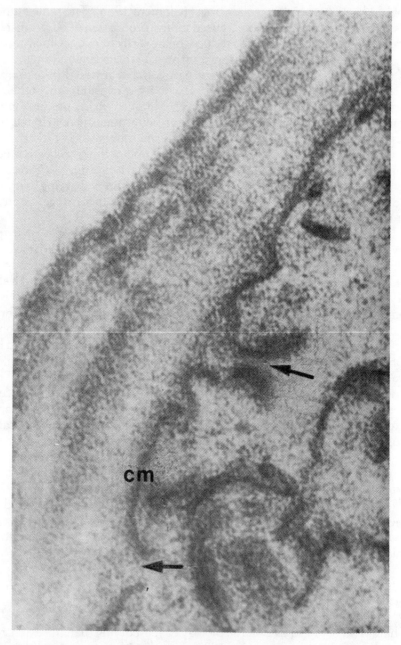

**Figure 6-15.** Electron microscopy of ultrathin sections of *Saccharomyces cerevisiae*. Ruptures of the cytoplasmic membrane are shown by arrows. Magnification 145,000 times. (*From Beker and Rapoport 1987*)

(Fig. 6-15). It also documents the changes in composition of nucleic acids, proteins, lipids, carbohydrates, and polyphosphates. Such changes cannot yet be directly related to the properties of the dried yeasts, although attempts have been made to relate changes in the degree of unsaturation of fatty acids during drying and rehydration to the metabolic pathways of S. cerevisiae (Zikmanis et al. 1983). The concentration of ergosterol that ensures the structural integrity of the cell membranes increases during drying and decreases again upon rehydration of the ADY (Zikmanis et al. 1985).

Drying Processes

The rate at which yeast press cake may be dried varies greatly with the temperature and the moisture content of the drying air stream, as well as with air velocity and the size and shape of the granulated yeast. However, in each case the general scheme of the drying curve is that shown in Figure 6-16 (Beker and Rapoport 1987). The fast rate of drying shown in period I is attributed to the loss of free water; the slower rate of drying in period II is attributed to the loss of bound water. The point of inversion of the drying curve usually takes place at a moisture content of about 20%. Up to this point, changes in cell viability and cell wall permeability are minor. At moisture levels below 20%, respiration stops and cell constituents are leached if the yeast is rehydrated in water. Drying is continued until moisture levels of ADY are 7.5-8.5% and those of IADY are 4-6%.

Early drying methods consisted generally of mixing the press cake with an edible low-moisture food. For instance, 1 kg of press cake could be

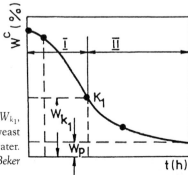

Figure 6-16. Schematic drying curve of yeast. $W_{k_1}$, bound water content, %; $W_p$, moisture content in yeast after drying, %. Period I indicates removal of free water. Period II indicates removal of bound water. (From Beker and Rapoport 1987)

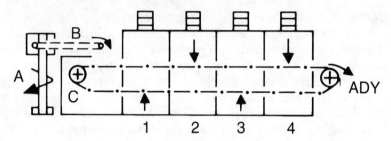

**Figure 6-17.** Tunnel drier (continuous): A, oscillating feeder; B, feeder belt; C, endless wire mesh belt; 1, 2, 3, 4, drying chambers. Arrows indicate direction of air flow. (*Courtesy Proctor and Schwartz Inc.*)

mixed with 5 kg of low-moisture starch followed by air drying (Rupprecht and Popp 1970). However, for commercial production of ADY such methods are not suitable. Continuous belt tunnel driers are generally used for the production of ADY. In this method, the press cake is extruded in strands with a circular cross-section of a diameter of 0.2-2 mm. The strands are cut off at the extruder to a length of 1-2 cm. They are then deposited on the wire mesh screen of the continuous belt that carries the yeast layer through three to six drying chambers.

The drying air is directed alternately in a downward and upward direction through the yeast bed as shown in Figure 6-17. Drying times vary from 2-4 hours and air inlet temperatures vary from 28°-42°C (Belokon 1962). Tunnel drying may also be carried out as a batch process in a single chamber, but this process requires consecutive changes in the temperature of the drying air. It is suitable for smaller installations.

Batch drying can also be carried out in a rotolouver drier. The extruded yeast strands are fed into a large, hollow cylinder equipped with baffles on the inside. The cylinder rotates slowly so that the yeast particles tumble constantly during the drying period. Warm air, up to 60°C, is blown into the cylinder and passes through the tumbling yeast particles.

Fluid-bed (air lift) driers are preferred for the production of IADY. The process can be carried out batchwise (Fig. 6-18) or in continuous fashion (Fig. 6-19). The extruded strands of yeast with a diameter of 0.2-0.5 mm are deposited on a metal screen or a perforated plate of the drier. Air is blown from the bottom through the yeast layer at velocities that suspend the yeast particles in the fluid bed.

Generally, fluid-bed driers can be operated with shorter drying times than belt-tunnel driers because they permit the use of more finely granulated yeast particles. Drying times of 1 to 2 hours are satisfactory, but much shorter drying times have been reported in the patent literature. Langejan (1972) used an air stream at 100°-150°C at the beginning of the drying

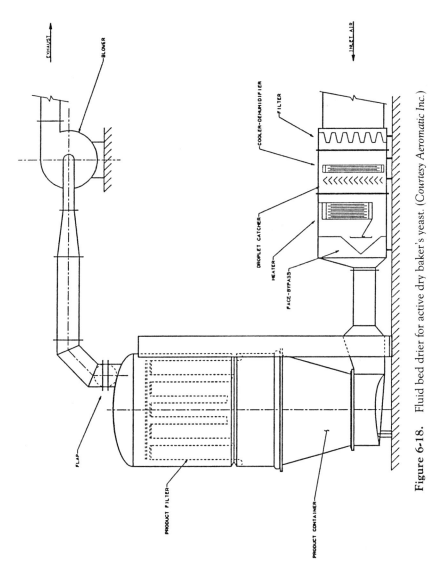

Figure 6-18. Fluid bed drier for active dry baker's yeast. (Courtesy Aeromatic Inc.)

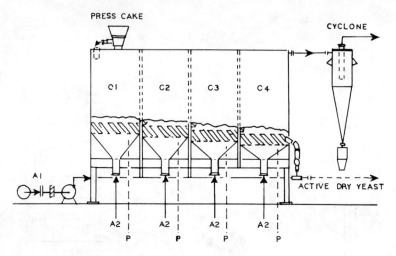

**Figure 6-19.** Continuous multichamber fluid bed drier for active dry baker's yeast: A1, air for pneumatic transport of ADY; A2, air for yeast drying; P, steam; C1, C2, C3, C4, drying chambers. (*Courtesy Pressindustria, Milan*)

period. The drying time was 10-30 minutes, and the temperature of the yeast particles was kept within the 25°-42°C range. Continuous fluid-bed driers are also in commercial use.

Other methods of drying yeast have been proposed, but are not in general use. The drying of yeast cream in oil has already been mentioned. Spray drying is highly productive and economical but results in a considerable loss of activity. Spray drying with agglomeration of the yeast particles followed by more conventional belt drying has also been suggested. Vacuum drum drying has been described by Shishatskii and Bocharova (1973). Freeze drying results in heavy losses in viability and fermentation activity. It may be used for culture preservation but not for the commercial production of ADY. Hartmeier (1977) has reported a process for drying of liquid yeast cream on an endless belt in a vacuum chamber. This process would eliminate the need for pressing of the yeast cream and for extrusion of the press cake. It has not been used on a commercial scale.

*Processing Aids and Additives*

The patent literature lists many compounds that may be added to the yeast cream or to the press cake and that are claimed to produce a more active ADY or IADY. For instance, Langejan (1980) lists methyl cellulose or carboxymethyl cellulose as suitable "swelling" agents; suggested level is 1-2% of the yeast solids. Hill (1987) lists emulsifiers such as monoglycerides, soya lecithin, glycerol polyesters, and sorbitan esters. In practice, sorbitan esters are widely used at levels from 0.2-1%. The added emulsifiers facilitate processing and lighten the color of the ADY or IADY, but the main reason for use of these compounds is the reduction of leached solids when ADY is rehydrated.

Active dry yeast loses some activity upon storage when exposed to the oxygen of the air. This storage stability of the yeast in air can be increased by addition of an antioxidant such as butylated hydroxyanisole (0.1%, based on yeast solids) to the yeast cream before pressing or filtration (Chen and Cooper 1962; Chen, Cooper, and Gutmanis 1966). The antioxidant is dispersed with the aid of one of the mentioned emulsifiers, preferably with a sorbitan ester. The treatment is effective for ADYs with a moisture content of less than 6%. The storage stability of such "protected" ADY is considerably enhanced. It does not match the extended stability that can be achieved with complete exclusion of air during storage.

*Leached Solids*

For use in baking ADY must be rehydrated in water at 30°-40°C. Figure 6-20 shows the effect of the temperature of the rehydration water on the

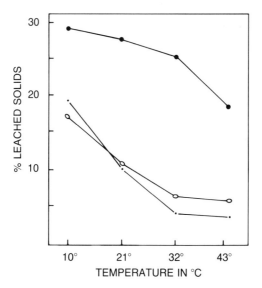

**Figure 6-20.** Leached solids of ADY as a function of rehydration temperature, yeast moisture content, and presence of emulsifier:

- •—•  ADY, 8.1% moisture
- ●—●  ADY, 5.3% moisture
- ○—○  ADY, 4.0% moisture plus 2% sorbitan monostearate

(*After Chen, Cooper, and Gutmanis 1966*)

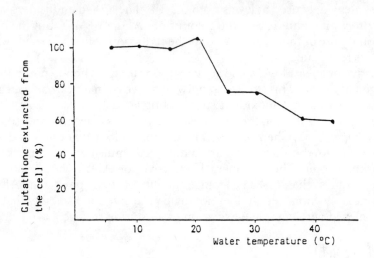

**Figure 6-21.** Leaching of glutathione from baker's yeast upon rehydration at different water temperatures. (*From Hill 1987*)

percent of solids that are leached from the yeast. Rehydration at lower water temperatures lessens the fermentation activity of the ADY. The material leached from the yeast cells has a slackening effect on doughs due to the extraction of reducing compounds such as glutathione or cysteine.

Figure 6-21 shows the effect of water temperature on the extraction of glutathione from ADY (Hill 1987). In many dough systems slackening is not desired. For bucky doughs made from very strong flours slackening may be desired, but it is better achieved in a controlled fashion with the addition of proteolytic enzymes or of cysteine to the dough. IADY need not be rehydrated in water, but may be mixed in with the other dry dough ingredients. Its activity is better upon direct incorporation than with separate rehydration (Langejan 1980; Bruinsma and Finney 1981), possibly due to the difference in moisture content between ADY and IADY.

*Storage Stability and Packaging of ADYs*

The storage stability of ADY is decreased by higher temperatures, a higher moisture of the yeast, and by the presence of air in the package. ADY loses about 7% of its fermentation activity per month if stored without a protective inert atmosphere at ambient temperature. For storage under nitrogen, or when vacuum packed, the loss is about 1% per month and generally less than 10% per year.

Table 6-8. Storage Stability of Active Dry Baker's Yeast in Various Atmospheres

| Atmosphere* | Residual activity† |
|---|---|
| Nitrogen | 79.5% |
| Carbon dioxide | 81.9% |
| Carbon monoxide | 85.2% |
| Argon | 80.7% |
| Hydrogen | 88.9% |
| Air | 43.3% |
| Vacuum‡ | 70.3% |

Source: Data from Hill 1987.
*400 Torr internal pressure in sealed glass ampoules.
†After storage for 3 days at 55°C.
‡10 Torr internal pressure in sealed glass ampoules.

IADY is always distributed in hermetically sealed pouches in vacuo or flushed with an inert gas. ADY adsorbs $CO_2$ readily (Amsz et al. 1956). If this gas is used with larger packages, a sufficient vacuum is created to give the appearance and performance of vacuum packaging. Hill (1987) has compared the effect of various atmospheres in an accelerated storage test at 55°C (Table 6-8). Such accelerated tests are useful for purposes of quality control. They do not provide reliable information on commercial conditions unless they have been correlated with long-term storage under more normal temperatures.

The packages may be nitrogen-flushed tins or aluminum foil pouches with low permeability to water vapor and gaseous oxygen. The aluminum foil, which must be free from pin holes, is laminated to a heat-sealable plastic film such as Saran or Pliofilm. Hill (1987) has indicated the composition of a four-layer composite film as follows: polyester, $20\,\mu$; aluminum, $12\,\mu$; polyester, $12\,\mu$; polyethylene, $8\,\mu$. Such films are suitable for larger packages in 500-g and 10-kg sizes vacuum-packed. Smaller 2-lb (907-g) tins may also be vacuum packed. Larger 10-kg tins must be nitrogen-flushed to prevent collapse under vacuum.

Once the package has been opened the ADY should be used quickly. In particular, it should not be exposed to an atmosphere with high humidity. Figure 6-22 shows the equilibrium moisture content of ADY at various relative humidities (Dobbs, Peleg, and Mudget 1982). It suggests rapid take-up of moisture by ADY in the atmosphere of a wholesale bakery and a concomitant loss of activity upon further storage.

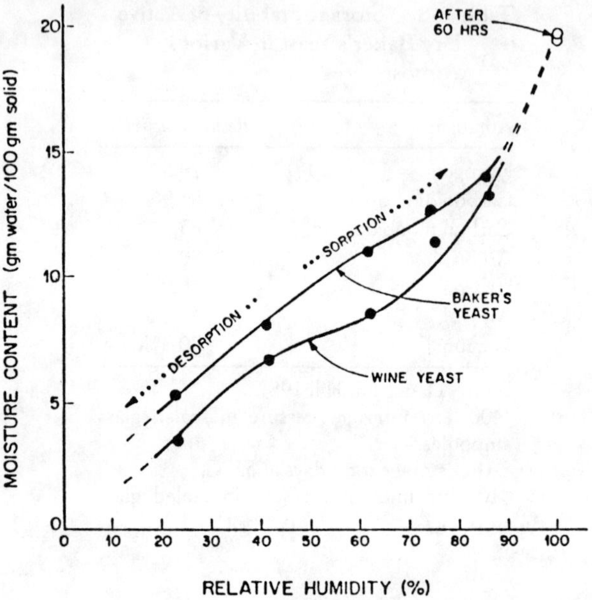

**Figure 6-22.** Equilibrium moisture content of industrial yeasts exposed to different relative humidities at ambient temperature. (From *Dobbs, Peleg, and Mudget 1982*)

Table 6-9. Waste Water Volume and Specific COD Load for Three Processing Operations per 1000 kg of Yeast Solids*

| Process Waste Liquid | Volume of Waste ($m^3$/1000 kg yeast solids) | COD Load as % of Total |
|---|---|---|
| Spent fermenter beer | 12.6 | 80% |
| Centrifuging wash water | 11.8 | 16% |
| Filtration waste water | 1.9 | 4% |
| Total process waste† | 26.3 | 100% |

*Source:* Data from Bronn 1985.
*For a harvest of 337 kg of yeast solids or 270 kg of newly formed yeast solids in a volume of 6 $m^3$ (growth to 5.6% yeast solids).
†Does not include water from rinsing, cleaning, and wort preparation.

Table 6-10. Average Composition of Total Waste Water*

| Component | Molasses-Grown† | Corn-Syrup-Grown‡ |
|---|---|---|
| COD, g $O_2$/L | 20.0 | 4.4 |
| $BOD_5$, g $O_2$/L | 15.2 | 4.0 |
| $BOD_5$ as % of COD | 76% | 90% |
| Total solids, g/L | 60 | 20.1 |
| Minerals, g/L | 7.5 | 3.7 |

Source: Modified after Bronn 1985.
*Conditions as in Table 6-9.
†Beet molasses, 80° Brix, 47% sugar.
‡Corn syrup, 70% solids, DE 96.

## Waste Water of Baker's Yeast Plants

The principle source of high gallonage-high BOD effluents is the fermenter beer after separation of the yeast. Table 6-9 shows the liquid volume of this beer as well as that of the effluent resulting from washing of the yeast during separation and the liquid from the filtration of the yeast cream. It also shows the distribution of the COD load between these three fractions. The actual composition of these effluents is shown in Table 6-10 for growth on beet molasses or on a high-DE corn syrup (Bronn 1985). The BOD in effluents of these processing steps with high-DE corn syrups is only 20-25% of that of effluents from molasses-grown yeasts. Methods of disposal of these effluents vary from plant to plant and from country to country. A fairly common method of disposal is through local municipal sewage plants with charges based on gallonage as well as on BOD load.

## FURTHER READINGS

The following books or book chapters contain extensive reviews on the production of baker's yeast: White (1954); Peppler (1960); Butschek and Kautzmann (1962); Sato (1966); Burrows (1970); Harrison (1971); Reed and Peppler (1973); Reed (1982); Oura (1983); Chen and Chiger (1985); Trivedi, Jacobson, and Tesch (1986); Beudeker et al. (1990).

## REFERENCES

Aiba, S., S. Nagai, and Y. Nishizawa. 1976. Fed-batch culture of S. cerevisiae. A perspective of computer control to enhance the productivity in baker's yeast cultivation. Biotech. Bioeng. **18**:1001-1016.

Amsz, J., R. F. Dale, and H. J. Peppler. 1956. Carbon dioxide sorption by yeast. *Science* **123**:463.
Anon. 1982. Corning-Kroger combine technology to exploit lactose hydrolyzed whey. *Food Develop.* Jan. 1982:34-35.
Bach, H. P., W. Woehrer, and M. Roehr. 1978. Continuous determination of ethanol during aerobic cultivation of yeasts. *Biotech. Bioeng.* **20**:799-807.
Barnett, J. A. 1976. The utilization of sugars by yeast. *Adv. Carbohydrate Chem. Biochem.* **32**:125-234.
Beker, M. J., and A. I. Rapoport. 1987. Conservation of yeasts by dehydration. *Adv. Biochem. Eng./Biotechnol.* **35**:127-171.
Belokon, V. N. 1962. Yeast drying on a belt drier (in Russian). *Spirt. Prom.* **1**:40-42.
Bergander, E. 1969. The effect of various fermentation inhibitors in molasses (in German). *Lebensm. Industrie* **16**:219-221.
Beudeker, R. F., H. W. van Dam, J. B. van der Platt, and K. Vellenga. 1990. Developments in baker's yeast production. In *Yeast Biotechnology and Catalysis*, H. Verachtert and De Mot (eds.). Marcel Dekker, New York.
Blenke, H. 1987. Process engineering contributions to bioreactor design and operation. In *Biochemical Engineering*, H. Chmiel et al. (ed.). Gustav Fischer, Stuttgart, West Germany.
Brauer, H. 1985. Stirred vessel reactors. In *Biotechnology*, vol. 2, H. Brauer (ed.). VCH Publishing Company, Weinheim, West Germany.
Bronn, W. K. 1985. Investigations of the technological and economical possibility of using raw materials other than molasses for yeast production (in German). Research Report T 85-117, Ministry for Science and Technology, German Federal Republic.
Bruinsma, B. L., and K. F. Finney. 1981. Functional (bread making) properties of a new yeast. *Cereal Chem.* **58**:477-480.
Burrows, S. 1970. Baker's yeast. In *The Yeasts*, vol. 3, A. H. Rose and J. S. Harrison (eds.). Academic Press, New York.
Butschek, G., and R. Kautzmann. 1962. Production of baker's yeast (in German). In *Die Hefen*, vol. 2, F. Reiff et al. (ed.). Verlag Hans Carl, Nuremberg, West Germany.
Carlin, G. T. 1958. The fundamental chemistry of bread making. *Proc. Am. Soc. Bakery Eng.* pp. 56-63.
Ceijka, A. 1983. Preparation of media. In *Biotechnology* vol. 3, H. Dellweg (ed.). VCH Publishing Company, Weinheim, West Germany.
Chen, S. L., and M. Chiger. 1985. Production of baker's yeast. In *Comprehensive Biotechnology*, vol. 3, M. Moo-Young (ed.). Pergamon Press, Oxford, U.K.
Chen, S. L., and E. J. Cooper. 1962. Production of active dry yeast. U.S. Patent 3,041,249.
Chen, S. L., E. J. Cooper, and F. Gutmanis. 1966. Active dry yeast: Protection against oxidative deterioration during storage. *Food Technol.* **20**(12):79-83.
Chen, S. L., and F. Gutmanis. 1976. Carbon dioxide inhibition of yeast growth in biomass production. *Biotechnol. Bioeng.* **18**:1455-1462.
Clement, Ph. 1983. Preparation of dried baker's yeast. U.S. Patent 4,370,420.
Cooney, C. L. 1981. Growth of microorganisms. In *Biotechnology*, vol. 1, H. J. Rehm and G. Reed (eds.). VCH Publishing Company, Weinheim, West Germany.
Cooney, C. L., H. Y. Wan, and D. C. I. Wang. 1977. Computer-aided material balancing for production of fermentation parameters. *Biotechnol. Bioeng.* **19**:55-67.
Cooper, C. M., G. A. Fernstrom, and S. A. Miller. 1944. Performance of agitated liquid contactors. *Ind. Eng. Chem.* **36**:504-509.

Dairaku, K., Y. Yamasaki, K. Kuki, S. Shioya, and T. Takamatsu. 1981. Maximum production in a bakers' yeast fed-batch culture by a tubing method. *Biotechnol. Bioeng.* **23**:2069-2081.
DeBeczy, G., and A. J. Liebmann. 1944. Aeration in the production of compressed yeast. *Ind. Eng. Chem.* **30**:882-890.
Dellweg, H., W. K. Bronn, and W. Hartmeier. 1977. Respiration rates of growing and fermenting yeast. *Kem. Kemi* **4**(12):611-615.
Dobbs, A. J., M. Peleg, and R. E. Mudget. 1982. Some physical characteristics of active dry yeast. *Powder Technol.* **32**:63-69.
Drews, B., H. Specht, and A. M. Herbst. 1962. Growth of baker's yeast in concentrated molasses wort. *Branntweinwirtschaft* **102**:245-247.
Ebner, H., K. Pohl, and A. Enenkel. 1967. Self priming aerator and mechanical defoamer for microbiological processes. *Biotechnol. Bioeng.* **9**:357-364.
Fiechter, A., M. Meriners, and D. A. Sukatsch. 1987. Biological regulation and process control. In *Fundamentals of Biotechnology*, P. Prave et al. (ed.). VCH Publishing Company, Weinhcim, West Germany.
Finn, R. K. 1967. Agitation and aeration. *Biochem. Biol. Eng. Sci.* **1**:69-99.
Fowell, M. S. 1965. The identification of wild yeast colonies on lysine agar. *J. Appl. Bacteriol.* **28**:373-383.
Fowell, M. S. 1967. Infection control in yeast factories and breweries. *Proc. Biochem.* **2**(12):11-15.
Frey, C. N. 1957. History and development of active dry yeast. In *Yeast, Its Characteristics, Growth and Function in Baked Products*. Proc. Symp. U.S. Quartermaster Food Container Inst., Chicago, pp. 7-32.
Fries, H. von. 1962. Peculiarities of yeast growth in aerated fermentations (in German). *Branntweinwirtschaft* **102**:442-445.
Grylls, S. M., S. D. Rennie, and M. Kelly. 1980. Process for producing active dry yeast. U.S. Patent 4,188,407.
Harrison, J. S. 1967. Aspects of commercial yeast production. *Proc. Biochem.* **2**(3):41-45.
Harrison, J. S. 1971. Yeast production. *Prog. Industr. Microbiol.* **10**:129-177.
Hartmeier, W. 1977. Active dry yeast and method of its production (in German). German Patent Appl. 25 15 029.
Hatch, R. T. 1975. Experimental and theoretical studies of oxygen transfer in the airlift fermenter. In *Single-Cell Protein II*, S. R. Tannenbaum and D. I. C. Wang, (eds.). MIT Press, Cambridge, Mass.
Hautera, P., and T. Lovgren. 1975. The fermentation activity of baker's yeast. Its variation during storage. *Baker's Digest* **49**(3):36-37, 49.
Hill, F. F. 1987. Dry living microorganisms—Products for the food industry. In *Biochemical Engineeering*, H. Chmiel, W. P. Hammes, and J. E. Bailey (eds.). Gustav Fischer Verlag, Stuttgart.
Hongisto, H. J., and P. Laakso. 1978. Application of the Finn-sugar-Pfeiffer and Langen desugaring process in a beet sugar factory. 20th General Meeting of the American Society of Sugar Beet Technology, San Diego, Aug. 6.
Hospodka, J., Z. Caslavaky, K. Beran, and F. Stross. 1962. The polarographic determination of oxygen uptake and transfer rate in aerobic steady-state yeast cultivation. In *Continuous Cultivation of Microorganisms*, I. Malik, K. Beran, and J. Hospodka, (eds.). Academic Press, New York.
Jakubowska, J., and M. Wlodarczyk. 1969. Observations on yeast growth and metabolism influenced by beta-indolylacetic acid. *Antonie van Leeuwenhoek J. Microbiol. Serol.* **35** (Suppl. Yeast Symp.), p. G17.

Johnston, W. R. 1959. Active dry yeast products and processes for producing same. U.S. Patent 2,919,194.

Kautzmann, R. 1969. Effect of amino acids on yield and quality of baker's yeast. *Branntweinwirtschaft* **109**:214-222.

Kiby, W. 1912. Handbook of production of compressed yeast (in German). Friedrich Vieweg & Son, Braunschweig.

Kirsop, B. E., and C. P. Kurtzman. 1988. *Yeasts.* Cambridge University Press, Cambridge, U.K.

Kowalski, S., I. Zander, and S. Windisch. 1981. Hybrid yeast strains capable of raising an extraordinarily broad range of dough types. *Eur. J. Appl. Microbiol. Biotechnol.* **11**:146-150.

Kuestler, E., and K. Rokitansky. 1960. Process of producing yeast of increased dry solids content and reduced plasticity. U.S. Patent 2,947,668.

Kula, M. R. 1985. Recovery operations. In *Biotechnology,* vol. 2, H. Brauer (ed.). VCH Publishing Company, Weinheim, West Germany.

Lagunas, R. 1986. Misconceptions about the energy metabolism of *Saccharomyces cerevisiae*. *Yeast* **2**:221-228.

Langejan, A. 1972. A novel type of active dry baker's yeast. In *Fermentation Technology Today,* G. Terui (ed.), pp. 669-671. Society of Fermentation Technology, Osaka, Japan.

Langejan, A. 1980. Active dry baker's yeast. U.S. Patent 4,217,420.

Large, P. J. 1986. Degradation of organic nitrogen compounds by yeast. *Yeast* **2**:1-34.

Liljestrom-Suominen, P., V. Joutsjoki, and M. Korhola. 1988. Construction of a stable alpha-galactosidase producing baker's yeast strain. *Appl. Environ. Microbiol.* **54**:245-249.

Linek, V., and P. Benes. 1978. Enhancement of oxygen absorption into sodium sulfite solutions. *Biotechnol. Bioeng.* **20**:697-707.

McCann, A. K., and J. A. Barnett. 1986. The utilization of starch by yeasts. *Yeast* **2**:109-115.

Mateles, R. I. 1971. Calculation of oxygen required for cell production. *Biotechnol. Bioeng.* **13**:581-582.

Meyenburg, H. K. von. 1969. Energetics of the budding cell of *S. cerevisiae* during glucose limited aerobic growth. *Archives Microbiol.* **66**:289-303.

Mor, J. R., and A. Fiechter. 1968. Continuous cultivation of *S. cerevisiae*. I. Growth on ethanol under steady state conditions. *Biotechnol. Bioeng.* **10**:159-176.

Moulin, G., and P. Galzy. 1984. Whey, a potential substrate for biotechnology. *Biotechnol. Gen. Eng. Rev.* **1**:347-373.

Notkina, L. G., I. M. Balyberdina, and L. D. Lavrenchuk. 1975. The effect of nitrites on baker's yeast manufacture (in Russian). *Khlebopek. Konditer Prom.* **2**:28-31.

Olson, A. J. C. 1961. Manufacture of baker's yeast by continuous fermentation. I. Plant and process. *Soc. Chem. Ind. (London) Monograph* **12**:18-93.

Oura, E. 1974. Effect of aeration intensity on the biochemical composition of baker's yeast. I. Factors affecting the type of metabolism. *Biotechnol. Bioeng.* **16**:1197-1212.

Oura, E. 1983. Biomass from carbohydrates. In *Biotechnology,* vol. 3, H. Dellweg (ed.). VCH Publishing Company, Weinheim, West Germany.

Panek, A. D. 1975. Trehalose synthesis during starvation of baker's yeast. *Eur. J. Appl. Microbiol. Biotechnol.* **2**:39-46.

Peppler, H. J. 1960. Yeast. In *Bakery Technology and Engineering*, A. Matz (ed.). AVI Publishing Company, Westport, Conn.

Podel'ko, A. D., et al. 1975. The effect of ultrasound on the microflora in yeast manufacturing plants (in Russian). *Khlebopek. Konditer Prom.* **9**:23-24.

Reed, G. 1982. Production of baker's yeast. In *Prescott and Dunn's Industrial Microbiology*, 4th ed., G. Reed (ed.). AVI Publishing Company, Westport, Conn.
Reed, G., and H. J. Peppler. 1973. *Yeast Technology*. AVI Publishing Company, Westport, Conn.
Rodney, P. J., and P. F. Greenfield. 1984. Review of yeast ionic nutrition. I. Growth and fermentation requirements. *Proc. Biochem.* **19**(2):48-60.
Rosen, K. 1977. Production of baker's yeast. *Proc. Biochem.* **12**(3):10-12.
Rupprecht, H., and L. Popp. 1970. Preparation of stable concentrate of baking flours containing yeast. U.S. Patent 3,510,312.
Sambuchi, M., et al. 1974. Filtration and extrusion characteristics of baker's yeast. I. Results of compression permeability test and constant pressure filtration (in Japanese). *Hakko Kogaku Zasshi* **49**:880-885.
Sanderson, G. W., and G. Reed. 1985. Fermented products from whey and whey permeate. IDF Seminar (New Dairy Products via New Technology), Atlanta, Ga., Oct. 1985.
Sato, T. 1966. *Baker's Yeast* (in Japanese). Korin-Shoin Publ., Tokyo.
Schaaf, I. 1988. The effect of overproduction of glycolytic enzyme on the rate of alcoholic fermentation. *Yeast Newsletter* **37**(1):8.
Schmidt-Kastner, G., and Ch. Gölker. 1987. Product recovery in biotechnology. In *Fundamentals in Biotechnology*, P. Prave et al. (eds.). VCH Publishing Company, Weinheim, West Germany.
Schreier, K. 1974. Bioreactors: stage of development and industrial application; especially with regard to systems for transfer of gas. *4th International Symposium on Yeasts*, Vienna.
Schuegerl, K. and W. Sittig. 1987. Bioreactors. In *Fundamentals of Biotechnology*. P. Praeve, ed. VCH Publ. Co., Weinheim, West Germany.
Sher, H. N. 1961. Manufacture of bakers' yeast by continuous fermentation. II. Instrumentation. *Soc. Chem. Ind. (London) Monograph* **12**:94-115.
Shishatskii, Y. I., and G. A. Bocharova. 1973. Vacuum drying of baker's yeast (in Russian). *Izv. Vyssh. Uchebn. Zaved Pishch. Tekhnol.* **5**:73-77.
Sikyta, B. 1983. *Methods in Industrial Microbiology*. Ellis Horwood Ltd., Chichester, Sussex, U.K.
Sreekrishna, K., and R. C. Dickson. 1985. Construction of strains of *Saccharomyces cerevisiae* that grow on lactose. *Proc. Natl. Acad. Sci. USA* **82**:7909-7913.
Stineman, T. L., J. D. Edwards, and J. C. Grosskopf. 1980. Production of baker's yeast from acid whey. U.S. Patent 4,192,918.
Strohm, J. A., and R. F. Dale. 1961. Dissolved oxygen measurement in yeast propagation. *Ind. Eng. Chem.* **53**:760-764.
Stros, F., Z. Caslavsky, and I. Tomisek. 1968. The development of turbine aerators for the aerobic growth of yeast in Czechoslovakia (in Czech). *Kvasny Prum.* **14**(5):109-112.
Suomalainen, H. 1963. Changes in cell constitution of baker's yeast in changing growth conditions. *Pure Appl. Chem.* **7**:634-654.
Suomalainen, H. 1975. Some enzymological factors influencing the leavening activity and keeping quality of baker's yeast. *Eur. J. Appl. Microbiol.* **1**:1-12.
Trivedi, N. B., G. K. Jacobson, and W. Tesch. 1986. Baker's yeast. *CRC Crit. Rev. Biotechnol.* **24**:75-109.
Van Horn, D. R. 1989. Cream yeast. *Proc. Am. Soc. Bakery Eng.* pp. 144-153.
Volkova, G. A., V. I. Drobot, and I. M. Roiter. 1974. Use of liquid yeast concentrate in bread baking. *Kharchova Prom.* **6**:32-34.

Volkova, G. A., and I. M. Roiter. 1973. Changes in the quality of yeast cream during storage (in Russian). *Khlebopek. Konditer Prom.* **11**:13-16.

Wang, H. Y., C. L. Cooney, and D. I. C. Wang. 1977. Computer aided baker's yeast fermentation. *Biotechnol. Bioeng.* **19**:69-86.

Wang, H. Y., C. L. Cooney, and D. I. C. Wang. 1979. Computer control of baker's yeast production. *Biotechnol. Bioeng.* **21**:975-995.

Whaite, P., S. Aborhey, E. Hong, and P. L. Rogers. 1978. Microprocessor control of respiratory quotient. *Biotechnol. Bioeng.* **20**:1459-1463.

White, J. 1954. *Yeast Technology.* Chapman and Hall, London.

Williams, D., P. Yousefpour, and E. M. H. Wellington. 1986. On-line adaptive control of a fed-batch fermentation of *Saccharomyces cerevisiae*. *Biotechnol. Bioeng.* **28**:631-645.

Woehrer, W., and M. Roehr. 1981. Regulatory aspects of baker's yeast in aerobic fed-batch cultures. *Biotechnol. Bioeng.* **23**:567-581.

Zikmanis, P. B., S. I. Auzane, R. V. Kruce, L. P. Auzina, and M. J. Beker. 1983. Interrelationship between the fatty acid composition and metabolic pathways upon dehydration-rehydration of the yeast *Saccharomyces cerevisiae*. *Eur. J. Appl. Microbiol. Biotechnol.* **18**:298-302.

Zikmanis, P. B., S. I. Auzane, L. P. Auzina, M. V. Margevicha, and M J. Beker. 1985. Changes of ergosterol content and resistance of population upon drying-rehydration of the yeast *Saccharomyces cerevisiae*. *Appl. Microbiol. Biotechnol.* **22**:265-267.

CHAPTER
7

# USE OF YEAST IN BAKING

Cereal grains were an important source of food before the art of bread baking was known. Gruel or porridge can be made from mixtures of coarsely ground wheat, corn, oats, or other grain and water. Such pastes can be cooked or they can be baked on hot stones or in hot ashes. The Mexican tortilla, the Scandinavian flat breads, Jewish matzoh, and all kinds of pancakes are modern versions of these earlier products.

If such cereal mashes are kept for several days they are likely to ferment naturally, generally with an alcoholic fermentation by yeast and a lactic acid fermentation by bacteria. However, these fermentations do not proceed as regularly or as rapidly as the spontaneous alcoholic fermentation of crushed grapes. Probably for this reason a process was invented that improved the regularity and rapidity of the fermentation. It consists of the retention of a portion of a well-fermenting dough so that the portion can be mixed with more ground wheat or flour and water for a succeeding dough. In this manner each dough was inoculated with the active principle of the preceding dough. During the fermentation period the yeasts and bacteria multiplied so that one-third to one-sixth of a dough was sufficient for inoculation of the next dough. This type of inoculation from dough to dough is still practiced today, for instance, in China for the preparation of steam bread in the home or in the United States in the commercial production of San Francisco sour dough bread. Obviously, this type of propagation of a ferment does not depend on the recognition of the microbial nature of the active principle.

Such natural fermentations provided a baked product with improved flavor and texture: *bread*. Doughs made from any cereal can undergo such fermentation, but only ground wheat or wheat flour doughs have the cohesion that permits retention of the leavening gas: carbon dioxide. Such pieces of leavened dough cannot be baked on hot stones but require construction of an oven. It is likely that the art of baking originated in Egypt where the following three conditions were met: the availability of wheat; the discovery of the art of fermentation; and the ability to construct baking ovens. The baking of bread was certainly known four thousand to six thousand years ago in Egypt, but the invention of the art of baking may well have been made in prehistoric times. The history of baking has been described by Jacob (1944) in a book that is as charming as it is informative, and more recently by Darby, Grivetti, and Ghalionngui (1972).

## FUNCTION OF YEAST IN BAKING

### Leavening Gases

Leavened baked goods are preferred in all countries in which wheat is available as a diet staple. There are many ways in which such leavening can be obtained. Foam in the form of whipped egg whites may be mixed directly with the dough, as in the production of angel food cake. Water vapor (steam) that develops during baking at high temperatures has considerable leavening action in the production of Scandinavian flat breads and soda crackers. Other methods have been suggested but are not presently used. These are the forcing of air or carbon dioxide into the dough by mixing under pressure and by expansion of the gas when the pressure is removed, or the combined use of catalase and hydrogen peroxide to use oxygen as the leavening gas (Selman 1953).

In practice, yeast and baking powder are the important leavening agents for baked goods. Commercial baking powders for use in bakeries contain 30% sodium bicarbonate. The acid components of the powder that release the carbon dioxide are sodium acid pyrophosphate, monocalcium phosphate, sodium aluminum phosphate, and glucono-delta-lactone: 100 g of baking powder release about 15 g or 340 mM or 8.2 L of carbon dioxide gas. The rate of $CO_2$ release can be regulated by the choice of the leavening acid as well as by the temperature at which the acid is released. A portion of the gas evolves at the temperature of the dough (bench action) and a portion is released during baking.

Table 7-1 compares the leavening action of yeast with that of baking powder at commonly used percentages (Reed and Peppler 1973). Evolution of $CO_2$ by yeast is shown for a 1-hour period. The formation of the leavening gas may, of course, proceed much longer if sufficient ferment-

Table 7-1. Leavening Action of Yeast and Baking Powder

|  | Yeast | Baking Powder* |
|---|---|---|
| Leavener based on flour | 2.5% | 6.0% |
| Leavener based on dough weight | 1.47% | 3.42% |
| $CO_2$ evolved per g leavener | 0.5 g[†] | 0.15 g[‡] |
| $CO_2$ evolved per 100 g dough | 0.735 g[†] | 0.513 g[‡] |
| $CO_2$ evolved per 100 g dough | 350 ml[†] | 214 ml[‡] |

*A double-acting baking powder containing 30% $NaHCO_3$.
[†]$CO_2$ evolution per hr.
[‡]Total $CO_2$ evolution.

able sugar is available. Release of the gas by baking powder is a great deal faster during baking, but once it has been released there is no further leavening action.

Yeasts are not efficient in some baked goods. These are products that have a very high osmotic pressure (cakes, cookies) or that are leavened for a very short period (muffins, pancakes, waffles). For some baked goods either yeasts or baking powder may be used. These are doughnuts, coffee cakes, and pizza doughs. Most pizza doughs are now leavened by yeast.

Bread may be leavened with baking powder, which is sometimes necessary where yeast fermentation is not practicable, for instance, during military operations or for use by sportsmen. In such instances, glucono-delta-lactone should be used as leavening acid. The baked loaf will have a grain and texture resembling yeast-raised bread, but the flavor will be flat and cereal-like, and the bread will lack the aroma characteristic of yeast fermentations.

At 30°C and a pressure of 760 mmHg about 125 mg of $CO_2$ is soluble in 100 ml of water. The amount actually soluble in 100 g of dough can be only roughly estimated because dough ingredients such as salt and ethanol affect its solubility. This amount corresponds to about 7.5% of the total $CO_2$ produced in 1 hour. The dissolved gas becomes available for leavening during baking. It and the expansion of water vapor trapped in the dough contribute to the expansion of the dough during baking, the so-called oven spring.

## Yeast Strains and Their Evaluation for Use in Baking

Yeasts have three functions in baking. They produce the leavening gas that expands the dough; they affect the rheological properties of the dough;

and they contribute the typical fermentation flavor of yeast-raised products. The ability to produce leavening gas under practical conditions (dough composition, pH, temperature, osmotic pressure, etc.) is the most important function. Usually it is the only one considered in strain selection. The effect of yeasts on the rheological properties is assumed to depend on a lowering of the pH by $CO_2$, on ethanol formation, and on the mechanical action of expanding gas bubbles. It is difficult to assess these factors, and there is no reason to believe that yeast strains differ in these properties. Therefore, they are not considered in strain selection. Flavor formation is an important characteristic, but with some notable exceptions strains do not differ materially in the production of compounds that contribute flavor. Thus, the ability to ferment sugars anaerobically is the major criterion for strain selection.

The development of suitable strains is based largely on traditional methods of hybridization or mutation. Hybridization via spores has been largely replaced by protoplast fusion, which avoids the problem of poor sporulation and poor spore viability of industrial yeast strains. Construction of new strains by the techniques of genetic engineering has been attempted but presents problems. There is no detailed knowledge of the genes responsible for the particular properties of industrial strains, on their copy number or allelic forms (Trivedi and Jacobson 1986). In addition the ability of a strain to ferment glucose anaerobically through the glycolytic cycle depends on a multiplicity of genes.

Difficulties in methods of selection are common to all methods of strain modification. Strains cannot be selected on the basis of a single or on only a few characteristics in Petri dishes or shake flasks. They require evaluation in model bake systems. Thus, they have to be grown aerobically to a minimum of at least 1 gram of yeast solids for each test with 100 grams of dough.

A fairly large number of patents and publications describe the use of specific yeast strains suitable for the production of active dry yeasts. One of the earliest is a patent by Langejan and Khoudokormoff (1976) protecting use of two specific strains obtained by protoplast fusion for the production of instant active dry yeast. Other specific strains also obtained by protoplast fusion have been patented by Jacobson and Trivedi (1987). Legman and Margalith (1983) reported on interspecies fusion of *Saccharomyces cerevisiae* and *Saccharomyces mellis* to produce osmotolerant yeasts. Japanese workers obtained artificial diploids from the haploid yeast *Torulospora delbrueckii* (syn. *Saccharomyces rosei*) for use as a frozen dough leavener (Sasaki and Oshima 1987). These are just a few examples of strain improvement work. They will not be discussed further because it is usually impossible to determine whether commercially available yeasts are produced from one of the strains described in the literature or in patents, and much of the usefulness of a particular strain depends on the manner in which it is grown.

## Fermentable Sugars in Flour and Doughs

In grape musts and in brewer's worts all of the fermentable sugars are already available in the raw material. In contrast, flour contains only 1-2% of readily fermentable sugar (Friedemann, Witt, and Neighbors 1967). These sugars are glucose, fructose, sucrose, maltose, raffinose or melibiose, glucodifructose, and a series of polysaccharides composed of fructose and glucose, the glucofructosans. The monosaccharides and disaccharides are fermented by baker's yeast, but only some of the polysaccharides are fermentable. The concentrations of the individual sugars have been widely reported in the literature, but there is little agreement between authors either because of differences in flours or because the analytical methods are not sufficiently precise at concentrations of 0.01-0.2%. D'Appolonia et al. (1971) and Reed and Peppler (1973) compiled lists of such literature values. It need only be recognized that the total amount of fermentable sugar preformed in flour is generally no more than 1%.

Additional fermentable sugar becomes available as soon as the dough is mixed through the action of $\alpha$- and $\beta$-amylases on "damaged" starch, which refers to the portion of starch that is mechanically ruptured by milling so that it is subject to enzymatic hydrolysis. Flour contains a sufficient concentration of $\beta$-amylase, and the hydrolysis yields maltose. Flour milled from sound wheat, that is, flour that has not suffered sprout damage, does not contain enough $\alpha$-amylase. Therefore, $\alpha$-amylase preparations are generally added to flour at the mill or to doughs at the bakery in the form of barley malt or fungal enzymes (Kruger and Reed 1988).

The rate of maltose formation in doughs is difficult to judge because this sugar is fermented by yeast at rates that vary depending on the concentration of other sugars. Figure 7-1 (*top*) shows the presence of maltose, fructose, and glucose during the fermentation of a sponge and dough. At the end of the 4-hour sponge fermentation the maltose concentration is near zero. The very small concentrations of glucose and fructose are exhausted after about 1 hour. Glucose is fermented somewhat faster than fructose, which indicates that baker's yeast is a glucophilic yeast just as most wine yeast strains are. During the remix of the dough 5% sucrose (based on weight of flour) has been added. Sucrose is hydrolyzed by yeast sucrase (invertase) almost instantaneously. Additional maltose accumulates because its rate of fermentation is inhibited by the monosaccharides.

Figure 7-1(*bottom*) shows the presence of sugars during a straight dough fermentation. Here 5% sucrose has been added together with all other

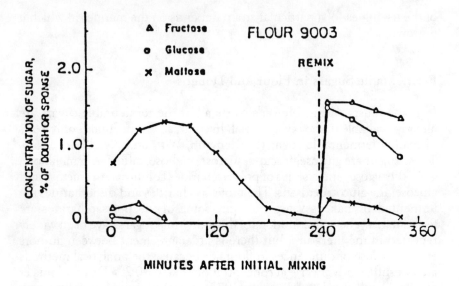
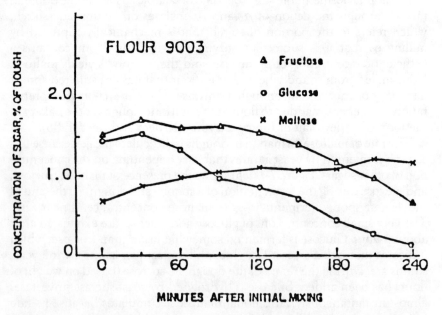

**Figure 7-1.** Concentration of sugars during dough fermentations. *Top:* Sponge dough with sucrose added at the remix stage. *Bottom:* Straight dough with sucrose added at the start. (*From Tang et al. 1972*)

ingredients in the dough. The increased accumulation of maltose is readily observed. The fast rate of glucose fermentation (compared to fructose fermentation) is again evident. Consequently, the concentration of residual fructose in bread is greater than that of glucose if sucrose is used as added sugar (Tang et al. 1972). The addition of $\alpha$-1,4-amyloglucosidase that is contained in almost all fungal amylase preparations increases the rate of maltose fermentation, presumably by the hydrolysis of maltose to glucose (Suomalainen, Dettweiler, and Sinda 1972). The rate of maltose formation can be observed in unyeasted doughs, that is, in doughs that do not ferment, as shown in Figure 7-2 (Oura, Suomalainen, and Viskari 1982).

It is instructive to relate the formation of maltose and the concentration of added sugars to the rate of carbon dioxide production. This relationship

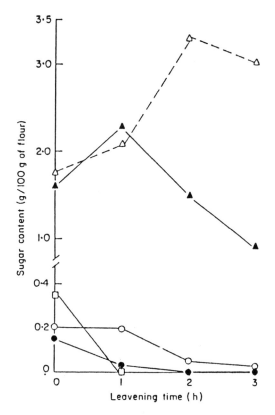

**Figure 7-2.** Concentration of sugars during straight dough fermentations: □ = sucrose; ● = glucose; ○ = fructose; and ▲ = maltose in a yeasted (fermenting dough), and △ = maltose in a nonyeasted (nonfermenting) dough. (*From Oura, Suomalainen, and Viskari 1982*)

## 322 USE OF YEAST IN BAKING

is shown for sponges and doughs in Figure 7-3. In actual bakery practice no sugar is added to the sponges shown in Figure 7-3. The rate of gas production increases rapidly as soon as the sponge is mixed. It reflects the fermentation of preformed flour sugars. The rate drops after about 1 hour when these sugars are exhausted. The rate increases again when maltose fermentation sets in. Continued maltose formation sustains the fermentation for at least the next two hours. But about 3.5 hours after mixing, maltose is not formed as fast as it can be fermented, and the rate of gas

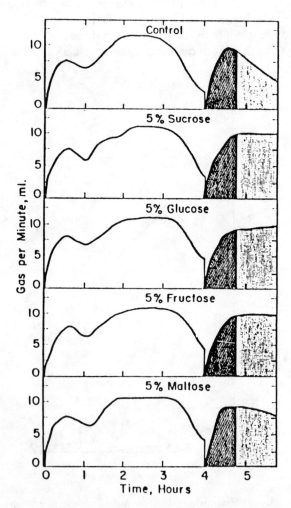

**Figure 7-3.** Rate of gas production in sponges (0-4 hr) and in doughs (4-5.7 hr) containing 2% yeast (total flour basis), fermented at 30°C. The sugars were added at the dough stage. (From Lee, Cuendet, and Geddes 1959)

production drops sharply. This time corresponds in the bakery to the time at which the volume of sponge in the trough shrinks rapidly.

During remix, 30% of the total amount of flour is added. It provides a new source of maltose that sustains the fermentation during the next 40 minutes (floor time), but it diminishes again during the next 60 minutes (proof time) as shown for the "control." Addition of sugars during remix as practiced for commercial sponge doughs permits adequate gassing rates during the floor time and proof periods. These rates are shown in Figure 7-3 for the addition of 5% glucose, sucrose, fructose, and maltose (Lee, Cuendet, and Geddes 1959). The above considerations apply to the relation of sugar concentration and gas production in sponge doughs and straight doughs. Their relationship in liquid ferments, lean doughs, and sweet doughs will be discussed when these processes are considered.

**Effect of pH**

The pH of the yeast cell protoplast is usually reported to be 5.8. There are differences in pH between different locations covering a pH range of 5.1 to 6.3. Yeasts maintain their internal pH quite well in solutions with pH ranges from 3 to 7. The rate of yeast fermentation between pH values of 4 and 6, that is, over a 100-fold range of H ion concentrations, is almost constant. Below a pH of 4 it drops off sharply, and it decreases more slowly on the alkaline side of this range (Garver, Navarine, and Swanson 1966; Franz 1961).

In traditional baking processes, such as sponge doughs and straight doughs (or lean and sweet doughs), the pH is generally between 4.8 and 5.5, that is, within the optimum range for fermentation. Figure 7-4 shows the changes in pH during a typical sponge dough fermentation. The increase in pH upon remix due to the addition of flour and during baking due to loss of volatile acids is marked. There are, however, some notable exceptions to this picture. In liquid preferments that are not well buffered or not buffered at all the pH may well drop to 4 or below and result in much lower fermentation rates. The role of yeasts in sour dough processes will be discussed separately in a later section of this chapter.

There are considerable differences in the effect of pH values below 4.2 on various strains of *S. cerevisiae*. Baker's strains of the species ferment doughs more slowly at pH values below 4.4 and cannot be used at pH levels below 4. In contrast, wine yeasts of the same species readily ferment grape juice of pH 3.2, although somewhat slower than at pH 3.8. Such wine yeasts can also leaven bread doughs at normal pH levels (4.5-5.5) with a fermentation rate only 20% slower than that of baker's yeast strains.

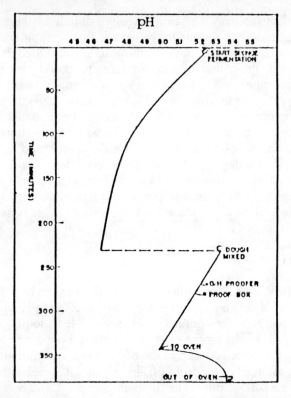

**Figure 7-4.** Changes in pH of a sponge dough fermentation. The pH drops due to the formation of $CO_2$ and organic acids. During baking the pH rises as $CO_2$ is driven off. (*From Selman 1953*)

### Effect of Temperature

The vegetative cells of baker's yeast are quickly killed at temperatures exceeding 50°C. Figure 7-5 shows that a 95% kill of cells occurs in 45 minutes at 48°C, in 18 minutes at 50°C, and in 6 minutes at 52°C. This fact is important because it affects the rate and extent of yeast fermentation during the early phase of baking. This phase is when oven spring occurs. The loaf expands rapidly, partly due to additional $CO_2$ formation by yeast, but also due to the expansion of gases ($CO_2$ and water vapor) and the driving out of dissolved $CO_2$ and alcohol. The role of the individual factors has not been clarified, and one can only guess at the contribution of yeast to oven spring. Figure 7-6 shows the internal temperature of a dough

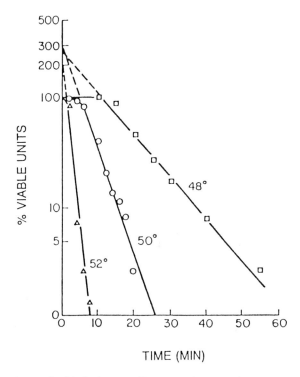

**Figure 7-5.** Survival of baker's yeast (*S. cerevisiae*) exposed to various temperatures (semilogarithmic scale). (*From Van Uden 1971*)

during baking. As the temperature of the dough increases, so does the rate of yeast fermentation as well as the rate of yeast kill.

The temperature of doughs or liquid preferments affects the maximal fermentation rate as well as the time period required to attain that rate. Table 7-2 shows data by Garver, Navarine, and Swanson (1966), one of the very few publications that expresses fermentation rate meaningfully on a basis of mM $CO_2$ production per g of yeast solids. By multiplying mM $CO_2$ by 0.088 one can readily calculate the amount of sugar (as glucose) fermented per hr/g of yeast solids. The data in Table 7-2 cover the range of temperatures commonly observed in fermenting doughs. It is a much narrower range than prevails in the wine or brewing industries. But it must be remembered that temperatures of doughs may increase by 4°C to 9°C during a 3-4-hour sponge fermentation. Table 7-2 shows an increase of 25% of the fermentation rate for an increase in temperature from 29°C to 33.5°C. At temperatures above 38°C the effect of yeast kill is overriding.

## 326  USE OF YEAST IN BAKING

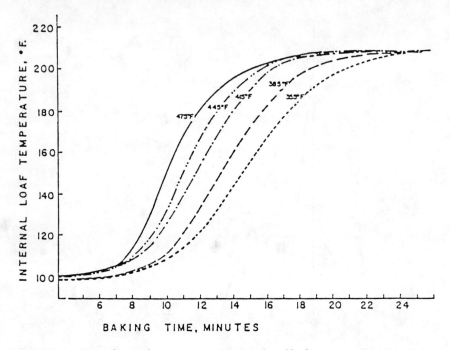

**Figure 7-6.** Rate of internal temperature increase in a bread loaf at various oven temperatures.

———— 475°F
●●——●● 445°F
●—●— 415°F
— — 385°F
- - - - 365°F

(*From Ponte, Titcomb, and Cotton 1963*)

Table 7-2.  Effect of Temperature on Gas Production Rate of Preferments

| Temperature in °C | Maximum Gas Production Rate in mM $CO_2$ per hr per g of Yeast Solids | Time Needed to Reach Maximum Gas Production Rate in min |
|---|---|---|
| 29 | 20 | 150 |
| 31 | 23 | 135 |
| 33 | 24.5 | 135 |
| 35.5 | 25 | 120 |
| 38 | 26 | 90 |
| 40 | 22.5 | 75 |
| 42 | 20 | 30 |

*Source:* Modified from Garver, Navarine, and Swanson 1966.

FUNCTION OF YEAST IN BAKING 327

Temperature effects during various processing techniques will be treated in the process section of this chapter.

## Osmotic Pressure

Yeast fermentation activity is inhibited at higher concentrations of sugars or salt. This fact was well known to bakers before various aspects of physical chemistry were considered important. It is the reason why sweet doughs require higher yeast concentrations and/or longer proof times than doughs with nominal levels of added sugars. Figure 7-7 shows the effect of variable sugar concentrations on fermentation rate, and Figure 7-8 shows the effect of variable salt concentrations. The inhibiting effect has traditionally been ascribed to the osmotic pressure exerted by solutes of low molecular weight. It is, however, quite difficult to determine the osmotic pressure in a dough.

The freezing point depression is most difficult to measure as shown in Figure 7-9. There is obviously an inflection in the freezing curve, but its exact location is indeterminate. An attempt could be made to calculate the osmotic pressure from the known composition of the bread dough; at least the contribution of salt and sugar to osmotic pressure could be calculated since these compounds are the principal low-molecular-weight solutes.

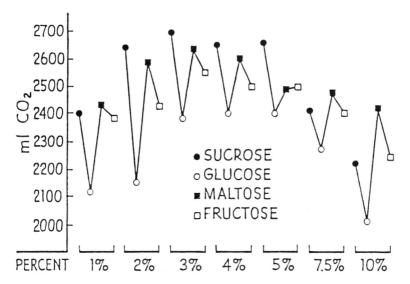

**Figure 7-7.** Effect of various sugar concentrations on the gassing activity of yeast fermenting sucrose, glucose, maltose, and fructose. (*From Schultz 1965*)

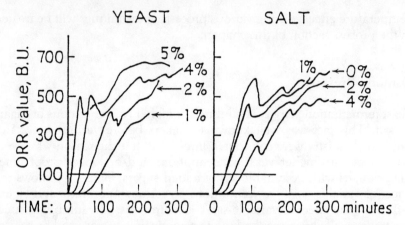

**Figure 7-8.** Gas production of doughs containing various concentrations of yeast and salt determined with the Oven-Rise recorder. (*From Marek and Buschuk 1967*)

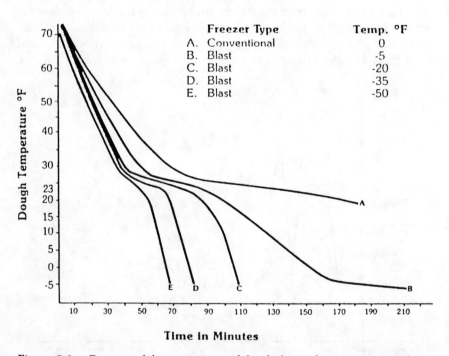

**Figure 7-9.** Decrease of the temperature of dough during freezing at various freezer temperatures and for two types of freezers. (*From Lehmann and Dreese 1981*)

FUNCTION OF YEAST IN BAKING 329

However, the concentration of free water in the dough is not known; some of the water is assumed to be bound to flour components. As expected, an increase in the water content of a dough leads to a greatly increased rate of fermentation.

Figure 7-10 shows the results of gassing power tests when the percentage of water was varied from 40% to 200% (based on the weight of flour) (Shogren, Finney, and Rubenthaler 1977). While the authors explain these results on the basis of additional yeast nutrients leached from the flour, it is fairly certain that the reduced osmotic pressure at higher water concentrations is responsible for the increased gassing power.

An attempt to relate the rate of yeast fermentation directly to varying osmotic pressures caused by four low-molecular-weight substances is shown in Figure 7-11. The straight lines for the two metabolized compounds, glucose and sucrose, are identical; and so are the lines for the two substances that are not metabolized, sodium chloride and xylose. One can certainly conclude that the inhibition of the fermentation at higher concentrations of metabolized and nonmetabolized solutes is indeed due to

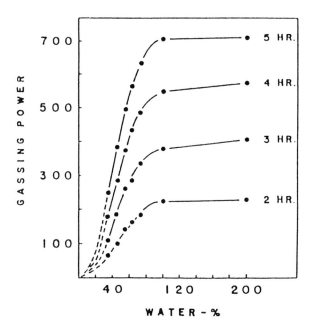

**Figure 7-10.** Gassing power of yeast (in mm Hg pressure) as a function of the percentage of water in the dough (2.75% yeast and 2-, 3-, 4- and 5-hr fermentation periods). (*From Shogren, Finney, and Rubenthaler 1977*)

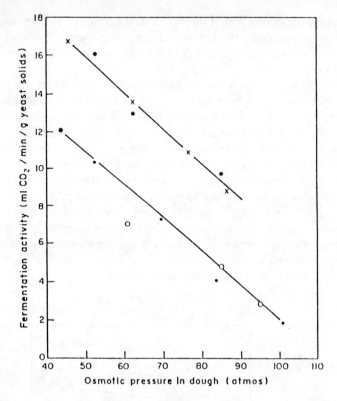

**Figure 7-11.** Yeast fermentation rate in doughs containing sucrose (●), glucose (x), xylose (•), or sodium chloride (○) as a function of the osmotic pressure (*From Chen and Chigger 1985*)

their osmotic pressure. The use of osmophilic yeasts for the leavening of doughs with high osmotic pressures will be taken up in the section on sweet doughs.

**Fermentation Activators and Inhibitors**

There is little if any growth of baker's yeast in doughs. However, the addition of low concentrations of a nitrogen source as a yeast nutrient reduces proof times. So-called yeast foods generally contain both yeast nutrients and oxidizing agents. Bakers often add them to doughs to satisfy the requirement for oxidants. The concentrations of ammonium salts that serve as yeast nutrients are then present in excess. Yeast foods added at a level of 0.5% based on flour contribute 0.05% of either ammonium chloride or ammonium sulfate. Such levels generally reduce proof times by about 5 minutes (Reed 1972).

Mold inhibitors are commonly added to bread doughs particularly if the bread is sliced before packaging. The commonly used inhibitor is calcium propionate, and an average rate of usage is 2 to 3 oz per 100 lbs. of flour (0.125 to 0.19%). Other mold inhibitors in use are sodium diacetate and vinegar. At nominal levels these additives have a mild but perceptible effect by extending proof times. The degree of inhibition reported in the literature is highly variable, probably because of differing pH values in the doughs. The effect of ethanol and of the carbon dioxide formed during the fermentation has already been discussed. The effect of low pH values in slowing the fermentation will be discussed in connection with the preferment process.

### Determination of Yeast Activity

The two ingredients in baked goods that show some variability are flour and yeast. Variability in yeast performance may be due to poor processing or to the perishable nature of compressed yeast. Bakers expect a high fermentation rate and good uniformity from delivery to delivery. Uniformity is by far the most important criterion of quality, although it is not always so treated.

There are two schools of thought on suitable methods of determining fermentation activity. The first demands that the method reflect actual bakery use as closely as possible. In one such test, two 1-pound loaves of bread were produced by the straight dough method (Peppler 1972). In this test, either the volume of the baked bread for a set proof time is measured, or the proof time required to attain a given height in the pan is measured.

Other bakery processes may, of course, also be carried out in the laboratory with the possible exception of the continuous dough-mixing process. Such tests have also been miniaturized by producing "pup" loaves that require only 100 grams of flour for each test. All of these bake tests have important disadvantages. They are imprecise. The results depend on the manual skill of the operator and on the proper functioning of all of the equipment, such as mixers, proof cabinets, molders, and ovens. Second, the tests reflect only a particular bakery operation, although the yeast may be used in a variety of processes.

The second school of thought suggests evaluation of the fermentation activity by a simple gassing test, that is, by measuring the amount of carbon dioxide produced in a given time period. Simple fermentometers for measuring fermentation rate in various sugar solutions have been described by Schultz, Atkin, and Frey (1942) for the evaluation of baker's yeast, and by Reed and Chen (1978) for the evaluation of active dry wine yeasts. With these and similar devices one can determine the formed $CO_2$ by titration, by volumetric determination, or by measurement of the pressure in a restricted volume. These simple tests have a serious disadvantage. They neglect the effect of osmotic pressure on yeast fermentation activity that can best be determined in a dough. In effect, neither bake tests that simulate bakery conditions nor measurement of gassing activity in solution are reliable.

Thus, it can be concluded that gassing power determinations with yeasted doughs must be used to determine the fermentation activity of baker's yeast. Such a procedure neglects any effect that the yeast may have on the gas-holding properties of doughs since total $CO_2$ production is measured. This problem is not a serious one with compressed yeasts since these yeasts do not differ in their effect on gas retention unless they have been autolyzed. With active dry yeast the effect of leached solids may have to be determined additionally (see Active Dry Yeast section later in this chapter).

The approved method of the American Association of Cereal Chemists for determining gas production is based on a yeasted lean dough. Shogren, Finney, and Rubenthaler (1977) have correctly pointed out that the test measures amylase activity of the flour. It is certainly not suitable for measuring yeast fermentation activity since availability of fermentable sugars may be limiting. The authors suggested use of a dough containing commonly used ingredients including 6% sugar. They evaluated gassing activity with pressure cups equipped with pressure gauges (National Manufacturing Co., Lincoln, Nebraska).

In 1980, Rubenthaler et al. reported on a newly designed instrument for measuring $CO_2$ production. The formula used contained sufficient sugar to sustain the fermentation. $CO_2$ was measured volumetrically. The instrument permits simultaneous determination of 12 samples in individual channels and automatic recording of the results. Figure 7-12 shows such a recording of six doughs containing 0%, 1%, 2%, 3%, 5%, and 10% yeast, respectively. Other instruments that permit automatic recording of gas evolution have been available for some time, for instance, the Swedish SJA Fermentograph.

Recently a method has been proposed for the estimation of yeast activity by measuring $CO_2$ formation in straight doughs, sweet doughs, and lean doughs. The method prescribes standard formulations for these doughs, including flour, water, yeast, sugar, salt, skim milk powder, and shortening, but without other additives. The yeast concentrations are adjusted so that 100 grams of dough contain 0.3509% of yeast solids (from compressed yeast or active dry yeast). Equal amounts of doughs are used in

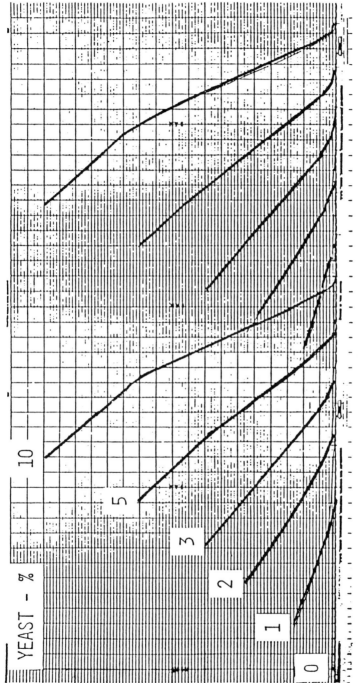

**Figure 7-12.** Replicated $CO_2$ production curves recorded with an automatic recording instrument (gasograph) for doughs containing 0, 1, 2, 3, 5 or 10% yeast. (*From Rubenthaler, Finney, and Demaray 1980*)

each test so that the quantity of yeast solids remains constant, which permits expression of the results as ml $CO_2$ per hr/g yeast solids, and cross-comparison between various lean, straight, and sweet dough formulae as well as cross-comparison between laboratories. The test is conducted during a 90-minute period at 30°C. Choice of the weight of the dough piece and choice of the instrument for measurement of $CO_2$ evolution are left to the discretion of the testing laboratory, but results are comparable since they are expressed on a common basis: the weight of the used yeast solids. The proposed method is subject to evaluation by collaborative studies (American Association of Cereal Chemists 1988).

## BREAD FLAVOR

The flavor and aroma of fermented yeast makes baked bread more popular than many wheat-derived products. Some individual components contributing to the overall flavor and aroma have been identified, but the total number of compounds contributing to flavor or thought to contribute to flavor is vast, and their interactions are not understood. These comments apply to the subtle fermentation flavor of bread, not to the perception of saltiness or sweetness that is readily regulated by appropriate formulation.

The odors of fermenting grape juice, fermenting beer wort, and fermenting dough are quite similar, which is not surprising in view of the similarity of the compounds formed by yeast fermentation. But the aroma of bread is formed during baking by thermal reactions between these compounds and more importantly between these compounds and other dough constituents such as sugars and amino acids. Baking forms the aroma that distinguishes bread from other products of the anaerobic alcoholic fermentation of sugars. Flavor formation in bread has been reviewed extensively by Maga (1974) and Rothe (1974) and more recently in a chapter by Pyler (1988). The subject will be treated here only in outline.

Yeast itself does not contribute much flavor to bread. Addition of a few percentages of inactive dried baker's yeast produces at most a slightly cooked flavor. At higher concentrations of yeast the content of thiamine and thiamine diphosphate may contribute a somewhat undesirable note. The real precursors to the fermentation aroma of bread are compounds formed during the anaerobic fermentation. Table 7-3 shows one such compilation by Magoffin and Hoseney (1974). These compounds are organic acids, alcohols, aldehydes, ketones, and carbonyl compounds. Esters not listed in the table are, for instance, ethyl acetate, ethyl lactate, ethyl succinate, and ethyl pyruvate (Coffman 1967).

Indeed a similar list of fermentation flavor compounds in wine would

Table 7-3. Compounds Reportedly Produced During Fermentation and/or Baking

| Organic Acids | | Alcohols | Aldehydes and Ketones | Carbonyl Compounds |
|---|---|---|---|---|
| Butyric | Acetic | Ethanol | Acetaldehyde | Furfural |
| Succinic | Lactic | n-Propanol | Formaldehyde | Methional |
| Propionic | Formic | Isobutanol | Isovaleraldehyde | Glyoxal |
| n-Butyric | Valeric | Amyl alcohol | n-Valeraldehyde | 3-Methyl |
| Isobutyric | Caproic | Isoamyl alcohol | 2-Methyl butanol | butanal |
| Isovaleric | Caprylic | 2,3-Butanediol | n-Hexaldehyde | 2-Methyl |
| Heptanoic | Isocaproic | 2-phenylethyl | Acetone | butanal |
| Pelargonic | Capric | alcohol | Propionaldehyde | Hydroxymethyl |
| Pyruvic | Lauric | | Isobutyraldehyde | furfural |
| Palmitic | Myristic | | Methyl ethyl ketone | |
| Crotonic | Hydrocinnamic | | 3-Butanone | |
| Itaconic | Benzylic | | Diacetyl | |
| Levulinic | | | Acetoin | |

*Source:* From Magoffin and Hoseney 1974.

be much longer and would include most of the mentioned compounds. It would identify the same categories and generally the same types of compounds. The odor of a fermenting bread dough can indeed be described as somewhat winey or fruity. *Bread* fermentation flavor is only created upon baking. The internal temperature of the bread crumb during baking does not exceed 99°C, whereas that of the crust reaches 150°-180°C.

At the higher temperatures both sugar caramelization and melanoidin formation by the Maillard reaction proceed. The typical bread flavor is definitely formed in the crust. Removal of the crust soon after baking prevents diffusion of the flavor from the crust into the crumb, and the crumb will have no fermentation flavor. Similarly, bread baked in a microwave oven or steamed bread (dumplings) will have little or no fermentation flavor.

There have been many attempts to produce fermentation flavors that could be used in chemically leavened products. One can indeed produce a fermentation-type flavor by drying a liquid preferment or a fermented dough, but such procedures are not economical since they merely replace one fermentation with another one. Some of the end products of the Maillard reaction have been described and/or identified as part of the bread flavor. Rothe (1974) listed the following: furfural, hydroxymethyl furfural, 2-methyl-propional, 3-methyl butanal, methional, and others. Amino acids can also be reacted with products of the yeast fermentation. Wiseblatt and Zoumut (1963) reacted proline with dihydroxyacetone and obtained a compound with a strong crackerlike flavor. Similar attempts involving free amino acids have been reported in the literature. So far none have led to a practical solution.

## WHITE PAN BREAD TECHNOLOGY

Several processes may be used for the production of white bread and closely related products such as buns and rolls. These processes will only be described in brief outline. This section will also serve as a reference to the processes used in the production of other yeast-leavened baked goods that will be treated in subsequent sections.

### Straight Doughs

Table 7-4 shows representative formulae for the production of white breads as they are used in different processes. The simplest process is the straight dough process in which all of the listed ingredients are mixed and developed in a mixer. All of the mixing processes are batch processes except for the continuous-mix process. However, subsequent operations are always or almost always continuous processes as discussed below.

Table 7-4. Formulations of Commonly Used White Bread Processes*

| Ingredients | Sponge Dough Process | | | Straight Dough Process | Preferment Process[†] | | |
|---|---|---|---|---|---|---|---|
| | Sponge | Dough | Total | | Brew | Dough | Total |
| Flour | 70.0 | 30.0 | 100.0 | 100.0 | 0 | 100.0 | 100.0 |
| Water | 40.0 | 24.0 | 64.0 | 64.0 | 32.0 | 32.0 | 64.0 |
| Yeast, compressed | 3.0 | 0 | 3.0 | 2.5 | 3.5 | 0 | 3.5 |
| Salt | 0 | 2.0 | 2.0 | 2.0 | 0.5 | 1.5 | 2.0 |
| Sugar or sweetener solids | 0 | 8.5 | 8.5 | 8.0 | 2.7 | 7.0 | 9.7 |
| Shortening | 0 | 3.0 | 3.0 | 2.75 | 0 | 2.75 | 2.75 |
| Nonfat dry milk or milk replacer | 0 | 2.0 | 2.0 | 2.0 | 0 | 2.0 | 2.0 |

Source: Modified from Kulp and Dubois 1982.
Note: Ingredients comprising less than 1% are not listed in the above table. All or some of the following may be used: yeast food, fungal protease, L-cysteine, potassium bromate, ascorbic acid, vinegar, monocalcium phosphate, monoglycerides and diglycerides, dough strengtheners, and calcium propionate.
*All ingredients are expressed in lb per 100 lb of flour.
[†]This is for a concentrated water preferment. For other preferment formulations see Table 7-5.

In *dividing* the mixed dough is divided mechanically (by volume) into pieces of suitable weight: generally 18-ounce balls for a 1-pound loaf of bread. In *rounding* the dough pieces are shaped into balls, an operation that stretches the surface of the piece to permit better gas retention. In *intermediate proofing* the rounded dough balls are permitted to "relax" for a rest period of 3-10 minutes to make them again pliable and more extensible. The *molding* process forms the loaf, usually by sheeting, curling, rolling, and sealing of the ends. In *panning* the loaves are deposited in metal baking pans. *Proofing* is the final fermentation stage. The pans are moved continuously through a proof cabinet in which both temperature and humidity are adjustable. Proof time is generally between 45 and 65 minutes; the temperature is between 35°C and 43°C; the humidity is 80-85% REH. During proofing the dough expands to a height often about .5 inch above the rim of the pan.

The pans enter the oven immediately for *baking*. Baking times and temperatures vary greatly depending on dough size and other parameters. For 1 pound white bread they may be 230°C for about 17 to 24 minutes. In the oven there is a further increase in the size of the loaf, the so-called oven spring, which has been mentioned previously. Subsequent operations are the depanning, cooling, slicing, and wrapping. Detailed descriptions of these processes have been published (Matz 1984; Kulp 1988).

The straight dough process is used by most retail bakers and in some wholesale plants. It is generally used in the production of variety breads.

### Sponge Dough Process

This process is commonly used for the production of white bread in the United States. It requires the mixing of a sponge composed of 70% of the total flour, about 62% of the total water, and all of the yeast and yeast food. Mixing is adequate to combine the ingredients into a coherent mass, but there is no attempt to develop the final rheological properties of the dough. This sponge is fermented for a period of 3 to 4 hours.

A sponge leaving the mixer at 25°C reaches a temperature of about 30°C at the end of the fermentation period. No sugar is added to the sponge so that fermentation proceeds with the preformed sugar of the flour and maltose, the latter being formed by enzyme action in the sponge. The leavened sponge collapses after these sugars are exhausted.

Soon thereafter the sponge is used for the mixing of the dough. In this second mixing step all of the remaining flour and water and all of the other ingredients (sugar, salt, milk solids, shortening, dough improvers, etc.) are added, and the dough is fully developed by high-speed mixing. The subsequent processing steps are those listed above for the straight dough process.

The second mixing step of a sponge dough, the so-called remix, is most important for the proper development of a dough. *Development* refers to the desired rheological properties of a dough. These are very important for the gas-holding properties of the dough, for the ability to process the dough through the mechanical equipment, and for the grain and texture of the bread. But they are most difficult to describe. Doughs show retarded elasticity, and terms such as buckiness, extensibility, and slackness are often used. They can be measured with various instruments, but have no precise physical meaning. Yeast fermentation affects these properties, but the effects of fermentation and the physical effects of mixing and shaping are not readily separated. Sponge dough processes provide bread with a fine grain, a desirable texture, and good flavor.

**Liquid Preferment Process**

The processing steps are similar to those of the sponge dough process except that the semisolid sponge is replaced by a liquid ferment that contains no flour or only a portion of the normal sponge flour. The process is often called the brew or the liquid sponge process. About 50% of the plants in the United States use this process for the production of bread. Formulations are more variable than those for the sponge dough process.

The advantages of the preferment process, compared to the sponge and dough process, are a shorter fermentation time and replacement of a solid mixing step with a liquid mixing step with a saving of capital and labor. Details of the process have been described by Dubois (1984), Martinez-Anaya and Kulp (1984), Kulp (1988), and many others. Some comments on the role of yeast in the preferment are needed since preferments provide less tolerance than sponges, which is particularly true for so-called water brews, that is, preferments that do not contain flour. Water brews that are not buffered reach low pH levels at which yeast fermentation activity is inhibited.

Figure 7-13 shows the drop in pH for an unbuffered water brew and for two levels of added buffer. At pH values approaching 4, fermentation activity is inhibited and buffering of a brew dough is essential to provide stability. With the introduction of the liquid preferment process more than 30 years ago buffering was provided by inclusion of high concentrations of skim milk solids in the preferment formula (McLaren 1954). Milk solids are a valuable ingredient in many baked goods, but they are a very expensive source of buffer capacity.

Good buffering can be provided by calcium carbonate. Acids produced by fermentation react with the carbonate to form the calcium salt of the acids and carbon dioxide. Calcium carbonate is only sparingly soluble. It

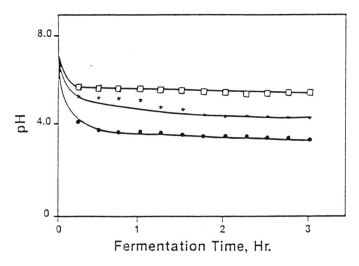

**Figure 7-13.** Change of pH in a liquid preferment during a 3-hr fermentation: ● = no buffer, * = 0.2% buffer, □ = 0.5% buffer. (*From Kulp et al. 1985*)

Table 7-5. Bread Characteristics of Breads Made from Water Preferments, Flour Liquid Preferments, and Sponge Doughs

| Process | Proof Time in min | Spec. Vol. cc/g | pH | TTA | Total Score |
|---|---|---|---|---|---|
| Preferment, water, unbuffered | 64 | 5.09 | 5.6 | 2.4 | 81.5 |
| Preferment, water, buffered | 59 | 5.17 | 5.5 | 2.1 | 82.8 |
| Preferment, 50% flour | 51 | 5.2 | 5.2 | 2.8 | 84.5 |
| Sponge and dough | 61.5 | 5.30 | 5.36 | 3.1 | 85 |

*Source:* From Martinez-Anaya and Kulp 1984.
*Notes:* Preferment temp. 28°C; preferment time 2 hr for water brews and 2.5 hr for flour brew. TTA = total titratable acidity. Total score: Perfect score = 100.

stabilizes the pH at about 4.5. Some of the acids produced during the liquid fermentation may be due to lactic acid bacteria. Robinson et al. (1958) reported the presence of about $10^7$ total bacteria per milliliter during the first 2 hours of a preferment. Most of the bacteria were lactobacilli introduced with the yeast, the flour, and the milk solids.

The addition of 20-50% of the total flour to the liquid preferment can serve as an adequate buffer. It reduces the proof time and has a slight positive effect on specific volume and the total bread score as shown in Table 7-5 (Martinez-Anaya and Kulp 1984). The fermentation time was 2 hours and the temperature was 28°C, which is representative of industrial practice. In bakeries fermentation times from 1 to 2 hours are common. The temperature of setting the preferment is often 28°C. Cooling is provided when the temperature exceeds 33°C. The addition of flour to preferments makes it more difficult to pump them into the mixer. More water is required in the preferment to keep it pumpable.

Another problem may arise from exhaustion of the fermentable sugar in a water preferment. Such ferments contain water, sugar, yeast, a brew buffer, and sometimes salt. For instance, for a preferment containing 7.75% sugar and 8% compressed yeast, only 1.2% sugar was left after a fermentation time of 1-1.5 hours (based on weight of the liquid ferment) (Martinez-Anaya and Kulp 1984). A slight extension of the fermentation time or a slight increase in the temperature would have led to a disappearance of the sugar and a consequent reduction in the fermenting power of the yeast.

Figure 7-14 shows the rapid disappearance of glucose in a water preferment (Kulp et al. 1985). Bakers are used to check pH and titratable acidity of preferments but not residual sugar levels. This situation may lead to the anomalous situation that a yeast that ferments faster in the preferment will lose fermenting power if the sugar is exhausted and will require a longer proof time of the dough.

Sometimes liquid preferments are prepared in concentrated form. That is, they contain only one-half of the total water but all of the yeast and sugar normally used in preferments. These so-called concentrated brews require only one-half of the tank space as single-strength preferments. Of course, they require the addition of a buffer and are subject to exhaustion of sugars just as single-strength preferments are.

Preferments may be preserved for later use in the preparation of the dough to provide flexibility for their use in the bakery. Cooling of the preferments to 4°-16°C permits their preservation for 16-24 hours. Again, with water preferments preservation requires a check of the residual sugar to assure that the yeast is maintained in an active state.

Flour-containing preferments have more tolerance because of the buffering action of the flour and because of the continued formation of maltose during the fermentation. However, they usually require a 2-3-hour fermentation time, whereas water preferments require 1-1.5 hours. It is more difficult to pump flour-containing preferments and to cool them for overnight storage. Table 7-6 shows formulations for liquid preferments with flour and without flour. The performance of yeast in liquid preferments has been treated in two excellent papers by Masselli (1959) and Thorn (1963), and more recently by Kulp et al. (1985) and the previously cited papers by Martinez-Anaya.

WHITE PAN BREAD TECHNOLOGY 341

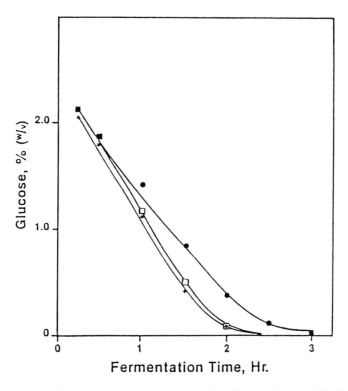

**Figure 7-14.** Depletion of sugar (glucose) in a liquid preferment during a 3-hr fermentation: ● = no buffer, * = 0.2% buffer, □ = 0.5% buffer. (*From Kulp et al. 1985*)

Table 7-6. Liquid Preferment Formulations (%)

| Ingredient | Flour Preferment (I) | Water Preferment (II) | Concentrated Water Preferment (III) |
|---|---|---|---|
| Flour | 40-70 | 0 | 0 |
| Water | 59-64 | 61-66 | 30 |
| Yeast, compressed | 2.5 | 3.5 | 3.0 |
| Sugar (sucrose) | 0.5 | 2.5 | 2.0 |
| Salt | 0 | 0.25 | 0.75 |
| Buffering agent | 0.1 | 0.19 | 0.25 |
| Yeast nutrient | 0.5 | 0.6 | 0 |

(I) and (II) from Uhrich 1975; (III) from Turner 1986.
*Note:* All values are based on flour = 100%.

## Processes with Accelerated Mechanical Dough Development

The development of a dough with suitable rheological properties, that is, suitable extensibility under pressure with a high work input, generally requires a mixing period of less than 1 minute. The very slack dough is extruded directly into pans, which eliminates the dividing, rounding, intermediate proofing and molding steps. Proofing and baking is done by conventional methods.

The development of the continuous mixer led to investigations of the optimum work input required for proper dough development. This input varies with the strength of the flour and formulation but also with the design of the impeller blades of the mixers (Fortmann, Gerrity, and Diachuk 1964). Power consumption during mixing can be measured on a recording watt meter that permits the operator to change the power input as required.

Formulations for the continuous mix process are similar to those for preferment processes, but require higher concentrations of yeast and oxidants. The continuous mix process saves time and labor and eliminates a great deal of equipment. As with all continuous processes it requires a dedicated equipment line and does not easily lend itself to production of a variety of bread types without serious interruption. It produces bread with a relatively weak sidewall and a very fine, almost cakelike, pore structure. At present, continuous processes do not have the prominence they attained in the 1960s and 1970s.

A shift to greater mechanical development of doughs has also occurred in the United Kingdom. The Chorleywood bread process was first described by Chamberlain, Collins, and Elton (1962) and Axford et al. (1963). It is a batch, mechanical dough development system. There is practically no fermentation prior to development in the mixer (no time dough principle), and fermentation takes place only during makeup and proofing. The process is based on the high, measured input of energy during mixing of about 11 W hr/kg. Mixing times are generally between 2 and 4 minutes. The mechanical work input is similar to that required by the continuous mix process (Redfern et al. 1968). The concentration of yeast has to be increased by about 50% above that used in conventional processes in order to achieve acceptable proof times. The process also requires high levels of oxidation (bromate plus iodate, ascorbic acid).

The Chorleywood process is used for at least 75% of the bread production in the United Kingdom, and it is used in many other European countries (Chamberlain 1984). The process has been particularly attractive in the United Kingdom because it permits utilization of flours of lesser strength. It produces a denser bread crumb with a specific volume of 3.8 to 4. North American breads are traditionally less dense with a specific volume of up to 10 (Kilborn and Tipples 1979). The Australian Brimec process was developed independently. It has some similarity with the Chorleywood process. It permits production of bread with normal grain by operating the mixing chamber at atmospheric pressure. Bread with a very fine grain is produced by mixing under pressure and extrusion of the dough directly into baking pans; the procedure resembles the continuous mix process (Marston 1967).

## No-Time Doughs and Short-Time Doughs

The terms *no-time doughs* and *short-time doughs* refer only to the period preceding the mixing step of a straight dough. Doughs start fermenting as soon as a yeasted dough is mixed, and fermentation continues throughout makeup and proofing until about 8 to 10 minutes after the proofed loaves enter the oven. The fermentation time of straight doughs can be considerably shortened by including a reducing compound, cysteine, in the formula. Short-time doughs with floor times (fermentation prior to makeup) of 20 to 30 minutes require increased mixing times and considerably higher concentrations of yeast and oxidants (Shirley 1977). The conditions for achieving optimum bread quality using fermentation time, yeast concentration, oxidant concentration, and proof time as variables have been investigated by Finney et al. (1976). As shown in Table 7-7, fermentation time of a straight dough can be greatly reduced, but at the cost of much higher concentrations of yeast and of oxidants.

## Lean Dough Process

Formulations for the sponge dough, straight dough, preferment, short-time, and continuous mix processes in North America always contain

Table 7-7. Conditions for the Production of Quality Straight Dough Bread by Short-Time Dough Procedures

| Fermentation Time (min) | Yeast Concn. (%) | $KBrO_3$ Concn. (ppm) | Proof Time (min) |
|---|---|---|---|
| 180 | 2 | 0-30 | 55 |
| 120 | 3.5 | 0-45 | 36.5 |
| 70 | 7.2 | 0-90 | 21.5 |
| 45 | 12.0 | 0-180 | 12 |

Source: From Finney et al. 1976.
Note: The range of bromate concentration indicates varying requirements for seven hard red winter and seven hard red spring wheat flours.

Table 7-8. Chorleywood Bread Process Formula

| | |
|---|---|
| Flour | 100 |
| Water | 61 |
| Yeast | 1.8-2.0 |
| Salt | 1.8-2.0 |
| Fat | 0.7 |
| Ascorbic acid | 75 ppm |

*Source:* From Ponte 1971.
Ingredients based on 100 parts flour.

sufficient added sugar to sustain yeast fermentation, generally in the range of 6-8%. Under these circumstances baker's yeasts perform equally well in any of these processes. Thus, there is no yeast that performs well in a straight dough and poorly in a sponge dough or vice versa.

In contrast, lean doughs are characterized by the absence of added sugar in the formula. In the United States, these doughs may be for the production of Vienna, Italian, or French breads as well as for some pizza doughs. In Europe, most of the bread is produced without the addition of sugar or with minimal amounts of added sugar. For instance, a formula used in the Chorleywood process is shown in Table 7-8. For all of these doughs the substrate for yeast fermentation is at first the preformed sugar of the flour, and then maltose formed by amylase action on damaged starch. Figure 7-15 shows the rate of $CO_2$ formation in such a lean dough for various temperatures (Harbrecht and Kautzmann 1967). This figure has frequently been reproduced because it shows clearly the drop in gassing rate after exhaustion of the preformed sugar after about 60 minutes. Obviously, this depletion of the preformed sugar is faster at the highest temperature.

Yeasts that show essentially the same fermentation activity with glucose, fructose, mannose, and sucrose may have quite different rates of maltose fermentation. The monosaccharides readily enter the yeast cell by a process called *facilitated diffusion*. Sucrose is immediately hydrolyzed by yeast invertase outside of the yeast cell membrane, and its constituent monosaccharides then enter the cell readily. But maltose enters the cell by an active transport process by a maltose permease. It was established quite early that these induced transport enzymes are probably the limiting step in maltose fermentation (Robertson and Halvorson 1957). This belief is convincing since there was no correlation between the internal maltase ($\alpha$-glucosidase) activity of the yeast cell and maltose fermentation rate (Suomalainen, Dettweiler, and Sinda 1972).

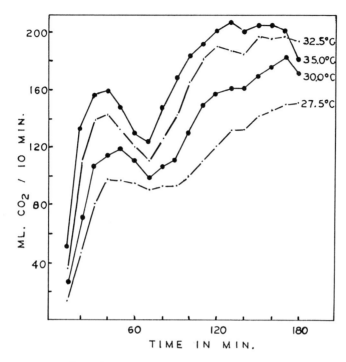

**Figure 7-15.** Rate of yeast fermentation in lean doughs at various temperatures (no added sugar). The indigenous sugars are fermented during the first 40 min. Thereafter maltose is fermented as it becomes available through enzymatic hydrolysis of the starch. (*From Harbrecht and Kautzmann 1967*)

Some baker's yeast strains have constitutive maltose fermenting systems; in others fermentation must be induced by maltose in the medium. Whereas low concentrations of glucose in maltose media induce maltose fermentation, glucose concentrations above 0.1-0.2% repress it.

A consideration of the gassing rate curves in Figure 7-15 shows indeed that substantial maltose fermentation does not set in until the preformed monosaccharides of the flour have been fermented. Induction of the maltose fermenting system in the absence of glucose occurs only slowly as shown in Figure 7-16.

The fermentation of maltose is controlled by a polymeric gene system (Winge and Roberts 1958). Six such genes have now been recognized: $M_1$ to $M_6$. Their functions are to regulate (MAL regulator), to form maltase, and to form maltose permease. The synthesis of these various enzymes is induced by maltose and inhibited by the monosaccharides glucose, fructose, and mannose. Busturia and Lagunas (1985) have shown that the carrier

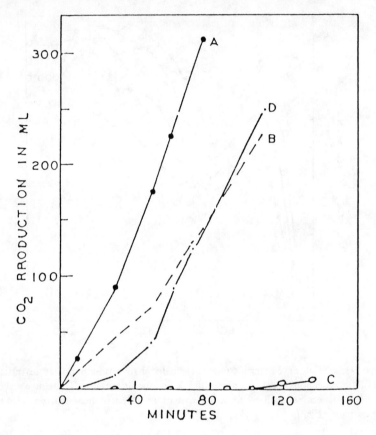

**Figure 7-16.** Effect of added glucose on maltose fermentation: A = 3.0 g glucose; B = 2.8 g maltose plus 0.2 g glucose; C = 3.0 g maltose; D = 3.0 g maltose plus 40 ml maltase extract of yeast. (*From Atkin, Schultz, and Frey 1946*)

system that transports maltose across the cell membrane, the maltose permease, exists in two forms that differ in their maximal velocity and in their affinity for the substrate ($K_m$). Osinga et al. (1989) announced the transformation of baker's yeast by a homologous DNA construct to strains that have increased maltase and maltose permease activity and that indeed show a higher maltose fermentation rate compared with the untransformed strain.

Figure 7-15 has clearly shown the drop in fermentation activity after about 60 minutes when the yeast shifts to maltose fermentation. The same picture can also be seen in Figure 7-8. Here the drop occurs after more than 100 minutes with 2% compressed yeast, and after only 30 minutes with 5% yeast, obviously reflecting the earlier exhaustion of preformed sugars of the flour at the higher yeast concentration. However, at least in the United States, small percentages of sugar are often added to lean dough formulae, which results in a sustained rate of fermentation. Thus, the need has decreased for baker's yeasts with strong constitutive maltose fermenting activity, and in the United States special lean dough yeasts are not marketed at the present time. There is, of course, always some incentive to improve maltose fermenting ability of yeasts because maltose formed from flour is a cheaper source of fermentable substrate than added sugars.

### Sweet Goods

The great variety of so-called sweet goods includes Danish pastry, yeast-raised doughnuts, coffee cakes, sweet buns, and puff paste shells. The important common characteristic, as far as yeast fermentation is concerned, is the relatively high sugar content. It may range from a moderate 6-12% sugar for doughnuts requiring 4-6% compressed yeast in the formula to Danish pastry doughs with 18% to more than 20% sugar requiring 8-10% yeast. The higher concentrations of shortening or eggs in some sweet goods do not require increased yeast concentrations. But high concentrations of sugar that raise the osmotic pressure of the dough inhibit fermentation and hence require higher yeast concentrations.

It has previously been said that yeast strains vary considerably in their sensitivity to osmotic pressure. Typically Japanese compressed yeasts have shown good tolerance to higher sugar concentrations as required by the high sugar levels in some Japanese breads. In the United States, the same compressed yeasts are used in lean, regular, and sweet doughs. However, special "sweet goods" yeast strains are available as instant active dry yeasts. It is a logical development since the good stability of vacuum-packaged active dry yeasts lends itself better to the production, storage, and distribution of several strains.

### Pizza Shells and Pretzels

Doughs for pizza shells and pretzels warrant a short description of their fermentation. Table 7-9 shows a formula for pizza shells that indicates the fairly wide ranges for the concentration of major ingredients as they are encountered in practice (Lehmann and Dubois 1980). Most pizza shells are yeast-leavened, but some bakers prefer chemical leavening. The pliable dough is brought out of the mixer rather warm (32°-37°C). There is no special fermentation time, and the doughs are merely given a rest period of 10-15 minutes. Of course, fermentation begins already during mixing,

Table 7-9.  Pizza Shell Formulae

| Basic | | Optional | |
|---|---|---|---|
| Ingredient | Flour Basis (%) | Ingredient | Flour Basis (%) |
| Flour | 100 | Protease enzyme | as recommended |
| Water | 35.0-70.0 | L-cysteine or | 45.0-90.0 ppm |
| Salt | 1.0-2.0 | Na bisulfite | |
| Sugar | 1.0-5.0 | Corn meal | 10.0-20.0 |
| Shortening or oil | 3.0-14.0 | Sours | 1.0-3.0 |
| Yeast | 0.5-5.0 | Vinegar, 100 grain | 0.5-1.0 |
| Baking powder | 0.5-4.0 | Na stearoyl lactylate | 0.25-0.50 |
| Ca propionate | 0.1-0.3 | Vital wheat gluten | 1.0-2.0 |

*Source:* From Lehmann and Dubois 1980.

and processing of the batch-mixed dough provides additional time. Following the rest period pizza doughs are either stamped out or sheeted and cut. Particularly for the stamping operation the doughs must be soft and pliable. Therefore, reducing agents such as cysteine or sulfite are added. Active dry yeasts that leach some cysteine during rehydration contribute to the pliability of the doughs.

Pretzel doughs require a lean dough formula and only small concentrations of yeast. After mixing and shaping into the well-known pretzel configuration the fermentation proceeds from a few minutes to as long as 30 minutes. The pretzels are then immersed on a continuous wire mesh belt into a hot (90°-93°C) lye bath (about 1.25% NaOH). The time of immersion is very short, about 10-15 seconds. The pretzels are then salted and baked. For the production of the more popular hard pretzels they are kiln-dried to a moisture content of 2-3%, which provides for the crispness of the product (Reed 1974).

### Growth of Yeast in Doughs

It is common practice in the wine and beer industries to determine the number of yeast cells and to follow their growth. This procedure is readily done in the liquid medium by microscopic cell count or by actual plating to determine the number of live cells. It is difficult to count yeast cells in doughs, and few data are available in the literature. To date the best information on sponge doughs and straight doughs was reported by Thorn and Ross (1960). Table 7-10 shows that for sponge doughs there is no significant increase in the number of yeast cells during the sponge

Table 7-10. Budding of Yeast in Doughs

| | Sponge | | Dough | |
|---|---|---|---|---|
| Sponge and Dough Process | Zero-time | 4 hr | Zero-time | 4 hr |
| Active dry yeast | 329(1.2%) | 323(37%) | 214(38%) | 223(38%) |
| Compressed yeast | 434(1.6%) | 456(50%) | 273(51%) | 295(42%) |
| | | | | |
| Sweet Dough Sponge | Zero-time | 3.5 hr | | |
| Active dry yeast | 800(3.4%) | 790(2.6%) | | |
| Compressed yeast | 1090(3.0%) | 1060(5.0%) | | |
| | | End of | End of | |
| Straight Dough Process | Zero-time | Third Rise | Proof | |
| Active dry yeast | 150(2.0%) | 140(16%) | 146(39%) | |
| Compressed yeast | 230(2.0%) | 224(12%) | 232(39%) | |

Source: Adapted from Thorn and Ross 1960.
Note: Cells in millions per gram of dough and percent of budding cells in parentheses.

fermentation, but the number of budding cells increases greatly. During dough mixing and to the end of the proof period there is a small but measurable increase in the total number of cells. (The drop in the number of cells from the end of the sponge period to the beginning of the dough period is entirely due to the dilution effect of the added dough ingredients.) In sweet dough sponges there is no measurable increase in the number of cells or the percentage of budding cells. These conclusions are, of course, based on doughs inoculated with a normal amount of compressed or active dry yeast at cell counts of $150\text{-}1000 \times 10^6$ cells per gram of dough.

At low concentrations of yeast there is considerable growth. For the millenia-old method of inoculating a new dough from a portion of the preceding dough it is obvious that growth of yeast must occur. For instance, there must be a 10-fold increase in the number of yeast cells if 10% of a fermented dough is retained for inoculation of the next dough.

## Frozen Doughs, Yeast-Leavened and Unbaked

Baked breads or cakes are often frozen to preserve their freshness for prolonged storage periods. The freezing preservation of yeast-leavened, unbaked doughs poses problems that are immediately related to yeast activity and yeast preservation in the frozen state.

The installation of bake-off sections in supermarkets permits the sale of freshly baked bread with a minimum of processing and capital investment if frozen doughs are used. The doughs are delivered frozen to the store. There they are thawed in the refrigerator (usually overnight), proofed, and baked. The frozen doughs are generally prepared in a central bakery for delivery to supermarkets or to institutional users such as schools, restaurants, and prisons. This commissary-type operation requires stability of frozen doughs for a period of 4-12 weeks. There is also a market for consumer products requiring a frozen shelf life of 3-6 months. This shelf life is difficult to achieve in practice, and consequently the consumer market has not developed as expected.

Frozen doughs deteriorate upon frozen storage. Proof times increase, loaf volumes decrease, the grain gets coarser, and the texture worsens. These changes are definitely caused by a loss of gassing power of the yeast as can be readily ascertained by a quantitative determination of the $CO_2$ production rate of thawed dough pieces. Many authors have also reported a change in the rheology of the doughs since thawed and reshaped doughs have shown better loaf volumes than doughs that were not reshaped before proofing. For instance, Bruinsma and Giesenschlag (1984) report that remolding of doughs after thawing greatly improves the grain, but contrary results have also been reported.

It must be emphasized that the loss of fermenting activity of the yeast occurs during frozen storage, not during freezing of the dough. Yeast can be frozen and thawed without loss of fermenting activity or with only minimal losses. It can also be stored as frozen compressed yeast for weeks without such loss (Godkin and Cathcart 1949). Wolt and D'Appolonia (1984a) stored commercial compressed yeast in frozen form for 6 weeks without significant loss of gassing power. This early work has been confirmed by commercial practice for shipment of compressed yeast in frozen form (although it is not a routine operation).

In 1968, Mazur and Schmidt showed in a well-documented study that yeasts could be killed by fast freezing. Their tests were carried out in capillaries with freezing rates of several degrees centigrade per second. Commercial doughs are frozen very much slower as shown in Figure 7-9 (Lehmann and Dreese 1981). Depending on the type of freezer, it takes from 1 to 2.5 hours before the dough reaches $-18°C$. Some bakers use cryogenic systems (liquid $N_2$ or $CO_2$) that freeze the dough faster, but most bakers use mechanical refrigeration and strong air blasts to freeze the doughs. The temperature in the primary freezer is usually kept at $-29°C$ to $-40°C$, and the doughs are frozen until the core temperature of the loaf reaches at least $0°C$. At that point, a thick outer layer of the loaf has formed a completely frozen, hard shell. The temperature is then permitted to equilibrate. The loaves are protected against loss of moisture by packaging in plastic bags and kept in frozen storage at $-20.6°C$ to $-23.3°C$.

There are some practical steps that can be taken to minimize the loss of yeast activity during frozen storage of the doughs. Trivedi et al. (1989) have summarized the suggestions of earlier authors as well as practical experience in bakeries as follows:

Increase yeast level by 2-3% above normal.
Minimize fermentation prior to freezing by using a short-time straight dough method, by producing cool doughs, and by reducing the time between mixing and freezing.
Use strong oxidation (30 ppm $KBrO_3$ plus 100 ppm ascorbic acid); use an efficient dough conditioner (Na stearoyl lactylate); increase the level of shortening to 4%; and use a fairly strong bread flour.
Freeze immediately in blast freezers and store between $-15$ and $-21°C$; avoid freeze/thaw cycles during storage and delivery.
Thaw the doughs for at least 8 hours in a refrigerator (retarder).

It has been vigorously debated whether the deterioration of frozen doughs is exclusively due to the deterioration of the yeast. Varriano-Marston, Hsu, and Mahdi (1980) and Wolt and D'Appolonia (1984b) have shown that the rheology of doughs changes upon frozen storage for yeasted as well as for nonyeasted doughs. Figure 7-17 shows the decrease in extensibility of yeasted doughs as the result of frozen storage and as the result of freeze/thaw cycles.

The figure also shows the same phenomenon for the nonyeasted counterparts of these doughs. More specifically, Wolt and D'Appolonia (1984b) demonstrated that the leaching of reducing compounds (glutathione) from yeast was not responsible for these rheological changes. In view of the rheological changes in frozen doughs, it is not surprising that the use of increased levels of oxidants and dough conditioners has a desirable effect on the shelf life of frozen doughs. The use of sodium stearoyl lactylate was recommended by Davis (1981) and Wolt and D'Appolonia (1984a). An increase of the fat concentration to 4% was suggested by Marston (1978).

The major cause of deterioration occurs as the result of fermentation prior to freezing. It is recognized by all authors beginning with Merritt (1960) and by succeeding authors: Kline and Sugihara (1968), Lorenz (1974), Tanaka and Miyatake (1975), and many others. Thus, short-time straight-dough methods give better results than regular straight doughs and sponge doughs, doughs should be mixed as cool as possible, intermediate proof should be minimized, and the dough pieces should be frozen as quickly as possible.

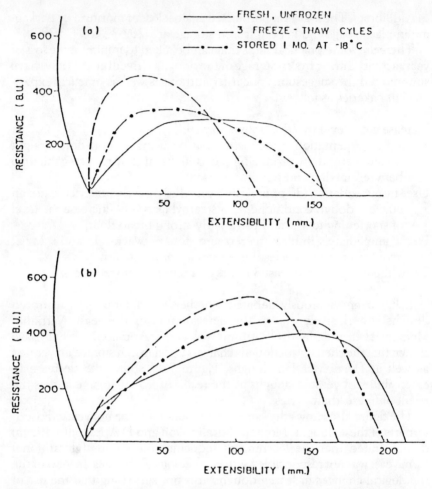

**Figure 7-17.** Effect of freezing and of frozen storage on the extensibility of doughs: extensigrams of yeasted (a) and non-yeasted (b) doughs. (*From Varriano-Marsten, Hsu, and Mahdi 1980*)

At present, there is no satisfactory explanation for the deleterious effect of fermentation prior to freezing. Hsu, Hoseney, and Seib (1979a, 1979b) ascribed the deleterious effect to the end products of fermentation. They assessed yeast damage upon freezing in the presence and absence of a fermentation broth containing the end products of yeast fermentation, and concluded that "When the yeast was frozen together with the fermentation products, damage was major and was particularly severe if the yeast had been activated." Similar results have been reported by Tanaka,

Kawaguchi, and Miyatake (1976). Unfortunately, this promising lead has not been pursued.

Finally, the role of yeast production and yeast strain on performance in frozen dough systems must be assessed. Gelinas (1988) determined the influence of growth conditions of baker's yeast on its cryoresistance. Cryoresistance in this instance was the proof height of a frozen dough as a percentage of that of an unfrozen control. Yeast grown by the fed-batch process had better cryoresistance than yeast grown in a batch process. Even partial oxygen starvation during a fed-batch process lowered the cryoresistance of the yeast.

Hino, Takano, and Tanaka (1987) isolated two yeast strains from banana peel that appear to have excellent cryoresistance and that were not damaged by fermentation prior to freezing. However, tests with frozen doughs were not extended beyond a 7-day storage period. Oda, Ono, and Ohta (1986) screened 300 strains of *Saccharomyces cerevisiae* and selected 11 strains for their improved cryoresistance. All of the selected strains had higher trehalose concentrations than commercial baker's yeasts. These yeasts performed exceptionally well in sweet doughs (30% sugar) after 7 days of frozen storage, but poorly in 5% sugar doughs, always in comparison with unfrozen controls.

It is likely that sooner or later a yeast strain will be found that shows better stability during the storage of frozen doughs. Proper selection of such strains requires extended storage in frozen doughs, not just an evaluation of the resistance of the yeast to freezing.

The recommended increase in the concentration of yeast used for frozen dough production does not require further comment. Such higher concentrations compensate for the longer proof periods of frozen doughs and extend the shelf life to some extent. They do not solve the problem of the decline in yeast activity.

It is often difficult to judge the practicability of work reported in the literature. In the laboratory it is possible to minimize fermentation time drastically, that is, to bring the dough into the freezer 5 to 10 minutes after the end of mixing. In the bakery this time period is likely to be at least 30 to 40 minutes, depending among other factors on the size of the batch.

### Rye and Wheat Sour Doughs

Rye can be grown in colder climates than wheat. It is a major source of bread flour in northern and eastern Europe, and rye breads are baked extensively in Scandinavia, the USSR, and Central Europe. In the United States, rye is grown in the North Central region. Rye flour doughs do not develop the cohesive gluten structure that is the distinguishing characteristic of wheat flour doughs.

The gas-holding properties of rye doughs depend largely on rye starch. Rye starch gelatinizes at a lower temperature than wheat starch, but there are differences in the gelatinization temperature of European and American varieties, and the gelatinization point of rye starch increases with extended storage of rye flour. Rye flours have a higher concentration of preformed sugars. The concentration of pentosans (gums) in rye flours is also much higher. The chemistry and technology of rye and rye flours has been reviewed by Rozsa (1976).

Rye flours are classified as light, medium, and dark flours depending on the degree of extraction. The light flours are lightest in color and lowest in ash content. Most rye bread formulations call for a mixture of rye and wheat flours. In the United States, the percentages of rye flour are generally quite low. For instance, Rozsa (1976) reports 20-40% rye flour and 60-80% wheat flour in typical U.S. rye breads. In contrast, Spicher and Stephan (1982) show that rye bread made in Germany has the following percentages of rye flour: rye bread, 90-100%; rye mixed grain bread, 50-89%; and wheat mixed grain bread, 10-49%.

Doughs containing more than small percentages of rye flour require acidification. It is necessary to improve the handling characteristics of the doughs, the quality of the bread crumb (elasticity, grain, slicing, and shelf life), and the flavor and aroma that characterize true rye bread (Spicher 1983). Table 7-11 shows the recommended pH and titratable acidity values for rye breads.

Acidification may be carried out in the United States by the addition of lactic acid or mixtures of lactic and acetic acids. Such chemical additions will not be discussed further. The traditional and preferred method is a fermentation by lactic acid bacteria that proceeds at the same time as the yeast fermentation of the doughs. The lactic acid bacteria may be introduced in the process in three ways: (1) by a spontaneous fermentation of rye flour for inoculation of a water sponge; (2) by retention of a portion of a

Table 7-11. Required pH and Acidity in Breads Made with Rye Flour

| Bread Type | Recommended pH | Specification Acidity* |
|---|---|---|
| Rye bread | 4.2-4.3 | 8.0-10.0 |
| Rye mixed grain bread (50-89% rye) | 4.3-4.4 | 7.0-9.0 |
| Wheat mixed grain bread (10-49% rye) | 4.65-4.75 | 6.0-8.0 |
| Whole rye bread | 4.0-4.6 | 8.0-14.0 |

Source: From Spicher and Stephan 1982.
*ml 0.1 N NaOH per 10 g of dough.

preceding sponge for inoculation of a succeeding sponge; or (3) by inoculation of the sponge with a selected bacterial culture.

A spontaneous fermentation can be started by incubating rye flour with twice its weight of water at 26°-28°C for 24 hours. This starter can then be carried through succeeding stages by addition of more rye flour and lesser amounts of water. This method can be compared to the spontaneous fermentation of grape musts in wine making. Like other spontaneous fermentations it lacks reproducibility.

A common method of producing sour doughs is the retention of a piece of a fully fermented dough piece for inoculation of succeeding sponges. This process may be carried out in several stages. In a four-stage process these stages may be called fresh sour, basic sour, full sour, and dough. A piece of the full sour is retained and used as a starter culture for the fresh sour. Simpler and faster methods are also used. The Berlin short sour dough process uses only two stages: the seed sour and the final dough stage. European bakers refer to it as a single-stage process since the final dough is not considered one of the sours.

The important bacteria of sour dough processes belong to the genus *Lactobacillus*, although *Pediococcus*, *Leuconostoc*, and *Streptococcus* are sometimes encountered. Among the lactobacilli the following species have been encountered: *L. plantarum, L. casei, L. farciminis, L. acidophilus* (all homofermentative) and *L. brevis, L. brevis* var. *lindneri, L. fermentum, L. buchneri*, and *L. fructivorans* (all heterofermentative). The homofermentative lactic acid bacteria produce principally lactic acid; the heterofermentative organisms produce lactic acid and considerable quantities of acetic acid, $CO_2$, and some ethanol.

Both lactic and acetic acids are required for the development of the typical sour dough flavor and their ratio is important. The range of lactic acid concentrations in fully fermented doughs is 0.8-1% of the dough and that of acetic acid is 0.25-0.3%. The ratio of acids is also affected by the firmness of the dough and by temperature. A softer dough produces relatively more lactic acid and so does a warmer dough (Fig. 7-18). The number of bacteria in various stages may reach several 100-1000 × $10^6$ per gram. Their growth is limited when the pH drops to the neighborhood of 3.7.

The indigenous species of yeast in rye flour are *Candida crusei, Pichia saitoi, Saccharomyces cerevisiae*, and *Torulopsis holmii*. The yeasts may multiply and reach 20-100 × $10^6$ cells per gram of dough. They are inhibited at pH values below 4 as shown in the often-reproduced graph of Rohrlich and Stegeman (1958) (Fig. 7-19). Yeasts are also greatly inhibited by the acetic acid formed by heterofermentative bacilli. Yeasts and bacteria may inhibit each other or there may be a symbiotic effect (Spicher and Schroeder 1978). It is difficult to elucidate because in practice a mixed bacterial flora

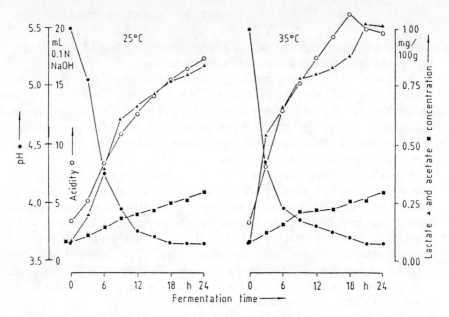

**Figure 7-18.** Effect of temperature on the formation of lactic and acetic acids by a sour dough starter. (*From Spicher and Stephan 1982*)

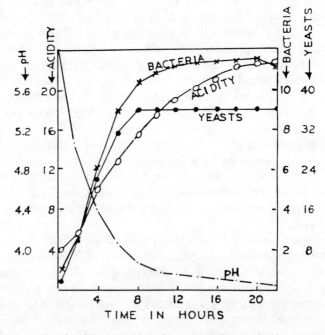

**Figure 7-19.** Growth of bacteria (x) and yeasts (●) during the fermentation of a rye sour dough, and changes in pH (●—●) and acidity (○). (*From Rohrlich and Stegemann 1958*)

and a mixed yeast flora grow in a medium that is constantly transformed by the metabolic activity of the microorganisms. In this respect, the fermentation of sour doughs also resembles the malolactic wine fermentation that has been described in Chapter 4.

Finally, sour doughs for seeding may be prepared by inoculation with commercial preparations of dried bacterial cultures. These are usually mixtures of cultures of the most important sour dough organisms: *L. plantarum*, *L. brevis*, and *L. fermentum*. Freeze-dried preparations of a fermented sour dough (Ulmer full sour) are also commercially available.

The processing of sour doughs after the fermentation follows in most respects the processing of wheat breads. Commercial baker's yeast is often used in the final dough to increase the rate of fermentation. It is essential in some processes, for instance, in the Berlin two-stage process, where the high temperature of the seed sour fermentation (35°C) inhibits the growth of the indigenous yeasts.

The production of rye sour doughs in Europe has been described in considerable detail by Spicher (1983), in the Soviet Union by Kosmina (1977) and in the United States by Pyler (1988).

Wheat sour doughs are well known, but they are not produced on a larger commercial scale. However, there are some interesting aspects that warrant description. The San Francisco French sour dough process produces a white hearth bread made entirely from wheat flour. The bread has a decided flavor and aroma due to the development of lactic and acetic acids in addition to the normal fermentation flavor of white bread (Galal, Johnson, and Varriano-Marston 1978). The process is based on the perpetuation of the microflora by retention of a portion of the sponge for seeding subsequent sponges. This process is the only one practiced by wholesale bakers that uses the age-old method of culture propagation.

The yeast was first described as *Saccharomyces exiguus*, but is now designated *Torulopsis holmii*, its nonsporulating form. It is capable of fermenting doughs at low pH values and has been reported to occur in grape musts and in rye sour doughs (see above). The yeast does not ferment maltose, which makes this sugar available for the bacterial fermentation. It is the most likely explanation for the truly symbiotic relationship between the two species.

The acid-producing bacterium has been designated *Lactobacillus sanfrancisco* (Sugihara, Kline, and Miller 1971). It can be produced in submerged culture. However, it is difficult to grow *T. holmii* because the yeast grows very slowly. Therefore, the percentage of the sponge that has to be retained for addition to the next dough is large: about 40%. Some baker's yeast, *S. cerevisiae*, is generally added on the dough side. The practical aspects of the process in the bakery have been described by Kline (1970) and Sugihara (1985). The Italian Pannetone is a traditional festive cakelike

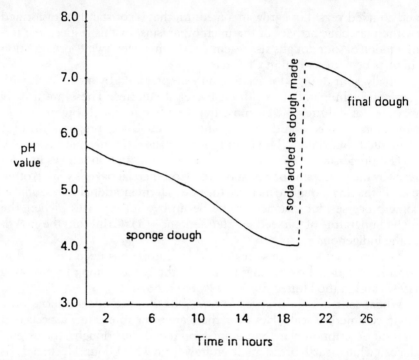

**Figure 7-20.** Changes in pH during a cracker sponge and dough fermentation. (*From Manley 1983*)

bread. It is made from a dough rich in eggs. The sweet/sour taste is due to fermentation by *T. holmii* and *L. brevis*.

Soda crackers are also produced by a mixed fermentation by yeasts and lactic acid bacteria. Again it is a sponge dough process characterized by a very long sponge fermentation of 18 hours. The sponge containing 70% flour and 30% water is very stiff. It is inoculated with 0.2-0.3% baker's yeast and may in addition contain a small percentage of a preceding sponge. The yeasts multiply rapidly during the first 10 hours of the fermentation, but the lowering of the pH to about 4 inhibits the yeast. The active bacteria are lactic-acid producers with a strong presence of *L. plantarum*.

Figure 7-20 shows the changes in pH during the entire fermentation period, and Table 7-12 shows the changes in bacterial and yeast counts. The dough is neutralized by the addition of sodium bicarbonate so that the final pH of the baked saltine (soda cracker) is above 7.5. The fermentation provides flavor and the rheological properties that permit machining of the final doughs. However, leavening takes place mostly in the oven by steam during the short baking period in the very hot oven. The microbiol-

Table 7-12. Changes in Bacterial and Yeast Counts, pH, and Temperature in Cracker Sponges and Doughs Made with 0.75% Compressed Yeast

|  | Time (hr) | Bacteria × 10⁶/ml Total | Acid Formers | Yeasts × 10⁶/ml | pH | Temp. (°C) |
|---|---|---|---|---|---|---|
| Sponge | 0 | 0.025 | 0.018 | 17.9 | 6.0 | 22 |
|  | 5 | 8.3 |  | 25.4 | 5.5 | 24 |
|  | 10 | 57.5 |  | 68.7 | 5.2 | 27 |
|  | 15 | 62.7 |  | 83.6 | 5.2 | 29 |
|  | 20 | 64.2 | 4.2 | 86.4 | 5.2 | 29 |
| Dough | 0 | 46.5 | 3.5 | 59.5 | 6.9 | 29 |
|  | 5 | 41.8 | 3.2 | 54.3 | 6.8 | 29 |
| Cracker (pH only) |  |  |  |  | 8.9 |  |

Source: From Micka 1955.

ogy of the cracker dough fermentation has not been explored as well as that of sour doughs. The following references provide some further insight: Micka 1955; Jensen 1974; Fields, Hoseney, and Varriano-Marston 1982; Manley 1983; and Rogers and Hoseney 1989.

Flat breads are consumed in most countries in the Middle East and in India, Pakistan, and many parts of Africa. They may be unleavened as the crisp rye wafers of northern Europe and Russia, the Mexican tortilla, or the Jewish matzoh. The leavened, fermented flat breads are sour dough products usually produced by the indigenous mixed flora of the doughs. *Saccharomyces, Hansenula, Candida,* and *Torulopsis* are some of the genera participating in these fermentations.

In recent years, pita bread has become popular in the United States. The dough for pita bread is a simple flour/water/yeast/salt dough (1% yeast). The flat, pancakelike dough is baked in a very hot oven (500°C, 60-90 sec). The large pocket inside the bread is formed by steam (Faridi and Rubenthaler 1984). Flat breads and other fermented cereal foods have been reviewed by Faridi (1988), Pomeranz (1987), and Beuchat (1983).

## ACTIVE DRY YEAST

The production of active dry yeasts as well as important properties have been described in the preceding chapter. The following discussion deals with their performance in baked goods, particularly in comparison with compressed yeast. In the mid-1970s, active dry yeasts had only 75-80% of

Table 7-13. Composition and Activity of Commercial Yeast Products

|  | Compressed Yeast | Active Dry Yeast | | |
| --- | --- | --- | --- | --- |
|  |  | Regular | Protected | Instant |
| Moisture | 67.0-72.0% | 7.5-8.3% | 4.5-6.5% | 4.5-6.0% |
| Nitrogen (a) | 8.0-9.0% | 6.3-7.3% | 6.5-7.3% | 6.3-8.0% |
| $P_2O_5$ (a) | 2.5-3.5% | 1.7-2.5% | 2.2-2.5% | 1.8-2.8% |
| Ash (a) | 4.0-6.5% | 4.0-6.5% | 4.0-6.5% | 4.0-6.5% |
| Fermentation activity (b) |  |  |  |  |
| in regular doughs (c) | 24.5-26.1 | 15.8-17.4 | 15.8-17.4 | 17.9-20.5 |
| in sweet doughs (d) | 10.9-12.5 | 9.2-10.0 | 9.2-10.0 | 9.2-10.0 |
| in lean doughs (e) | 25.9-28.8 | 13.6-14.3 | 13.6-14.3 | 20.5-21.9 |

Source: From Sanderson et al. 1983.
(a) On a dry-weight basis.
(b) In mM $CO_2$ produced per g of yeast solids per hr.
(c) 4-12% sugar added.
(d) 15-25% sugar added.
(e) No sugar added.

the gassing activity of compressed yeast if the two forms were compared on an equivalent moisture basis.* In the intervening years the gap has narrowed but it has not disappeared. It is difficult to quantify because at present 3 forms of active dry yeasts are available: ADY (regular active dry yeast), IADY (instant active dry yeast), and PADY (protected active dry yeast). Several strains are used for the production of these dry yeasts, and the relative performance of the various yeasts depends on the application, for instance, use in sweet doughs versus use in lean doughs.

Dry yeasts are available in several forms. IADY is dried in fluid bed driers. It is most sensitive to loss of activity if it is stored in air and the concentration of oxygen exceeds 0.5% in the gas. Therefore, it is always packaged under vacuum or in an inert atmosphere. The name IADY suggests that it may be added directly to flour and other dry ingredients before the formation of the dough. This method is actually preferred as shown in Table 7-13 (Bruinsma and Finney 1981). Separate rehydration resulted in lower gassing activity and consequently longer proof times. It also resulted in a less satisfactory crumb grain unless 3% sucrose was added to the rehydration water.

ADY is dried in continuous belt driers. Its friable strands may be sold in fiberboard drums. It may be stored for several weeks, preferably refrigerated.

---

*All values and comparisons will be expressed on a dry solids basis: CY, 30% solids; ADY, 92% solids; IADY, 96% solids.

ADY should be rehydrated in warm (35°-40°C) water before addition to the dough. This process minimizes leaching of reducing compounds from the yeast into the rehydration water (reduced glutathione, GSH) as well as minimizing loss in fermenting activity. Such reducing compounds produce slacker, more extensible and more relaxed doughs. Ponte, Glass, and Geddes (1960) have shown that the effect is indeed due to the leached compounds.

Dough slackening does not occur if the rehydration water is discarded, a procedure that is not practical in the bakery. Rehydrated ADY reduces mixing time of doughs approximately 20%. It is similar to the effect of added cysteine or $NaHSO_3$. It is beneficial for doughs made from strong bucky flour, or when a well-relaxed dough is needed, as for some pizza doughs or bun doughs. ADY may be counteracted by additional oxidation with bromate or azodicarbonamide where it is not desired. IADY also exerts a slackening effect but to a lesser extent.

ADY is also available in ground form. This product is sold in aluminum foil pouches ($N_2$ flushed) to consumers. It may be added directly to the dry ingredients if the dough water is quite warm (hot tap water of 45°-55°C). This procedure cannot be used in bakeries because doughs would become too hot. The consumer yeast packages are subject to mechanical shock during shipment and handling in grocery stores, which may lead to minor leaks that expose the ADY to air, and a consequent loss of activity during the long shelf life of these products. Therefore, the integrity of the package is more important for consumer yeast than the activity of the yeast. The development of a fast-fermenting-consumer IADY has been described by Trivedi, Cooper, and Bruinsma (1984).

PADY is produced with the addition of an antioxidant (about 0.1% BHA) to the press cake before drying. The antioxidant is dispersed in warm water with an emulsifier, usually 1% of a sorbitan ester (percentages expressed on a yeast solids basis). The antioxidant slurry may be added to the yeast cream before pressing or filtration. PADY is more stable than ADY with an extended shelf life of 3 months. It is not as stable as ADYs packaged under vacuum (Chen and Cooper 1962; Chen, Cooper, and Gutmanis 1966). The yeast is suitable for inclusion into complete bakery mixes, that is, mixes that contain all dough ingredients except water, but the use of a low moisture flour (8-9% moisture) is required. Such mixes have a shelf life of at least 3 months. With flour of a normal moisture content of about 13% the yeast picks up additional moisture and loses activity.

PADY is also suitable for use in premixes of dry ingredients. Such premixes containing PADY, sugar, salt, and sometimes shortening have become quite popular for use in small pizzerias.

Table 7-13 shows the composition and the fermentation activity of various active dry yeasts in comparison with compressed yeast (Sanderson

Table 7-14. Composition and Shelf Life of Yeast Products

| | Compressed Yeast | Cream Yeast | Active Dry Yeasts | | |
|---|---|---|---|---|---|
| | | | Regular | Protected | Instant |
| Moisture, % | 70 | 82 | 7-8 | 5-6 | 4-5.5 |
| Protein, dry basis, % | 60 | 60 | 38-48 | 40-42 | 39-41.5 |
| Shelf life | | | | | |
| refrigerated (2°-4.5°C) | 3-4 wk | 3-4 wk | 6 mo* 1 yr† | 9 mo* | 1 yr plus† |
| room temp. (21°C) | perishable | perishable | 3 mo* 1 yr† | 6 mo* | 1 yr† |

*Source:* From Trivedi et al. 1989.
*In drums or bags, not packaged under vacuum or inert atmosphere.
†Packaged under vacuum or inert atmosphere.

et al. 1983), and Table 7-14 shows the approximate shelf life of these yeasts (Trivedi et al. 1989). The fermentation activity in the various doughs is expressed as millimoles of $CO_2$ produced per gram of yeast solids per hour. The values given can be converted to milliliters by dividing by 22.4.

Table 7-13 indicates that all of the yeasts tested, including the CY, have only about one-half of the fermentation activity in sweet doughs. Thus, it is necessary to use 2.5-3% of CY in most sponge doughs and straight doughs, and about 6% of CY in sweet doughs. ADY is less active than CY on an equivalent solids basis, and IADY falls somewhere between the activities determined for CY and ADY.

Oszlanyi (1980), who worked with lean doughs, reported the following fermentation activities: IADY, 18.5 mM and CY, 21.1 mM $CO_2$ per g of yeast solids per hr. These values are somewhat lower than those shown in table 7-13. They also show a lesser fermentation activity for IADY than CY. It must be added that such values cannot be strictly compared unless the temperatures are also given.

Bruinsma and Finney (1981) show values for fermenting activity in Gasograph (arbitrary) units that precludes calculation of mM $CO_2$ evolved. With a straight dough system with 6% sugar, they report that samples of IADY had 86-91% of the activity of CY on an equivalent solids basis.

## FURTHER READINGS

The following books or book chapters published since 1980 are useful references for the technology of baking and the role of yeasts in baking: Oura, Suomalainen, and Viskari 1982; Ponte and Reed 1982; Dubois

1984; Chen and Chiger 1985; Trivedi and Jacobson 1986; Trivedi, Jacobson, and Tesch 1986; Nagodawithana 1986; Hoseney 1986; Pomeranz 1987; Kulp 1988; Pyler 1988; Beudeker et al. 1990.

## REFERENCES

American Association of Cereal Chemists. 1988. Yeast activity, $CO_2$ production. Proposed AACC method 89-01. American Association of Cereal Chemists, St. Paul, Minn.
Atkin, L., A. S. Schultz, and C. N. Frey. 1946. Yeast fermentation. In *Enzymes and Their Role in Wheat Technology*, J. A. Anderson (ed.). Interscience, New York.
Axford, D. W. E., N. Chamberlain, T. H. Collins, and G. H. A. Elton. 1963. The Chorleywood process. *Cereal Sci. Today* **8**:265-270.
Beuchat, L. R. 1983. Indigenous fermented foods. In *Biotechnology*, vol. 5, H. J. Rehm and G. Reed (eds.). VCH Publishing Company, Weinheim, West Germany.
Beudeker, R. F., H. W. van Dam, J. B. van der Plaat, and K. Vellenga. 1990. Developments in baker's yeast production. In *Yeast Biotechnology and Biocatalysis*, H. Verachtert and R. De Mot (eds.). Marcel Dekker, New York.
Bruinsma, B. L., and K. F. Finney. 1981. Functional (bread-making) properties of a new dry yeast. *Cereal Chem.* **58**(5):477-480.
Bruinsma, B. L., and J. Giesenschlag. 1984. Frozen dough performance. *Baker's Dig.* **58**(6):6-7, 11.
Busturia, A., and R. Lagunas. 1985. Identification of two forms of the maltose transport system in *S. cerevisiae* and their regulation by catabolite inactivation. *Biochim. Biophys. Acta* **820**:324-327.
Chamberlain, N. 1984. The Chorleywood bread process. *Cereal Foods World* **29**:656-658.
Chamberlain, N., T. H. Collins, and G. H. A. Elton. 1962. The Chorleywood bread process. *Baker's Dig.* **36**(5):52-53.
Chen, S. L., and M. Chiger. 1985. Production of baker's yeast. In *Comprehensive Biotechnology*, M. Moo-Young (ed.). Pergamon Press, Oxford, United Kingdom.
Chen, S. L., and E. J. Cooper. 1962. Production of active dry yeast. U.S. Patent 3,041,249, 6-26.
Chen, S. L., E. J. Cooper, and F. Gutmanis. 1966. Active dry yeast. Protection against oxidative deterioration during storage. *Food Technol.* **20**(12):79-83.
Coffman, J. R. 1967. Bread flavor and aroma: A review. *Baker's Dig.* **41**(1):50-51, 54-55.
D'Appolonia, R. L., K. A. Gilles, E. M. Osman, and Y. Pomeranz. 1971. Carbohydrates. In *Wheat Chemistry and Technology*. 2nd. ed., Y. Pomeranz (ed.). American Association of Cereal Chemists, St. Paul, Minn.
Darby, W., L. Grivetti, and P. Ghalioungui. 1972. *Food: Gift of Osiris*. Academic Press, London, United Kingdom.
Davis, E. W. 1981. Shelf life studies on frozen doughs. *Baker's Dig.* **55**(3):12-13, 16.
Dubois, D. K. 1984. Processes and ingredient trends in U.S. breadmaking. In *International Symposium Advances in Bakery Science and Technology*. Department of Grain Science, Kansas State University, Manhattan, Kan., pp. B-1-B-14.
Faridi, H. 1988. Flat breads. In *Wheat Chemistry and Technology*, 3rd ed., Y. Pomeranz (ed.). American Association of Cereal Chemists, St. Paul, Minn.
Faridi, H. A., and G. L. Rubenthaler. 1984. Effect of various flour extractions, water absorption,

baking temperature and shortening level on the physical quality and staling of pita breads. *Cereal Worlds Food* **29**:575-576.

Fields, L., R. C. Hoseney, and E. Varriano-Marston. 1982. Microbiology of cracker sponge fermentation. *Cereal Chem.* **59**:23-26.

Finney, P. L., C. D. Magoffin, R. C. Hoseney, and K. F. Finney. 1976. Short time baking systems. I. Interdependence of yeast concentration, fermentation time and oxidation requirements. *Cereal Chem.* **53**:126-134.

Fortmann, K. L., A. B. Gerrity, and V. R. Diachuk. 1964. Factors influencing work requirements. *Cereal Sci. Today* **9**:268-271.

Franz, B. 1961. Kinetics of the alcoholic fermentation during the propagation of baker's yeast (in German). *Die Nahrung* **5**:458-481.

Friedemann, T. E., N. F. Witt, and B. W. Neighbors. 1967. Determination of starch and soluble carbohydrates. I. Development of method for grains, stock feeds, cereal foods, fruits and vegetables. *J. Assoc. Offic. Anal. Chem.* **50**:945.

Galal, A. M., J. A. Johnson, and E. Varriano-Marston. 1978. Lactic and volatile ($C_2$-$C_5$) organic acids of San Francisco sourdough French bread. *Cereal Chem.* **55**:461-468.

Garver, J. C., I. Navarine, and A. M. Swanson. 1966. Factors influencing the activation of baker's yeast. *Cereal Sci. Today* **11**:410-418.

Gelinas, P. 1988. Conditions of growth and cryoresistance of baker's yeast, *S. cerevisiae*, incorporated into frozen dough. Ph.D. thesis. Laval University, Sainte-Foy, Quebec.

Godkin, W. J., and W. H. Cathcart. 1949. Fermentation activity and survival of yeast in frozen fermented and unfermented doughs. *Food Technol.* **3**:139-146.

Harbrecht, A., and R. Kautzmann. 1967. Comparative investigation of the determination of the leavening power of baker's yeasts (in German). *Branntweinwirtsch.* **1078**:21-23.

Hino, A., H. Takano, and Y. Tanaka. 1987. New freeze-tolerant yeast for frozen dough preparations. *Cereal Chem.* **64**:269-275.

Hoseney, R. C. 1986. *Principles of Cereal Science and Technology.* American Association of Cereal Chemists, St. Paul, Minn.

Hsu, K. H., R. C. Hoseney, and P. A. Seib. 1979a. Frozen dough. I. Factors affecting stability of yeasted doughs. *Cereal Chem.* **56**:419-424.

Hsu, K. H., R. C. Hoseney, and P. A. Seib. 1979b. Frozen dough. II. Effects of freezing and storage conditions on the stability of yeasted doughs. *Cereal Chem.* **56**:424-446.

Jacob, H. E. 1944. *Six Thousand Years of Bread*. Doubleday, Doran & Co., Garden City, N.J.

Jacobson, G. K., and N. B. Trivedi. 1987. Yeast strains, method of producing and use in baking. U.S. Patent 4,643,901, 2-17.

Jensen, O. G. 1974. Biscuit and cracker technology. In *Encyclopedia of Food Technology*, vol. 2, A. H. Johnson and M. S. Peterson (eds.). AVI Publishing Co., Westport, Conn.

Kilborn, R. H., and K. H. Tipples. 1979. Sponge and dough type bread from mechanically developed doughs. *Cereal Sci. Today* **13**:25-28, 30.

Kline, L. 1970. San Francisco sour dough bread. *Proc. Am. Soc. Bakery Eng.* (Chicago), 83-91.

Kline, L., and T. F. Sugihara. 1968. Factors affecting the stability of frozen bread doughs. I. Prepared by the straight dough method. *Baker's Dig.* **42**(5):44-46, 48-50.

Kosmina, N. P. 1977. Biochemistry of bread making (in German). VEB Fachbuchverlag, Leipzig, East Germany.

Kruger, J. E., and G. Reed. 1988. Enzymes and color. In *Wheat Chemistry and Technology*, 3rd ed., Y. Pomeranz, (ed.). American Association of Cereal Chemists, St. Paul, Minn.

Kulp, K. 1988. Bread industry and processes. In *Wheat Chemistry and Technology*, 3rd ed., Y. Pomeranz (ed.). American Association of Cereal Chemists, St. Paul, Minn.

Kulp, K., and D. K. Dubois. 1982. Breads and sweet goods in the U.S., *Am. Inst. Baking, Res. Dept. Tech. Bull.* **4**(6).

Kulp, K., H. Chung, M. A. Martinez-Anaya, and W. Doerry. 1985. Fermentation of water ferments and bread quality. *Cereal Chem.* **62:**55-59.

Langejan, A., and B. Khoudokormoff. 1976. High protein active dry baker's yeast. U.S. Patent 3,993,782, 11-23.

Lee, J. W., L. S. Cuendet, and W. F. Geddes. 1959. The fate of various sugars in fermenting sponges and doughs. *Cereal Chem.* **36:**522-533.

Legman, R., and P. Z. Margalith. 1983. Interspecific protoplast fusion of S. cerevisiae and S. mellis. *Eur. J. Appl. Microbiol. Biotechnol.* **18:**320-322.

Lehmann, T. A., and P. Dreese. 1981. Frozen bread dough. *Am. Inst. Baking, Res. Dept. Tech. Bull.* **3**(7).

Lehmann, T. A., and D. K. Dubois. 1980. Pizza crust: Formulation and processing. *Cereal Foods World* **28**(5):589-592.

Lorenz, K. 1974. Frozen dough. *Baker's Dig.* **48**(2):18-19, 22.

McLaren, L. H. 1954. Practical aspects of the stable ferment process. *Baker's Dig.* **28**(3):23-24, 30.

Maga, J. A. 1974. Bread flavor. *CRC Crit. Rev. Food Technol.* **5:**55-142.

Magoffin, C. D., and A. C. Hoseney. 1974. A review of fermentation. *Baker's Dig.* **48**(6):22-23, 26-27.

Manley, D. J. R. 1983. *Technology of Biscuits and Cookies*. Ellis Horwood Ltd., Chichester, United Kingdom.

Marek, C. J., and W. Buschuk. 1967. Study of gas production and retention in doughs with a modified Brabender oven-rise recorder. *Cereal Chem.* **44:**300-307.

Marston, P. E. 1967. The use of ascorbic acid in bread production. *Baker's Dig.* **41**(6):30-33, 70.

Marston, P. E. 1978. Frozen dough for bread making. *Baker's Dig.* **52**(5):18-20, 37.

Martinez-Anaya, M. A., and K. Kulp. 1984. Fermentation of liquid ferments and bread quality. In *International Symposium Advances in Baking Science and Technology*. Department of Grain Science, Kansas State University, Manhattan, Kan., pp. D-1-D-12.

Masselli, J. A. 1959. The fundamentals of brew fermentation. *Proc. Am. Soc. Bakery Eng.* (Chicago), 160-167.

Matz, S. A. 1984. Modern baking technology. *Sci. Am.* **25**(5):123-134.

Mazur, P., and J. J. Schmidt. 1968. Interactions of cooling velocity, temperature and warming velocity on the survival of frozen and thawed yeast. *Cryobiology* **5**(1):1-17.

Merritt, P. P. 1960. The effect of preparation on the stability and performance of frozen, unbaked yeast-leavened doughs. *Baker's Dig.* **34**(4):57-58.

Micka, J. 1955. Bacterial aspects of soda cracker fermentation. *Cereal Chem.* **32:**125-131.

Nagodawithana, T. W. 1986. Yeasts: Their role in modified cereal fermentations. In *Cereal Science and Technology*, vol. 8, Y. Pomeranz (ed.). American Association of Cereal Chemists, St. Paul, Minn.

Oda, Y., K. Ono, and S. Ohta. 1986. Selection of yeasts for breadmaking by frozen dough method. *Appl. Environ. Microbiol.* **52**(4):941-943.

Osinga, K. A., R. F. Beudeker, J. B. Van der Plaat, and J. A. De Hollander. 1989. New yeast strains providing for an enhanced rate of the fermentation of sugars. A process to obtain such yeasts and the use of these yeasts. Eur. Patent Appl. 0 306 107, 3-8.

Oszlanyi, A. G. 1980. Instant yeast. *Baker's Dig.* **54**(4):16, 18-19.

Oura, E., H. Suomalainen, and R. Viskari. 1982. Bread-making. In *Economic Microbiology*, vol. 7, A. H. Rose (ed.). Academic Press, New York.

Peppler, H. J. 1972. Yeast. In *Bakery Technology and Engineering*, 2nd ed., S. A. Matz (ed.). AVI Publishing Co., Westport, Conn.

Pomeranz, Y. 1987. *Modern Cereal Science and Technology*. VCH Publishers, New York.

Ponte, J. G., Jr., 1971. Bread. In *Wheat Chemistry and Technology*, 2nd ed., Y. Pomeranz (ed.). American Association of Cereal Chemists, St. Paul, Minn.

Ponte, J. G., Jr., R. L. Glass, and W. F. Geddes. 1960. Studies on the behavior of active dry yeast in bread making. *Cereal Chem.* **37**:263-279.

Ponte, J. G., Jr., and G. Reed. 1982. Bakery foods. In *Prescott and Dunn's Industrial Microbiology*, 4th ed. AVI Publishing Co., Westport, Conn.

Ponte, J. G., Jr., S. T. Titcomb, and R. H. Cotton. 1963. Some effects of oven temperature, flour levels and malted barley on bread making. *Baker's Dig.* **37**(3):44-48.

Pyler, E. J. 1988. *Baking Science and Technology*. Sosland Publishing Co., Merriam, Kan.

Redfern, S., H. Gross, R. L. Bell, and F. Fischer. 1968. Research with a pilot scale continuous bread-making unit. *Cereal Sci. Today* **13**:324-326.

Reed, G. 1972. Yeast food. *Baker's Dig.* **46**(6):16-17, 60

Reed, G. 1974. Pretzels. In *Encyclopedia of Food Technology*, vol. 2, A. H. Johnson and M. S. Peterson (eds.). AVI Publishing Co., Westport, Conn.

Reed, G., and S. L. Chen. 1978. Evaluating commercial active dry wine yeasts by fermentation activity. *Am. J. Enol. Vitic.* **29**(3):165-168.

Reed, G., and H. J. Peppler. 1973. *Yeast Technology*. AVI Publishing Co., Westport, Conn.

Robertson, J. J., and H. O. Halvorson. 1957. The components of maltozymase in yeast, and their behavior during deadaptation. *J. Bacteriol.* **73**:186-198.

Robinson, R. J., T. H. Lord, J. A. Johnson, and B. S. Miller. 1958. The aerobic microbiological population of pre-ferments and the use of selected bacteria for flavor production. *Cereal Chem.* **35**:295-305.

Rogers, D. E., and R. C. Hoseney. 1989. Effects of fermentation in saltine cracker production. *Cereal Chem.* **66**(1):5-10.

Rohrlich, M., and J. Stegeman. 1958. Multiplication of sour dough organisms and acid formation during dough fermentations in a discontinuous process. *Brot Gebaeck* **12**:41-63.

Rothe, M. 1974. Aroma of bread. In *Handbook of Aroma Research* (in German). Akademie Verlag, Berlin.

Rozsa, T. A. 1976. Rye milling. In *Rye: Production, Chemistry and Technology*, W. Buschuk (ed.). American Association of Cereal Chemists, St. Paul, Minn.

Rubenthaler, G. L., P. L. Finney, D. E. Demaray, and K. F. Finney. 1980. Gasograph: Design, construction and reproducibility of a 12 channel gas recording instrument. *Cereal Chem.* **57**(3):212-216.

Sanderson, G. W., G. Reed, B. Bruinsma, and E. J. Cooper. 1983. Yeast fermentation in bread making. *Am. Inst. Baking, Res. Dept. Tech. Bull.* **5**(12).

Sasaki, T., and Y. Oshima. 1987. Induction and characterization of artificial diploids from the haploid yeast *Torulaspora delbrueckii*. *Appl. Environ. Microbiol.* **53**(7):1504-1511.

Schultz, A. S., L. Atkin, and C. N. Frey. 1942. Determination of vitamin $B_1$ by a yeast fermentation method. *Ind. Eng. Chem. (Anal. Ed.)* **14**:35-39.

Schultz, A. 1965. Investigations of the leavening activity of baker's yeast (in German). *Brot Gebaeck* **19**(4):61-65.
Selman, R. W. 1948. The significance of pH in baking. *Baker's Dig.* **22**(1):28-29, 34.
Selman, R. W. 1953. Leavening by oxygen is key to new baking process. *Baker's Dig.* **27**(4):67-68, 73.
Shirley, E. H. 1977. The Canadian concept of a no-time dough system. *Proc. Am. Soc. Bakery Eng.* (Chicago), 36-43.
Shogren, M. D., K. F. Finney, and G. L. Rubenthaler. 1977. Note on the determination of gas production. *Cereal Chem.* **54**(3):665-668.
Spicher, G. 1983. Baked goods. In *Biotechnology*, vol. 5, H. J. Rehm and G. Reed (eds.). VCH Verlagsgesellschaft, Weinheim, West Germany.
Spicher, G., and R. Schroeder. 1978. The microflora of sour dough, IV. The species of rod shaped lactic acid bacteria of the genus Lactobacillus occurring in sour doughs. *Z. Lebensm. Unters. Forsch.* **168**:397.
Spicher, G., and H. Stephan. 1982. *Handbuch Sauerteig* (*Handbook on Sour Dough*; in German). BBV Wirtschaftsinformationen, Hamburg.
Sugihara, T. F. 1985. Microbiology of bread making. In *Microbiology of Fermented Foods*, vol. 1, B. J. B. Wood (ed.). Elsevier, New York.
Sugihara, T. F., L. Kline, and N. W. Miller. 1971. Microorganisms of the San Francisco sour dough process. I. Yeasts responsible for the leavening action. *Appl. Microbiol.* **21**:456-458.
Suomalainen, H., J. Dettweiler, and E. Sinda. 1972. Alpha-glucosidase and leavening of baker's yeast. *Process Biochem.* **7**(5):16-19, 22.
Tang, R. T., J. Robert, J. Robinson, and W. C. Hurley. 1972. Quantitative changes in various sugar concentrations during bread making. *Baker's Dig.* **46**(4):48-55.
Tanaka, Y., W. Kawaguchi, and M. Miyatake. 1976. Studies on the injury of yeast in frozen dough. II. Effect of ethanol on the frozen storage of baker's yeast (in Japanese). *J. Food Sci. Technol.* **23**(9):419.
Tanaka, Y., and M. Miyatake. 1975. Studies on the injury of yeast in frozen dough. I. Effect of prefermentation before freezing on the injury of yeast (in Japanese). *J. Food Sci. Technol.* **22**(8):366.
Thorn, J. A. 1963. Yeast performance in liquid ferments. *Baker's Dig.* **37**(3):49-51.
Thorn, J. A., and J. W. Ross. 1960. Determination of yeast growth in doughs. *Cereal Chem.* **37**:415-421.
Trivedi, N. B., and G. Jacobson. 1986. Recent advances in baker's yeast. In *Progress in Industrial Microbiology*, vol. 23, M. R. Adams (ed.). Elsevier, New York.
Trivedi, N. B., E. J. Cooper, and B. Bruinsma. 1984. Development and applications of quick-rising yeast. *Food Technol.* **38**(6):51, 54-55, 57.
Trivedi, N. B., G. K. Jacobson, and W. Tesch. 1986. Baker's yeast. *CRC Crit. Rev. Biotechnol.* **24**:75-109.
Trivedi, N., J. Hauser, W. T. Nagodawithana, and G. Reed. 1989. Update on baker's yeast. *Am. Inst. Baking, Res. Dept. Tech. Bull.* **11**(2).
Turner, S. E., Sr. 1986. Liquid pre-ferments. *Proc. Am. Soc. Bakery Eng.* (Chicago), 176-181.
Uhrich, M. G. 1975. Formulation of liquid pre-ferments. *Proc. Am. Soc. Bakery Eng.* (Chicago), 42-51.
Van Uden, N. 1971. Kinetics and energetics of yeast growth. In *The Yeasts*, vol. 2, A. H. Rose and J. S. Harrison (eds.). Academic Press, New York.

Varriano-Marston, E., K. H. Hsu, and J. Mahdi. 1980. Rheological and structural changes in frozen dough. *Baker's Dig.* 54(1):32-34, 41.

Wiseblatt, L., and H. F. Zoumut. 1963. Isolation, origin and synthesis of a bread flavor constituent. *Cereal Chem.* 40:162-169.

Winge, O. and C. Roberts. 1958. Life history and cytology of yeasts. In *The Chemistry and Biology of Yeasts.* A. H. Cook (ed.). Academic Press, New York.

Wolt, M. J. and B. L. D'Appolonia. 1984a. Factors involved in the stability of frozen dough. I. The influence of yeast reducing compounds on frozen dough stability. *Cereal Chem.* 61:209-212.

Wolt, M. J. and B. L. D'Appolonia. 1984b. Factors involved in the stability of frozen dough. II. The effect of yeast type, flour type, and dough additives on frozen dough stability. *Cereal Chem.* 61(3):213-221.

CHAPTER
8

# YEAST-DERIVED PRODUCTS

Although yeast is perhaps considered the oldest microbial associate of humankind, the role it played in shaping the lives of past civilizations was not recognized until the discovery of the microscope by van Leeuwenhoek two centuries ago. Many authorities now believe that complicated beverages like beer originated in Egypt around 6000 B.C. (Corran 1975). By 3000 B.C., bread making and brewing of beer were closely allied arts. Likewise, Assyrian and Egyptian historical documents dating as far back as 3500 B.C. mention grapes and wine. Although these civilizations were unaware of the chemical changes induced by yeast, they had sufficient empirical knowledge to modify their food products to make them more palatable, nutritious, and in some instances, intoxicating. The knowledge they acquired through trial and error was also transmitted through the ages to succeeding generations.

Initial steps of brewing were similar to those of baking. The germinated barley was first crushed and the powdered meal was wetted, and the resulting dough was flattened and baked on a heated surface. With further experience in baking, bakers began to realize that the final baked product was more palatable with respect to stability, digestibility, and texture if the dough was left undisturbed for some time prior to baking. The next important development in baking was the recognition of the importance of yeasty beer as a leavening agent. This new development resulted in the expansion of the baking industry utilizing the excess yeast that was generated in the brewing industry.

Toward the end of the eighteenth century, with the expansion of the brewing industry, the spent yeast generated was far in excess of what the bakers could use for their baking. The brewers then began to experience the problem of excess yeast disposal, which accounted for a substantial proportion of their waste load. As a means of disposal, the spent yeast was often sprayed on agricultural lands as a source of fertilizer.

Toward the end of the nineteenth century, several studies demonstrated the importance of yeast as a nutritional supplement for both animals and humans, primarily due to its high level of protein. More recent studies have indicated the importance of brewer's yeast as an excellent source of the vitamin B complex. Many major brewers who saw an opportunity following these findings began an extensive search to produce products of greater commercial value to maximize their overall profitability.

Pioneering efforts made in England and Germany led to the development of a variety of nutritional and flavoring agents derived from brewer's yeast to meet the needs of the food industry. Continued research in this area during the last few decades led to the commercialization of several other products derived from both baker's and brewer's yeast. The major yeast-derived products of commercial importance are listed as follows: (1) flavor products and flavor enhancers, (2) products of nutritional value, (3) colorant, (4) enzymes, and (5) products of pharmaceutical and cosmetic value.

## FLAVOR PRODUCTS AND FLAVOR ENHANCERS

### Yeast Extracts

Yeast extracts are commercially available as a powder or paste and have been used extensively by the food industry as a flavoring agent. It is a concentrate of the soluble fraction of yeast generally made following autolysis, which is essentially a degradation process carried out by activating the yeast's own degradative enzymes to solubilize the cell components found within the cell. Yeast extracts can also be made by hydrolysis using external reagents or enzymes capable of releasing the cell content in a highly degraded form. Yeast extracts have received wide acceptance in the food industry during the last few decades as a natural flavoring agent. They are also cost-effective in relation to other flavoring agents (which are often not natural) on the basis of equivalent flavor intensity. They are used extensively in the fermentation industry and as a complex nutritional source in the preparation of microbiological growth media for plate assays. The production of yeast extracts worldwide amounts to approximately 35,000 tons.

In general, since the surplus yeast in the brewing industry is relatively cheap, it is utilized extensively in the production of extracts to meet the needs of the food and fermentation industries. Extracts made from the primary-grown yeast such as *Saccharomyces cerevisiae* (baker's), *Candida utilis*, or *Kluyveromyces marxianus* have flavor profiles somewhat different from those of brewer's yeast. Nevertheless, they have unique flavor characteristics and are widely used as flavoring agents in the food industry.

Although brewer's yeast is available in large quantities at a relatively cheap price, it is likely to contain undesirable flavor characteristics as a result of the carryover of hop resins and beer solids from the brewery fermentation. The hop resins, which are often bitter, are primarily humulones and isohumulones present as beer solids or as compounds firmly adsorbed to the surface of the yeast cells. Further processing of this yeast cream requires a debittering step that is sometimes wasteful and often costly. A partial debittering is less costly and makes the final extract less bitter, often providing a beneficial effect to the overall flavor profile. In contrast, primary-grown yeast is more costly compared to brewer's yeast, and the products are also priced higher in view of the higher raw material cost. Because of their unique flavor characteristics, such extracts are capable of imparting an overall flavor to a food formulation.

Different types of yeast extracts commercially available for food use differ from each other with respect to their flavor profiles. There are several factors that influence such quality differences and these may include the type and the condition of the yeast, presence of extraneous matter, conditions used in the extraction and processing of yeast solubles, and level of bacterial contamination during processing.

Autolysis, which is essentially an autodigestion of the yeast, requires the mediation of several endogenous enzymes. These enzymes, generally compartmentalized within the living cell, are located in the general matrix of the cell. Conditions that cause cell death also cause inactivation of the endogenous degrading enzymes, thereby making the cell less sensitive to further autolytic activity. Hence, a high viability in the yeast is a prerequisite for maximum autodegradation during an autolysis.

The flavor of the final product is also influenced by the extraneous matter present in the yeast slurry, which is often the case when the substrate used in the process is spent brewer's or distiller's yeast. Generally, well-washed baker's yeast produces an extract that is more bland than the one made from an inadequately washed molasses-grown yeast. This phenomenon is also seen in the production of extracts of *K. marxianus* grown on whey permeate. For this reason, some molasses is incorporated during processing of *K. marxianus* extracts to bring out the desired meaty flavor notes. Likewise, the level of bitter components introduced from brewer's spent yeast to the process can have a profound influence on the quality of the final product.

Some food manufacturers do not object to the presence of some bitterness in the extracts they wish to use in their formulations. The presence of such flavor compounds enable them to generate desirable reaction flavors so important for improving the product quality. The bitterness level in the final extract can generally be controlled by the level of efficiency of the initial cleaning of the spent brewer's yeast. An alkaline wash is sometimes applied to develop a less bitter flavor in the final product even though such a treatment reduces the final extract yield. Contamination is common during extract production, but it is generally controlled at a low level either by control of process conditions or by use of certain antimicrobial agents. Commercial extracts must meet certain microbiological specifications before they can be sold for food applications.

Detailed manufacturing procedures for the production of various types of extracts are not well documented due to obvious reasons. Nevertheless, the basic operational steps currently in use by most extract manufacturers are cleaning of the yeast if it is a by-product of another industry; solubilization; separation of solubles from the digested material; clarification; concentration; and drying.

Primary-grown yeast requires an appropriate wash to remove the mother liquor, especially if grown on molasses. Brewer's yeast, which is extensively used in the industry, must be highly viable at the time of processing and should contain a minimal level of microbial contaminants, hop resins, trub, and other undesirable soluble or particulate matter. The trub and other insoluble matter are separated by passing the cream through a 150-200 $\mu$ vibrating mesh. A variety of bitter components added as hops at the early stages of the brewing process can be detected in the spent yeast cream either in the soluble fraction as beer solids or as compounds adsorbed to the yeast cell wall. The latter, if not removed at the initial cleaning operation, may be released into the extract medium during autolysis, imparting a bitter flavor to the final product.

These hop resins present on the surface of the yeast cell can, however, be solubilized by an alkaline wash. Due to the variability in quality of brewer's yeast at different brewery locations as a result of their minor variations in processing, some level of blending of different-quality yeasts is necessary to maintain consistency of the final products. There are three distinct manufacturing practices for the production of yeast extracts: autolysis, plasmolysis, and hydrolysis.

*Autolysis*

All macromolecules present in viable yeast cells are held back within the cell by the semipermeable membrane. The cell wall provides the mechanical support necessary to maintain the size and shape of the cell. Autolysis is a process by which the cell components are solubilized by activation of the degradative enzymes inherently present within the cell. It is achieved by the application of carefully controlled conditions such as temperature, pH, time, and addition of certain enhancing agents. Such conditions bring death to the yeast population without inactivating the degrading enzymes that remain compartmentalized in a live cell. Under these conditions, important enzyme systems responsible for carrying out regular metabolism are no longer governed by the delicate control mechanisms that exist in a healthy living cell. Death causes disorderliness within the cell, allowing the free degradative enzymes to indiscriminately attack their specific substrates. This attack causes the breakdown of the corresponding macromolecules like proteins and nucleic acids to their basic units that are soluble by nature. Disorderliness within the cell results in a leaky cell wall and a loss in the integrity of the semipermeable membranes and permits the soluble components to leak out of the cell into the surrounding environment.

The final flavor profile and the extract yield in a well-controlled process are largely dependent on the temperature; pH; duration; viability and concentration of yeast; and type of solubilizing aids used in the process. Some conditions that favor higher extract yields may not necessarily favor the production of extracts with the desirable flavor. A substantial amount of research effort is thus necessary to optimize conditions so that the product has the desired flavor profile.

Common solubilizing aids used during autolysis are most often enzymes that are proteases, gluconases, nucleases, or phosphodiesterases. Ethyl acetate is often used to enhance autolysis and also to control the contaminants to the minimum. Proteases, such as papain, are known to increase extract yield only in longer autolysis procedures as practiced in the United States, United Kingdom, and Australia. Their effect is minimal when the duration of autolysis is shorter than 20 hours as practiced in Japan and France.

The autolysis procedure is well suited for the production of low-sodium yeast extracts that are gaining wide acceptance in the food industry. Although these extracts taste bland, they have the capability of improving the flavor profile by combining with other food ingredients. Low-sodium yeast extracts are becoming increasingly important components of convalescent and infant food formulations.

Autolysis, or self-digestion, of viable yeast cells takes place at an acceptable pace when the yeast slurry of approximately 15% solids is maintained at 45°-50°C for 24-36 hours at a pH of about 5.5. The lysis of these live whole cells is primarily due to the action of $\beta(1\text{-}3)$ gluconase and protease enzymes present within the cells. The $\beta(1\text{-}6)$ gluconase and mannanase enzymes participate in the further solubilization of the wall matrix.

More than 40 proteolytic enzymes have been identified to date in

*Saccharomyces cerevisiae* of which only a few enzymes can be described as important in the autolytic process. Four of the important proteolytic enzymes that play a major role during lysis are proteinase ysc A, proteinase ysc B, carboxypeptidase ysc Y, and carboxypeptidase ysc S. All four of these soluble proteolytic enzymes appear to be localized in the vacuole whereas their inhibitors are in the cytoplasm outside the vacuole (Table 8-1). Such a compartmentalization segregates biologically compatible components to allow the control and integration of the intracellular activities for the orderly functioning of a viable system.

Under autolytic conditions the yeast cells begin to die after all the cell reserves have been used up. This condition, then, can bring about a disorderliness within the yeast cell, thus marking the beginning of the autolysis process. There is a gradual breakdown of barriers in the highly compartmentalized cell matrix with the release of proteolytic enzymes from the vacuole into the degenerating cell matrix. These enzymes are, at first, inactivated due to the formation of complexes by reacting with corresponding inhibitors present in the immediate vicinity. This inactivation was evident when proteinase ysc A and proteinase ysc B were extracted by autolysis at pH 7. However, incubation of the yeast extracts at pH 5 caused reactivation of these proteinase enzymes (Hata, Hayashi, and Doi 1967) by the proteolytic inactivation of the firmly bound inhibitor (Lenney and Dalbec 1969; Saheki and Holzer 1975).

Addition of purified proteinase ysc A to fresh crude yeast extracts also accelerated the inactivation of the proteinase ysc B inhibitor and the appearance of maximal activity of proteinase ysc B and carboxypeptidase ysc Y. Their existence is thus reflected by a proteolytically degradable inhibitor of the proteinases that impart minimal influence in the long run on the proteolytic enzymes under conditions conducive for autolysis (pH 5.5; temperature 45°-55°C).

The vacuolar proteinases, like proteinase ysc A, proteinase ysc B, carboxypeptidase ysc Y, carboxypeptidase ysc S, and aminopeptidases are involved in highly nonspecific protein and peptide degradation. No other function has been attributed to vacuolar proteolytic enzymes to date. Under natural conditions, these enzymes most likely provide the living cell the ability to maintain an amino acid pool in the vacuole at the expense of the unneeded cell proteins for new protein synthesis. Under autolytic conditions, these enzymes provide the necessary proteolytic degradation by indiscriminately attacking the proteins present in the cytoplasmic matrix. Likewise, the nucleases present in the cell begin to act on the RNA and DNA, reducing them to polynucleotides, mononucleotides, and nucleosides. These soluble components of low molecular weight should finally be released to the surrounding medium.

The cell wall, which is generally rigid, provides the characteristic shape

Table 8-1. Properties of Yeast Proteases

| | Proteinase ysc A | Proteinase ysc B | Carboxypeptidase ysc Y | Carboxypeptidase ysc S |
|---|---|---|---|---|
| Type | Acid endopeptidase | Serine endopeptidase | Serine exopeptidase | Metallo exopeptidase ($Zn^{2+}$) |
| Optimum pH | 2-6 | 6-7 | 4-7 | 7 |
| Optimum T (°C)* | 35-40 | 45-55 | 45-55 | 60 |
| Cell location | Vacuole | Vacuole | Vacuole | Vacuole |
| Solubility | Soluble | Soluble | Soluble | Soluble |
| Molecular weight | 60,000 | 32,000-44,000 | 61,000 | Not determined |
| Isoelectric pt. | 3.8 | 5.8 | 3.6 | |
| Inhibitors | Protein $I_3^A$ Pepstatin, etc. | Protein $I_2^{13}$ Chymostatin, etc. | Protein $I^e$ $Hg^{2+}$ | EDTA |
| Cellular role | Protein degradation | Protein degradation | Protein degradation | Protein degradation |

*Source:* Based on data from Achstetter and Wolf 1985.
*Maddox and Hough 1969.

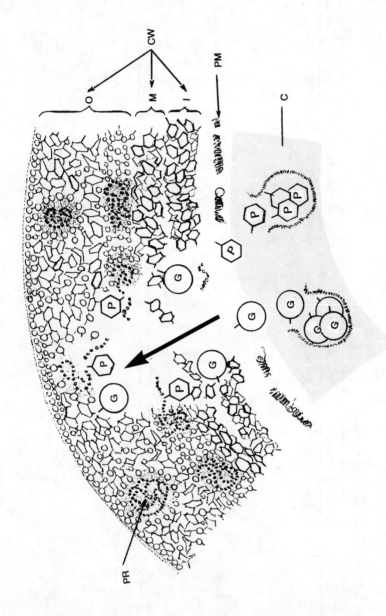

**Figure 8-1.** Schematic diagram of a lysing cell. C, cytoplasm; CW, cell wall; I, inner alkali-insoluble glucan layer; M, middle alkali-soluble glucan layer; O, outer glycoprotein layer; P, protease; G, gluconase; PM, plasma membrane; PR, protein.

to the cell. It is generally thought of as consisting of an alkali insoluble $\beta$-glucan inner layer, a middle layer of alkali-soluble $\beta$-glucan, and an outer layer of glycoproteins in which the carbohydrate is phosphorylated mannan (Fig. 8-8). The yeast cell wall also contains approximately 1% of chitin, a linear polymer of N-acetyl glucosamine that occurs only in the bud scars. Both alkali-soluble and alkali-insoluble glycan layers represent a mixture of polysaccharides, the major component (85%) consisting of predominantly $\beta$-(1-3)-linked glucan and a minor component (15%) made up of highly branched $\beta$(1-6) glucan with $\beta$(1-3) interchain linkages.

The gluconase enzymes have been isolated from yeast, and these enzymes are known to hydrolyze the $\beta$(1-3) and $\beta$(1-6) linkages of the glycan layers. The gluconase enzymes are known to participate in the budding process. Nonetheless, under autolytic conditions, $\beta$(1-3) gluconase enzymes, with the support from the proteinases, are capable of disrupting the cell wall, thereby facilitating the release of hydrolyzed components from within the cell. This process can, however, be enhanced by subjecting the yeast to an external protease attack initially like the one occurring with papain followed by treatment with a lytic gluconase (Fig. 8-1). These enzymes are now available at commercial scale to improve extract yields.

The yeast proteases previously described have a broad specificity toward their substrates. They are known to exhibit different pH and temperature optima. Therefore, the process conditions used by most extract manufacturers are not optimal for all these proteolytic enzymes. Likewise, the variation in the flavor profile of extracts made by autolysis under different pH and temperature conditions is understandable. Thus, it is implied that the control of conditions of autolysis is vital to maintain product consistency.

Yeast mutants with altered cell wall structure, highly sensitive to autolytic enzymes, can be of importance to an extract manufacturer. However, such mutants are generally unstable, grow slowly, and tend to lyse during growth. They also have an aberrant morphology that appears to be related to defects in cell wall construction. However, with the use of rDNA techniques, it will be possible to engineer yeasts capable of rapid autolysis with the release of desirable cell components that are of value to the food industry.

*Plasmolysis*

Plasmolysis is a phenomenon often used by extract manufacturers for rapid initiation of the cell degradation process. This procedure is highly cost-effective, and the most frequently used agent to achieve this effect is common salt. Organic solvents such as ethyl acetate and isopropanol can also facilitate this process. Yeast, in the presence of a high level of common salt in the medium, begins to loses it water content in an attempt to

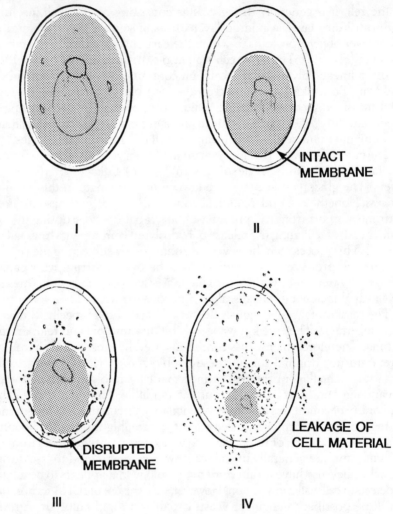

**Figure 8-2.** Plasmolysing yeast cell.

equilibrate its osmotic pressure with that of the surrounding medium. Under extreme conditions, as in the case of plasmolysis, cell content within the plasma membrane pulls away from the cell wall and tends to occupy only part of the cell volume within the cell wall. Prolonged plasmolysis results in cell death, marking the beginning of the cell degradative process (Fig. 8-2). Salt is often added to the pressed cake to achieve the best results.

Although common salt has been used as a processing aid to enhance plasmolysis, its beneficial effect as a bactericide or as a flavoring agent cannot be underestimated. Its bactericidal effect makes the process less susceptible to microbial contaminations. Because of this unique property of common salt, it has been possible to produce yeast extracts by plasmolysis with significantly lower microbiological counts. However, because of the high salt content, these products have limited use in the food industry. Low-sodium extracts are in higher demand and such extracts are made by autolysis rather than plasmolysis.

*Hydrolysis*

For the production of yeast extracts, this procedure requires the use of strong acids like hydrochloric acid to degrade the macromolecules such as proteins, carbohydrates, and nucleic acid within the cell to their corresponding subunits of higher solubility. Hydrolysis is considered the most efficient procedure for the solubilization of yeast. However, it is the least practical method for extract production on a commercial basis.

Unlike the autolytic procedure, this method does not require the use of yeast of high viability. In order to achieve maximum efficiency, a typical hydrolysis can be initiated with dried yeast, slurried to a solids concentration of 65-85%, followed by a treatment with concentrated hydrochloric acid. The optimal hydrolysis can be achieved at 100°C in a wiped film evaporator equipped with a reflux condenser. The degree of hydrolysis is controlled by the hydrolysis time. In commercial operations, 50-60% of the total nitrogen is converted to amino nitrogen in a 6-1 ratio under the conditions just described (Ziemba 1967). The highly acidic material is then neutralized with sodium hydroxide to pH 5-6, filtered and concentrated to a syrup, paste, or dried to a powder of approximately 5% moisture.

Although the hydrolytic procedure is well known for its higher yields and productivity, it has certain disadvantages that make it highly unattractive to certain extract manufacturers. A serious drawback is that the processing equipment has to be made of unreactive materials, or the surfaces have to be lined with glass because of the corrosive nature of hydrochloric acid. This modification would obviously make the process very cost-intensive. In addition, pressure vessels for hydrolysis have to be constructed to withstand required high pressures to meet the required safety standards.

Extensive use of concentrated acids can make the process more hazardous, and any accident in the plant could be dangerous and often costly. The high salt concentration in the final product could even make it less attractive for the food processor. Acid hydrolysis brings about the destruction of certain amino acids and vitamins, thereby reducing the nutritive value of the final extracts. There is also the possibility of generating chlorinated compounds (3-chloro-propanediol) that could be detrimental to human health. The presence of such chlorinated compounds in hydrolyzed vegetable protein (HVP) produced by a similar procedure have been reported recently. The disadvantages of acid hydrolysis seem to far outweigh the advantages of the process, thus making it highly unattractive for commercial extract manufacture.

## Autolysates

An autolysate is the total content of the yeast following autolysis. The product is in the form of a concentrated paste or dried without separation of the particulate matter. Hence, autolysates are less intense in flavor than an extract on an equal weight basis because of the dilution effect by the bland cell wall material.

Brewer's yeast cream used for autolysis is generally bitter due to the presence of beer solids and certain hop resins that remain attached to the surface of the yeast cell following a beer fermentation. Unlike in the extract manufacture where the cell walls are removed, the brewer's yeast cream intended for use in whole autolysate production requires an intense cleanup to minimize the carryover of hop resins into the process. Although the beer solids can be separated by use of a simple water wash, a complete debittering of the yeast cream entails a more rigorous treatment that often includes an alkaline wash. This treatment can loosen the hop components that are firmly attached to the yeast cell walls. A subsequent separation followed by a water wash may bring about a substantial reduction of the intensity of the bitterness in brewer's yeast. Few extract manufacturers rely on ion exchange technology for debittering of brewer's yeast.

At pH 5-6, the principal bitter components that include humulones and isohumulones present on the surface of the yeast cells are insoluble. These organic components are firmly bound to the cell surface, and very little can be removed by a simple water wash. At pH 5, only a small amount of humulone can be isomerized to form the more soluble isohumulone, owing to its low solubility. When treated with sodium hydroxide, the hop resins tend to form sodium salts that are readily soluble in water. Hence, by a simple adjustment of the brewer's yeast cream to pH 9 with sodium hydroxide, there is a high probability of solubilizing the bitter hop components. By use of a separation followed by a water wash, the undesirable bitter components in brewer's yeast cream can be considerably reduced prior to an autolysis. This treatment is vital for the production of autolysates, although it is of less significance in the extract production.

Unlike extracts, autolysates do not solubilize completely due to the presence of the insoluble cell wall material. Thus, they can be used in formulations where clarity is of less importance. The cell wall material present in a dry autolysate has an excellent water-binding capacity. Autolysates can also be used as flavor carriers. These unique features, combined with the characteristic flavor such autolysates impart, make them valuable ingredients in food formulations.

Flavor Enhancers

Although certain food products, like mushrooms and dried bonito, have often been used in the diets of early civilizations, mainly in the Orient, their significance was not fully recognized until their flavor-enhancing properties were discovered in Japan during the early years of the twentieth century. Monosodium glutamate (MSG) was isolated by Ikeda (1912) from sea tangle, a common sea weed, and Kodama (1913) identified the histidine salt of inosinic acid as the active flavor component of dried bonito. Almost 50 years later, Nakajima et al. (1961) and Shimazono (1964) reported that 5'-GMP was the main flavor component of the Japanese mushroom, shiitake. These flavor enhancers, which are now commercially available either as crude extracts or in the highly purified form, are particularly effective in improving the flavor of savory meats, gravies, sauces, and soups.

The old Japanese word *umami* is presently being used in the literature to describe the taste in foods elicited by glutamic acid (present in vegetables), disodium 5'-inosinate (present in fish and meat), and disodium 5-guanylate (present in mushrooms like shiitake). These unique components are known to stimulate the corresponding taste buds to bring about the full potential of the savory character in foods for maximum enjoyment. Recent studies in Japan have, in fact, shown that monosodium glutamate, which functions as an important component responsible for umami, has its special receptor sites in the olfactory and taste cells of aquatic animals separate from those sensory receptors that are specific for perceiving the four basic tastes: sweet, sour, salty, and bitter.

Recently, extensive research by sensory physiologists has strongly suggested that the flavor-enhancing property of 5'-GMP (guanosine monophosphate) is several times higher than that of 5'-IMP (inosine monophosphate) (Kuninaka 1986). A marked synergism has also been identified between glutamate and the flavor-enhancing nucleotides. The nucleotides that have been shown to best enhance the glutamate taste, as described above, are those that have a purine nucleus with a hydroxy group in the 6 position and a ribose moiety esterified in the 5' position with phosphoric acid. Both 5'-GMP and 5'-IMP meet these specific criteria.

These unique properties of both 5′-GMP and 5′-IMP offer the food processor the means to reduce or eliminate the use of MSG, thereby making the process cost-effective without distorting the flavor characteristics of the final product. Furthermore, use of relatively low levels of flavor-enhancing nucleotides in the recipe enhances the perceived flavor of sodium chloride (salt). Thus, the ability is provided for the food processor to reduce the salt level in the recipe without a major loss in salty taste.

Umami is a taste that has only been recognized since the turn of the century. It is now beginning to emerge as the fifth taste uniquely different from the other four basic tastes that characterize all food products (Yamaguchi 1987). Evidence in support of these findings also comes from feeding behavior (Torii, Mimura, and Yugari 1986) and basic research on taste receptors (Torii and Cagan 1980).

Recognizing the commercial potential of these flavor enhancers, the food industry began to focus its attention on the development of cost-effective methods to manufacture these products at a commercial level. Yeast was evaluated initially as a possible source of RNA because of its high RNA content (2.5-15%) (Nakao 1979). Yeast is also considered as an organism generally recognized as safe (GRAS), and it can be grown economically to produce large quantities of biomass containing high levels of nucleic acids. Further evaluations conducted with other microorganisms made it clear that yeast is the organism of choice for the extraction of RNA for the 5′-nucleotide production. Currently, there are microorganisms such as the mutants of *Brevibacterium* (e.g., *Brevibacterium ammoniagenes*) that can excrete large quantities of RNA into the medium, which can be further processed to 5′-nucleotides.

The RNA content of different yeasts is highly variable. *Candida utilis* has an RNA content of 10-15% on a dry solids basis. The actual level of RNA is dictated by the phase of the growth cycle in which the yeast is harvested for further processing. Because of the high RNA content, *Candida utilis* is often used for 5′-nucleotide production. Although baker's yeast *Saccharomyces cerevisiae* has a somewhat lower RNA level (8-11%), this strain can also be used for 5′-nucleotide-rich extract production. Control of process conditions is highly critical to maximize the RNA content in yeasts. Studies have indicated that to maximize RNA content the yeast has to be harvested during the logarithmic phase when the protein synthesis is at its peak (Katchman and Fetty 1955).

Maximum recovery of RNA is highly critical for process economics. The conventional method includes an extraction of RNA from dried yeast cells by treatment with hot alkaline sodium chloride (5-12%) solution. A soluble fraction is separated by centrifugation and the RNA is precipitated by the addition of an acid or ethanol (Nakao 1979). Following neutralization of the precipitate, the product can be spray-dried for use as a substrate for the production of 5′-nucleotides. One drawback in this procedure is due to the transformation of 3′-5′-phosphodiester linkages to 2′-5′-phosphodiester linkages as a result of isomerization reactions that occur when the extraction of RNA is conducted using a hot alkaline sodium chloride treatment. The 5′-phosphodiesterase enzyme is only specific for splitting 3′-5′-phosphodiester linkages. Every attempt must thus be made to limit the production of 2′-5′-phosphodiester linkages to maximize the 5′-nucleotide yields. Kuninaka et al. (1980) was able to achieve this goal by immediately neutralizing the RNA just at the time it was released from the yeast cell. This procedure included the treatment of the yeast packed in a column containing celite with dilute alkaline solution containing salt at a temperature below 40°C.

Another less complicated method used for the isolation of RNA from yeast includes an initial heat treatment of the yeast slurry followed by a selective precipitation of the released RNA by treating with acids. In general, most of the RNA within the cell is released to the environment by heating the yeast suspension at 90°-100°C for 1-3 hours. pH control at a 6-6.6 range prevents the nucleases in the yeast from degrading the RNA during the heating.

The production of 5′-nucleotides requires the hydrolysis of the crude RNA by means of an enzyme, 5′-phosphodiesterase. Cohn and Volkin (1953) first demonstrated the presence of enzyme 5′-phosphodiesterase activity in snake venom by identifying the presence of 5′-nucleotides in a reaction mixture containing RNA and snake venom. Despite its high efficacy, it cannot be used in the food industry for obvious reasons. Other important sources for 5′-phosphodiesterase enzyme include certain fungi such as *Penicillium citrinum* and certain *Actinomyces* such as *Streptomyces aureus*. In recent years, it has been possible to improve the 5′-phosphodiesterase level of these fungi several fold through genetic manipulations. The enzyme 5′-phosphodiesterase derived from *Penicillium citrinum* is now commercially available under the name Nuclease Pl. Due to its present high cost, this enzyme is currently used on an industrial scale in the immobilized form.

The presence of 5′-phosphodiesterase activity in certain germs of some plants was shown by Schuster (1957). The studies that followed led to a number of patents in France, Germany, and the United States describing the methods for the production of 5′-nucleotide-rich extracts from yeast using malt rootlet extracts. Malt rootlets are a by-product in the brewing industry, and hence could serve as a cheap enzyme source for the production of 5′-nucleotides from RNA or 5′-nucleotide-rich extracts directly from yeast.

Extracts prepared from malt rootlets generally contain at least two groups of competing enzymes for RNA. These enzymes, if present, could have an adverse effect on the final outcome of the process. The presence of nucleases and phosphatases will result in the degradation of the RNA to the basic units that do not have the flavor-potentiating characteristics.

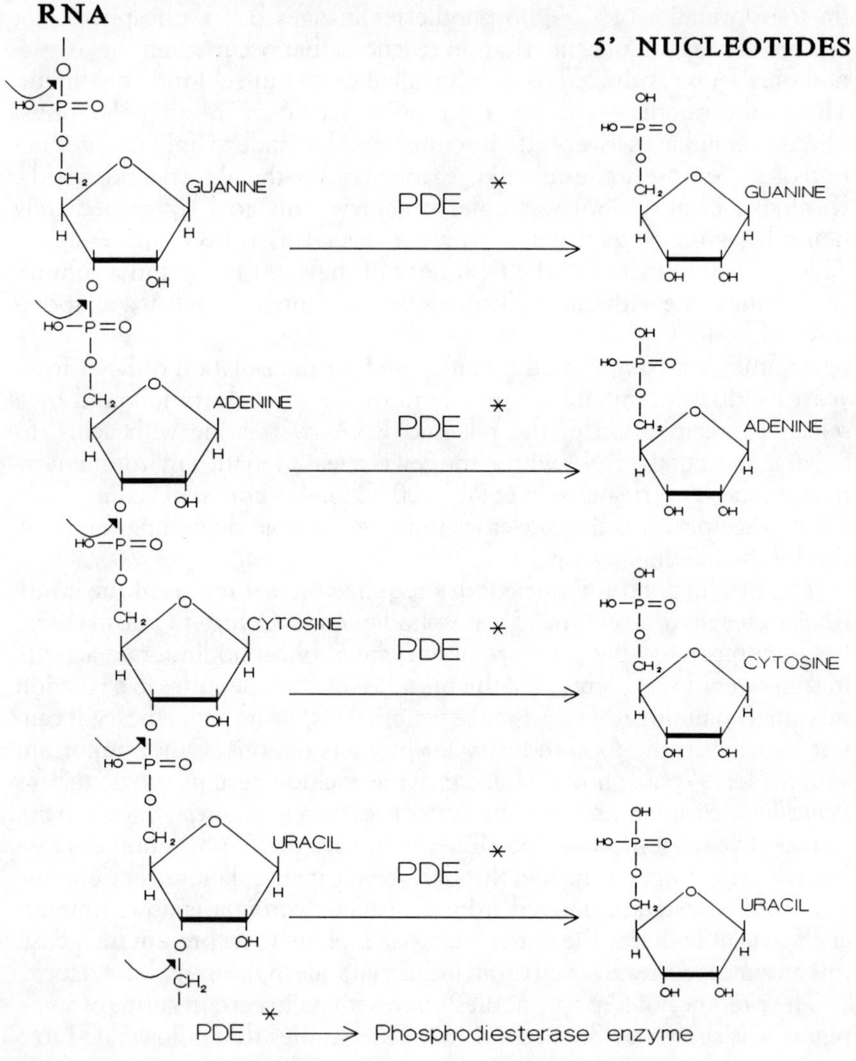

**Figure 8-3.** Degradation pattern of RNA in the presence of the enzyme phosphodiesterase.

Fortunately, these competing enzymes are thermolabile and can be destroyed at 60°C without seriously affecting the overall activity of the 5′-phosphodiesterase enzymes.

Extracts containing RNA can generate four types of 5′-nucleotides corresponding to the four bases present in the RNA when treated with the 5′-phosphodiesterase enzyme present in malt rootlet extracts (Figure 8-3). Only one of these 5′-mononucleotides, 5′-GMP, has the ability to impart

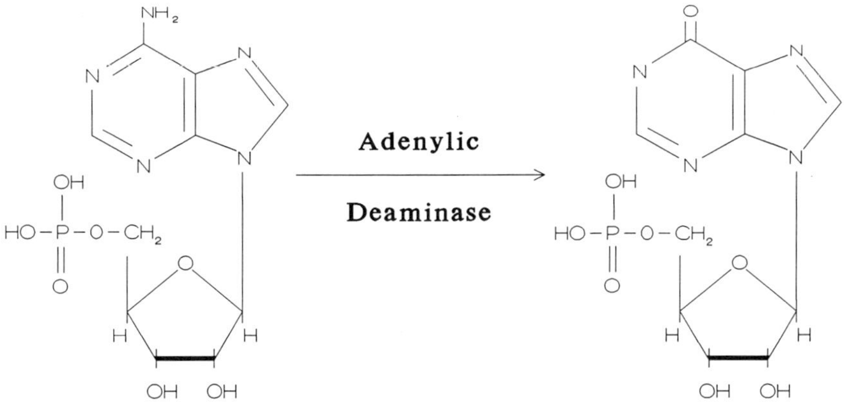

**Figure 8-4.** Conversion of 5′-AMP to 5′-IMP by the action of enzyme adenylic deaminase.

flavor-enhancing characteristics. Although 5′-AMP is incapable of imparting flavor enhancement, it is a precursor for another important flavor enhancer, 5′-IMP. This commercially valuable product can be made by the conversion of 5′-AMP to 5′-IMP through the mediation of the enzyme adenylic deaminase (Fig. 8-4). Although the two remaining 5′-nucleotides, 5′-CMP (cytidine monophosphate) and 5′-UMP (uridine monophosphate), are not important for the food industry, they have shown commercial value in the pharmaceutical industry for the preparation of antiviral drugs. The four 5′-nucleotides can be separated by use of ion exchange chromatography.

## NUTRITIONAL YEAST

Throughout the ages, yeasts have entered the human diet, often unwittingly, by way of their extensive use in the traditional fermented foods and beverages. Although the significance of fermented foods was not realized at the time, the nutritional value of such foods containing these microorganisms would have undoubtedly contributed to the good health and sustenance of these civilizations. Today yeast is generally recognized as an excellent source of proteins, B vitamins, fiber, and many other micronutrients. However, excessive intake of yeast has caused serious human health problems. These disorders were later identified as due to excessive consumption of nucleic acid present in yeast. This phenomenon will be discussed later in detail.

### Yeast Proteins

The crude protein value of yeast is generally expressed as $N \times 6.25$ where the nitrogen value (N) stays in the 7-9% range. The actual protein value is known to be lower than this amount due to the presence of 8-15% nucleic acid that accounts for part of the nitrogen in the yeast. The level of protein in yeast is generally dictated by the genetic constitution of the yeast and the conditions to which yeasts were exposed during the propagation.

Yeast proteins are generally high in lysine and deficient in sulfur amino acids. However, the protein efficiency ratio (PER) for baker's yeast is 1.8, whereas that for casein, a highly acceptable protein for food use, is 2.50. Accordingly, the nutritional value of baker's yeast is 72% that of casein. The addition of 0.5% by weight of methionine is known to increase the PER value of baker's yeast proteins to 2.77. Hence, although yeast proteins "as is" are inferior to animal proteins, such deficiencies can be minimized in mixed diets.

Although yeast contains approximately 50% crude protein, its consumption as a protein source for human nutrition is limited by the presence of a high level of nucleic acid. Yeast contains 6-15% nucleic acid as compared to 2% in meat products. It is known that high nucleic acid intake in the human diet could increase the uric acid level in blood at physiological pH as described in Chapter 9. When the uric acid level reaches a certain concentration in the bloodstream, it tends to precipitate or crystallize at the joints, causing gout, arthritis, or renal stones in the urinary tract. Research has shown that the safe level for nucleic acid intake for humans is 2 g/day (Anon. 1970), which amounts to 20 g yeast solids/day considering that the yeast has 10% nucleic acid. This amount corresponds to approximately one-sixth of the recommended daily allowance (RDA) of 65 g protein/day for a 70-kg adult male.

The amino acid composition of four industrially important yeasts expressed as percentage of protein is listed in Table 8-2. Although the yeasts have a high lysine content, a deficiency of methionine compared to that of a well-balanced protein is apparent.

The use of higher levels of yeast protein in food formulations for human consumption requires the reduction of the nucleic acid level in the product. The level of RNA in the yeast varies with the growth phase with minimal RNA found during the stationary phase. The introduction of certain inhibitors such as nitomycin C and actinomycin during the stationary phase has been recommended by Gatellier and Gilkamans (1972) to achieve low RNA-containing yeast. Yeast strains genetically low in RNA can also be selected for the production of low RNA yeast. These methods may still be inadequate to bring about a change in RNA within the yeast sufficient to make it acceptable as a direct source of protein for human nutrition. It thus may seem probable that, in order for yeast proteins to compete aggressively with vegetable proteins and to share the protein market, it is necessary that such proteins be further processed to minimize the level of nucleic acids.

## NUTRITIONAL YEAST

Table 8-2. Amino Acid Content of Selected Yeasts of Interest (% protein)

| Amino Acid | Candida utilis (Sulfite Waste Liquor) (Peppler 1965) | Candida utilis (Amoco Food Co. 1974) | Kluyveromyces marxianus (Bernstein and Plantz 1977) | Saccharomyces cerevisiae (Reed and Peppler 1973) |
|---|---|---|---|---|
| Alanine | 5.8 | 5.5 | | |
| Arginine | 5.4 | 5.4 | | 5.0 |
| Aspartic acid | 9.2 | 8.8 | | |
| Cystine | | 0.4 | | 1.6 |
| Glutamic acid | 15.6 | 14.6 | | |
| Glycine | 3.6 | 4.5 | | |
| Histidine | 1.2 | 2.1 | 2.1 | 4.0 |
| Isoleucine* | 3.8 | 4.5 | 4.0 | 5.5 |
| Leucine* | 7.6 | 7.1 | 6.1 | 7.9 |
| Lysine* | 4.8 | 6.6 | 6.9 | 8.2 |
| Methionine* | 1.1 | 1.4 | 1.9 | 2.5 |
| Phenylalanine* | 8.6 | 4.1 | 2.8 | 4.5 |
| Proline | 6.0 | 3.4 | | |
| Serine | 5.0 | 4.7 | | |
| Threonine* | 5.4 | 5.5 | 5.8 | 4.8 |
| Tryptophan* | 2.4 | 1.2 | 1.4 | 1.2 |
| Tyrosine | 6.2 | 3.3 | 2.4 | 5.0 |
| Valine* | 3.8 | 5.7 | 5.4 | 5.5 |

*Essential amino acid for human nutrition.

There are several methods published in the literature for the reduction of nucleic acid content in yeast and other microorganisms. A method proposed by Decker and Dirr (1944) requires a direct extraction of C. utilis cells with 10% NaCl and sodium acetate at 80°C for 72 hours. This procedure has been effective in reducing the purine nitrogen content by 80%. There have also been several attempts to reduce the RNA content by activation of the ribonuclease enzymes already present in the yeast cells. Tannenbaum, Sinskey, and Maul (1973) were able to activate the endogenous RNAase in yeast by a heat shock treatment at 68°C for a few seconds followed by an incubation period of 2 hours at 53°C. These methods have provided a suitable source of yeast with low nucleic acid for protein extraction to meet nucleic acid limits acceptable for human nutrition.

Reduction of nucleic acid is often carried out following the separation of yeast proteins in crude form from the whole cell. Yeast is ruptured by any of the methods described in the literature. These include high-pressure homogenization, colloid mill, sonic disintegration, freeze-thaw treatment, use of lytic enzymes, or a combination of these, for maximum recovery.

Rupture of the cells by any of the procedures listed above results in cellular debris and a soluble cytoplasmic fraction containing a high degree of colloidal matter. Any attempt to precipitate the protein fraction from the soluble fraction at pH 4.3-4.5 would also result in the coprecipitation of nucleic acids. Other practical methods are thus necessary to recover the yeast proteins with the least amount of nucleic acid so that the isolated protein can serve as an acceptable source of protein in the human diet.

Newell, Robbins, and Seeley (1975) observed the possibility of reducing the RNA in the isolated proteins from 11.7% to 1-2% by heating the soluble cytoplasmic fraction of cell homogenate at high-temperature-low-alkali conditions (e.g., 75°-85°C at pH 10-10.5 for 1-4 hours) or at low-temperature-high-pH conditions (e.g., 55°-65°C at pH 11.5-12.5 for 1-2 hours) prior to acid precipitation of the proteins. Newell, Seeley, and Robbins (1975) were also able to reduce the nucleic acid content in isolated proteins to less than 3% by treating the soluble fraction of yeast following homogenization with exogenous nucleases such as phosphodiesterase enzyme derived from malt roots.

A method developed by Robbins (1976) required the heating of the soluble cytoplasmic fraction of yeast to over 100°C for 10 seconds to 60 minutes at pH 6.8 to achieve a protein fraction containing approximately 1-2% RNA. Kinsella and Shetty (1982) found that by derivatizing the proteinaceous material with cyclic anhydrides, it was possible to recover the protein with low levels of nucleic acids. The method relies on the dissociation of protein-nucleic acid complexes by changing the ionic configuration by derivatizing with cyclic anhydrides. The derivatized protein separated from the mother liquor can then be regenerated to the original native protein. The protein-rich product made by this method is expected to contain less than 3% nucleic acid.

In the past two decades there has been much research on the extraction and characterization of yeast proteins. One important function is that these proteins can be spun into fibers and texturized (Daly and Ruiz 1974). However, such isolated yeast proteins can be more expensive than soy proteins mainly due to the high production costs. Unless some special functional property is identified in yeast proteins to make them attractive to the food manufacturer, the demand for yeast protein with the existing functional properties and high cost will be small.

## Vitamins

Vitamins are organic molecules present in all living systems commonly in trace quantities, yet they are vital for normal cell functions such as growth, maintenance, and reproduction. Most species are unable to synthesize one or more of these vitamins and are dependent on exogenous sources. Yeast contains predominantly the vitamins of the B complex, thus serving as an excellent source of B vitamins for human and animal nutrition. The B complex includes a group of water-soluble organic compounds derived from purines and pyrimidines ($B_1$, $B_2$, $B_6$, niacin, and folic acid), complexes like porphyrin-nucleotide ($B_{12}$), and amino acid-carboxylic acid complexes (biotin, pantothenic acid). These participate in enzymatic reactions as enzyme activators or coenzymes ($B_1$, $B_6$, $B_{12}$, niacin, biotin, folic acid, and pantothenic acid), in nucleic acid synthesis ($B_{12}$, biotin, and folic acid), and as activators of mitochondrial functions ($B_2$ and niacin). The presence of these important vitamins in yeast makes it an excellent source of these growth factors for human nutrition. Yeast can serve as a satisfactory starting material for making yeast extracts for use in microbiological growth media.

Inactive dry yeast is frequently used as a vitamin supplement rather than as a source of protein in food formulations. Some dry yeast products are even fortified with vitamins like $B_1$, $B_2$, and niacin to meet certain special requirements of vitamin tablet manufacturers. Table 8-3 lists the concentration of vitamins of selected yeasts, RDA, and the percentage of the RDA met by the supplementation of the safe limit of 20 grams of dry yeast in the human diet. Although the daily requirements for most of the B vitamins are not fully met by the recommended amount of yeast intake, their contribution to meeting the B complex vitamin requirements in the human diet is substantial. However, yeast does not provide vitamin C, vitamin $B_{12}$, and fat-soluble vitamins like A, E, K, and D. Nevertheless, dry yeast can function as a valuable vitamin source when used in combination with other food ingredients, as is generally the case with human diets.

Table 8-3. Percent of the RDA for Vitamins Satisfied by a Daily Intake of 20 Grams of Nutritional Yeast

| Vitamins | $\mu g/g$ Dry Yeast* | $mg/20\,g$ Dry Yeast | U.S. RDA (mg) | % of RDA in 20 g Dry Yeast |
|---|---|---|---|---|
| Thiamine ($B_1$) | 120 | 2.40 | 1.5 | 160 |
| Riboflavin ($B_2$) | 40 | 0.80 | 1.7 | 47 |
| Niacin | 300 | 6.00 | 20.0 | 30 |
| Pyridoxine ($B_6$) | 28 | 9.56 | 2.0 | 28 |
| Pantothenic acid | 70 | 1.40 | 10.0 | 14 |
| Biotin | 1.3 | 0.026 | 0.3 | 9 |
| Folic acid | 13 | 0.260 | 0.4 | 65 |
| Vitamin $B_{12}$ | 0.001 | 0.000006 | 0.006 | 0.3 |

*Primary grown dried nutritional yeast, Universal Foods Corporation.

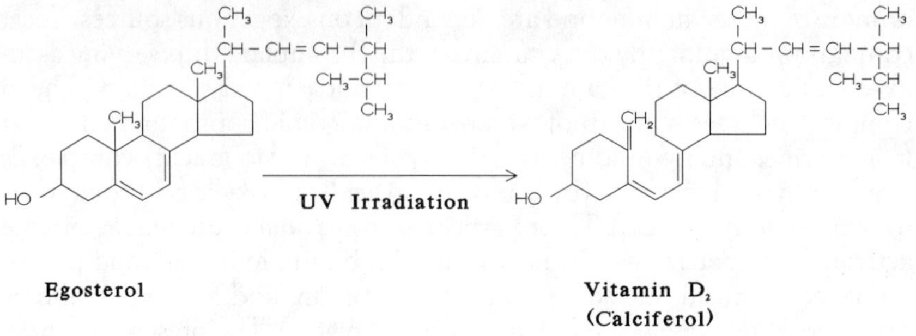

# 390 YEAST-DERIVED PRODUCTS

**Figure 8-5.** Effect of UV irradiation on ergosterol.

Baker's yeast contains 7-10% ergosterol located mainly in the membranes. It serves as the precursor for vitamin $D_2$, otherwise known as calciferol. This conversion is known to occur when ergosterol is irradiated with ultraviolet light (Fig. 8-5). Baker's yeast does not have the precursor 7-dehydrocholesterol; hence, it is unable to synthesize its own vitamin $D_3$.

## Mineral Yeasts

Yeast takes up a substantial quantity of macronutrients and micronutrients from the surrounding medium during growth. The mineral or ash content amounts to approximately 8% on a dry solids basis. Although the concentrations of potassium and phosphorus of dry yeast are higher than that of other minerals like Ca, Mg, and S, the mineral contribution made by an allowable daily serving of 20 grams of dry yeast is minor considering the high mineral content in the bulk of an average human diet.

Nevertheless, recent nutritional studies have shown that certain diets are deficient in some important trace elements. Supplementation of these diets with dry yeast has partially or wholly alleviated these dietary problems. Such inadequacies in the diet were later determined to be due to deficiencies of trace elements like chromium, selenium, and molybdenum. Brewer's and baker's yeast contain these elements in trace quantities. However, specially produced yeast products are now available in the marketplace with higher potencies for these micronutrients. A more detailed account of the significance of these special selenium- and chromium-containing yeasts is presented in Chapter 9.

## COLORANTS DERIVED FROM YEAST

In the last decade or so, there has been considerable interest worldwide to develop food colorants from sources that are of natural origin. Although

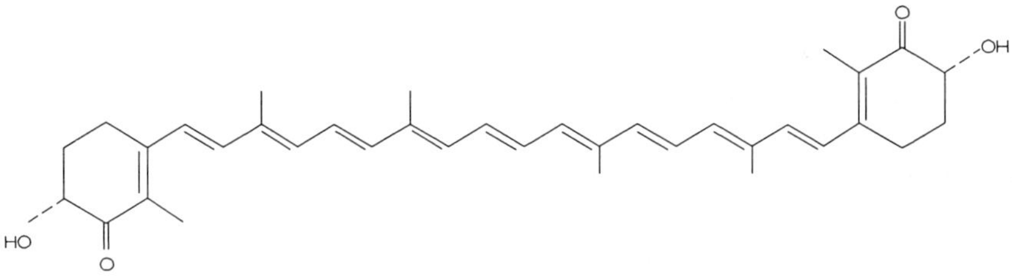

[Chemical structure diagram]

**ASTAXANTHIN (3R, 3R' ISOMER)**

**Figure 8-6.** Structure of astaxanthin.

different types of natural pigments are widely distributed in the plant and animal kingdoms, a variety of carotenoid pigments of various pink and yellow shades are also being produced by a number of yeast species. Yeast offers considerable advantages over other natural sources in that it is inexpensive to grow on an industrial scale. Considering this unique feature, yeast has now begun to emerge as a cost-effective source for the production of natural food colorants.

The best-known genera among the pigmented yeasts are *Rhodotorula, Rhodospondium, Cryptococcus, Sporidiobolus,* and *Sporobolomyces*. These genera are characterized by their ability to produce pigments like $\beta$-carotene, $\alpha$-carotene, torulene, torularhodin, plectaniaxointhin, and 2-hydroxyplectanisxanthin. Carotenoids produced by different *Rhodotorula* species contain primarily the two oxygenated compounds torulene and torularhodin, and the ratio of two major pigments determines the intensity of the shades of red in the different species. *Phaffia rhodozyma* is a basidiomycetous yeast that has been recently discovered (Miller, Yoneyama, and Soneda 1976) and is strikingly different from the other pigmented yeasts in producing the carotenoid pigment astaxanthin (3.3'-dihydroxy-$\beta$, $\beta$-carotene-4, 4'-dione; Fig. 8-6) (Andrews, Phaff, and Starr 1976).

Much of the interest in astaxanthin has arisen from the fact that this pigment is widely present in the animal kingdom. It is responsible for the characteristic color in the plumage of birds like flamingos; in the exoskeleton of marine invertebrates like lobsters, crabs, and shrimps; and in fishes like trout and salmon where their pink color is due to the storage of the pigment in the flesh. When these fish or crustaceans are raised in pens with limited mobility they often lack the characteristic pink color. Such products have limited acceptability. Studies conducted by Johnson, Conklin, and Lewis (1977) have shown that *Phaffia rhodozyma* is a potential source of astaxanthin to restore the pink color in the flesh of pen-reared salmonids. Their studies with rainbow trout indicated that pigments of intact yeast cells were not incorporated into the flesh of fish due to nonavailability. However, incorporation of *Phaffia rhodozyma* in the form of broken cells showed positive results (Johnson et al. 1978).

*Phaffia rhodozyma*, which is a basidiomycetous yeast, is significantly different from ascomycetous yeasts specifically with respect to the cell wall structure. *Phaffia*, for example, has a cell wall that is made up of regular sugars, glucose, and mannose and a unique sugar xylose that is absent in the genus *Saccharomyces*. Additionally, ultrathin sections of *Phaffia* cells viewed under electron microscopy show the presence of a capsule (Fig. 8-7). Certain viscous polymers form a sticky layer on the surface of the cells. Unlike *Saccharomyces*, *Phaffia* has been difficult to autolyze, probably due to these unique differences in cell structure.

A significant influence of environmental conditions on the growth rate and on pigment production has been observed in *Phaffia rhodozyma*. The optimum temperature for yeast growth and the pigment formation was found to be 20°-22°C at an optimal pH of 5, and a carbon source like D-cellobiose has shown improved pigment production. A high level of aeration during growth was found essential for optimal yields and astaxanthin production. Yields of astaxanthin under these conditions have been 0.5 mg/g dry yeast (Johnson and Lewis 1979; Johnson et al. 1978). Although

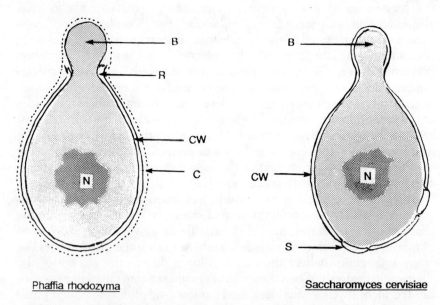

**Figure 8-7.** Schematic diagram of budding of *Phaffia rhodozyma* and *Saccharomyces cerevisiae*: B, bud; CW, cell wall; N, nucleus, R, collarlike remnants of cell wall formed by repeated budding at the same site; and S, bud scar.

this yeast strain is a slow grower, by engineering into the cell the ability to hyperproduce the pigment, and by constructing strains that have improved growth characteristics, a more economical production of astaxanthin will become possible under commercial conditions.

## YEAST-DERIVED ENZYMES

### Invertase

Invertase is an enzyme capable of splitting the sucrose molecule into glucose and fructose, a noncrystallizable mixture that is known as invert sugar. It is sweeter than sucrose on an equivalent weight basis due to higher sweetening power exhibited by fructose. Due to these characteristic properties, invert sugar is generally regarded as an important ingredient in the food and confectionary industries.

Sucrose inversion is largely effected at commercial scale by using the enzyme invertase ($\beta$-D fructofuranoside fructohydrolase; EC 3.2.1.26). Baker's yeast is very rich in invertase and represents the main source of almost all the commercially available invertase enzymes. The term *inversion* is commonly used to described the hydrolysis of sugar because of the change in optical rotation ($a_D$) that occurs from 66.7° to $-20$°C as a result of sucrose hydrolysis to glucose and fructose (invert sugar), in this case, by the action of the enzyme invertase.

There are two important applications for invertase enzyme in the food industry. In the manufacture of confectionaries, invertase is used mainly in the production of soft-centered candies. The use of this enzyme provides the plastic consistency to the center and prevents sucrose from crystallizing during storage. The enzymatic hydrolysis that brings about the creamy consistency to the center generally takes place after casting and subsequent to covering the hard center with chocolate. The other important application is in the inversion of syrups for the production of high-test molasses used extensively in the fermentation industry. The high-test molasses, if not inverted, would soon tend to crystallize due to the presence of high concentrations of sucrose. Inversion of sucrose to a mixture of glucose and fructose prevents crystallization, raises the osmotic pressure, and results in a microbiologically stable concentrate with a consistency that is easy to handle.

The production of invert syrups from highly refined sucrose solutions is generally carried out by acidification. However, this method is not applicable to unrefined sucrose solutions. These unrefined media are known to exhibit high buffering capacity so that the amount of acid required for inversion to occur would become cost-prohibitive. In that case, the enzyme process should provide an acceptable hydrolysis. Often, the enzyme source for such applications is baker's yeast dried under conditions that retain the invertase activity. However, for use in the candy industry more refined invertase preparations are preferred.

Yeast invertases have been extensively studied for over a century. The genetic analysis has established the existence of at least six genes, designated SUC, any one of which is sufficient for invertase formation and subsequent fermentation of sucrose or raffinose (Mortimer and Hawthorne 1969). Yeasts (*Saccharomyces*) carrying the SUC genes produce two forms of the sucrose-cleaving enzyme invertase: a secreted glycosylated enzyme and an intracellular, nonglycosylated enzyme. The external glycosylated form is located on the outside of the plasma membrane that is insoluble, and the internal soluble form is generally present in the vacuoles and vesicles. The external invertase consists of a mannoprotein with a molecular weight of 270,000, approximately half of which is due to the carbohydrate, predominantly mannose, with about 3% glucosamine. The smaller internal invertase contains no glycan and has a molecular weight of 135,000 (Table 8-4). The protein moieties of the two forms are very similar, each consisting of a dimer of approximately 60,000 dalton subunits.

External and internal invertases show similar enzymatic characteristics, although there are deficiencies in amino acid composition. Nucleotide sequence analysis of the 5' ends of the SUC genes have shown that a 1.9-kb mRNA encodes a signal peptide-containing precursor to the secreted invertase units, and that the 1.8-kb mRNA does not include the complete signal sequence and it is translated to yield intracellular invertase. The two mRNAs are differently regulated in response to glucose concentrations as is synthesis of the two forms of invertase. The properties of the two types of enzymes are very similar except for the wide pH stability of the external enzyme.

Table 8-4. Properties of External and Internal Invertases of Yeast

|  | External | Internal |
| --- | --- | --- |
| Molecular weight | 270,000 | 135,000 |
| Mannan content (%) | 50 | 3 |
| Glucosamine content (%) | 3 | 0 |
| Specific activity, units/mg protein | 2,700 | 2,900 |
| Michaelis constant (km) for sucrose | 26 mM | 25 mM |
| pH, optimum activity at 30°C | 3.5-5.5 | 3.5-5.5 |
| pH, optimum stability at 30°C | 3.0-7.5 | 6.0-9.0 |

*Source:* From Gascon, Neumann, and Lampen 1968.

The invertase enzyme is found stable even at 55°C, and inactivation of the enzyme is known to occur at 65°C (Arnold 1971). The stability of these enzymes to heat appears to be due to the protection role of the phosphomannan moities present in the external enzyme. It is known that there are at least 20 polysaccharide chains of varying lengths attached to the protein moiety (Tarantino, Plummer, and Miley 1974). It therefore seems likely that the protein tertiary structure will be stabilized by these carbohydrate-protein interactions as well as by possible cross-linking in the polypeptide chains. There are considerable variations between different strains in their ability to produce invertase. In general, the baker's yeast strains that are used for the production of active dry yeast have a lower invertase activity than those used for the production of compressed yeast.

The most common method of extracting and concentrating invertase from yeast is to autolyze the baker's yeast and to recover the invertase from the filtered autolysate by precipitation or concentration. The precipitating agents commonly used are ethanol (Ohashi and Kozutsumi 1966), acetone (Takei et al. 1967), and ammonium sulfate (Sotskaya, Smirnov, and Tikhomirov 1965). The precipitate thus formed readily dissolves in water. The filtered autolysate may be concentrated under vacuum. However, it is essential to remove the coloring matter by treating the filtered autolysate with activated carbon. The invertase enzyme can also be concentrated by ultrafiltration. Invertase is commercially available in the form of a concentrate containing about 50% by weight of polyols. Glycerine is commonly used to stabilize the solution. The pH is adjusted to between 4 and 5. Details of some of the commercial processes have been reported by Meister (1965).

The enzyme activity for commercially produced invertase enzyme is expressed in a number of ways. It is either expressed as the K value (unimolecular reaction velocity constant) or as inverton units (activity/unit weight of enzyme preparation). Commercially available sacarasa invertase preparations have an activity of 1200-1800 inverton units/g, which is equivalent to K values of 2-3 calculated according to the AOAC method 31.024 (published in 1970). One inverton unit represents the invertase activity that will hydrolyze 5.0 mg of sucrose per min at 25°C. Invertase activity can also be expressed in Sumner units. This unit represents the amount of invertase that forms 1.0 mg of invert sugar in 5 min in 6 ml of 5.4% sucrose solution at 20°C and at pH 4.7. In commercial preparations 1 K unit is equivalent to 570 invertons or 15,500 Sumner units. Specially prepared invertase-rich dry yeast yields a K value in the 3-5 range.

In the preparation of high-test molasses, 240 grams of sacarasa (1200-1800 inverton units/g) is rehydrated in water at 32°-38°C for the treatment of 1000 liters of cane syrup at 50°-55° Brix at a pH range of 4.5-5.5. The incubation period is about 9-12 hours at 54-63°C. The high-test molasses made this way is, however, more expensive than black strap molasses on a sugar basis.

## Lactase

Lactase or $\beta$-galactosidase (B-D-galactoside galactohydrolase; EC 2.2.1.23) catalyzes the hydrolysis of lactose milk sugar into its principal constituents, glucose and galactose. This enzyme is widely used in the food industry for the treatment of frozen whole milk concentrates to prevent crystallization during frozen storage, and most importantly to make the milk-based products acceptable for those who cannot tolerate lactose.

Although the lactase enzyme is not present in baker's yeast, it is present in a number of common yeasts such as *Candida pseudotropicalis* and *K. marxianus*. *Kluyveromyces* species appear to be one of the important species for $\beta$-galactosidase production. Methods for producing this enzyme by *K. marxianus* have been described (Stimpson 1954; Davies 1964; Young and Healy 1957). The yeast enzymes have an optimum pH between 6 and 7, and hence they are suitable for lactose hydrolysis in regular milk and skim milk and their concentrates. Molds, particularly *Aspergillus niger* or *A. awamori*, have been used as sources of the enzyme lactase. However, these mold-derived enzymes are active at pH 4 to 5, thus making them useful for hydrolyzing lactose in acid whey.

Among humans, the lactase activity within the large intestine remains high throughout life if milk is consumed on a regular basis. In the early 1960s, it had become clear that certain population groups were unable to tolerate the lactose level generally present in milk, and these groups happened to be those who did not consume milk in their adult life. In general, incidents of lactase deficiencies have been high among American Negroes, African Negroes, South American Indians, Asiatic Indians, Chinese, Japanese, Thais, and Australian Aborigines. In lactose-deficient individuals, the ingested lactose reaches the intestine without being affected and finally undergoes microbial fermentation within the lower intestine, causing abdominal bloating, rumbling, diarrhea, and other gastrointestinal disturbances. Hence, to improve the palatability of milk-based products for these populations, the lactose content must be reduced significantly. These problems have received widespread attention and are now being handled through the use of lactase enzyme.

The genetics associated with lactose utilization in certain yeasts have been presented in Chapter 2. There are a few species of yeast, for example, *K. marxianus* and *C. pseudotropicalis*, that are capable of utilizing lactose. The lactose-utilizing ability by these organisms is due to the induction of both lactose transport proteins, which are termed the lactose permease, and an intracellular β-galactosidase enzyme several folds above the standard basal level. The structural gene for β-galactosidase has been identified and designated LAC 4 (Sheets and Dickson 1981)

β-galactosidase enzyme has been isolated and purified from a variety of microbial sources such as bacteria, as in the case of *Escherichia coli* (Hu, Wolfe, and Reichel 1959), fungi (Borglum and Sternberg 1972), and yeasts (Biemaan and Glantz 1968). However, *Kluyveromyces* species has been considered a principal source for β-galactosidase production. Methods for producing this enzyme from *K. marxianus* have been described (Stimpson 1954; Davies 1964; Young and Healy 1957). The enzyme is available in a purified form or as vacuum-dried or spray-dried products. In the case of the latter, the chosen yeast in the live form is rapidly cooled to $-18°C$. Under these conditions, the fermentative system of the yeast is known to be destroyed while retaining the β-galactosidase activity. The product is then vacuum-dried or spray-dried to a moisture content of approximately 5-8%. Various other preparation methods of the enzyme are also described in the literature.

## Melibiase

The enzyme melibiase or α-galactosidase (EC 3.2.1.22) is capable of hydrolyzing the disaccharide melibiose to glucose and fructose. Because of the unique ability of this enzyme to degrade melibiose, there is likely to be a potential market for this enzyme, specifically to improve the production yield in the beet sugar industry.

In the production of beet sugar, raffinose is a trisaccharide (galactose-glucose-fructose) that is known to hamper the crystallization process and thereby reduce extract yields. The result is not only a high concentration of raffinose but also a higher proportion of sucrose in the final molasses. However, the attempts that have been made to eliminate raffinose have provided a means to improve extract yields. The enzyme that is likely to produce the most optimal results is melibiase, which splits the raffinose molecule to galactose and sucrose. This treatment not only increases the recoverable sucrose in the syrup but also minimizes the level of raffinose, thereby enhancing the crystallization process.

A potential application for melibiase enzyme is in the baker's yeast industry. Although the predominant sugar in both cane and beet molasses is sucrose, beet molasses contains raffinose in the 0.5-2.5% range. Only one-third of this trisaccharide molecule is utilized by the baker's yeast due to the presence of the enzyme invertase (β-fructosidase). The latter enzyme converts raffinose to fructose and melibiose. For complete utilization of the raffinose molecule the enzyme melibiase should either be added to the wort, wort should pass through a column of beads containing immobilized melibiase enzyme, or the baker's yeast strain must be genetically altered to produce the necessary enzyme, melibiase, that would hydrolyze the sugar melibiose. Although the first and second approaches seem uneconomical, efforts are currently being made to construct industrial strains with melibiose-utilizing ability.

Genes in *Saccharomyces* species that confer the ability to ferment melibiose have been designated MEL (von Borstel 1969), whereas strains of *S. italicus* var. *melibiosi* possess at least six apparently unlinked MEL genes (Roberts, Ganesan, and Haupt 1959). *S. carlsbergensis* has been variously reported to possess one (Gilliland 1956) or two (Lindegren, Speigelman, and Lindegren 1944) genes for melibiose utilization. Like most other exoenzymes of yeast, the $\alpha$-galactosidase is a glycoprotein. Even though this enzyme is not available at commercial scale, it could at least be a useful enzyme for the beet sugar industry.

## Other Enzymes

Yeast is considered a suitable source for the recovery of a variety of enzymes because it is available in large quantities and at a reasonable cost. Nevertheless, these enzymes are not recovered in quantities to the same extent as lactase or invertase because of their limited market potential. Some of these enzymes include alcohol dehydrogenase, hexokinase, L-lactate dehydrogenase, glucose 6-phosphate dehydrogenase, phenylalanine ammonia-lyase, and glyceraldehyde 3-phosphate dehydrogenase.

Baker's yeast contains a high level of constitutively formed alcohol dehydrogenase enzyme and the commercial recovery of this enzyme serves at least two principal objectives: (1) When the enzyme is of high purity it can be used as a biocatalyst in biochemical studies; and (2) in crude form, as in the form of yeast pellets, it is available in certain countries to be taken orally for suppressing alcoholic intoxication. Whether these products perform the intended objective is uncertain. It is assumed that the effect of consumed ethanol could be minimized perhaps by the conversion of the ethanol to acetaldehyde in the presence of the NAD within the cell. However, the conditions within the stomach may not be most conducive to such a reaction.

Phenylalanine ammonia-lyase (EC 4.3.1.5) which occurs in *Rhodotorula glutinis* is capable of catalyzing the deamination of L-phenylalanine to trans-cinnamic acid and ammonia. Phenylalanine ammonia-lyase besides playing an important role in phenylalanine metabolism, is of interest in the field of medicine as a possible treatment for phenylketonuria. At present, this disease is treated by eliminating phenylalanine from their diets. The new treatment can involve oral ingestion of phenylalanine ammonia-lyase which, on entering the small intestine, converts free phenylalanine produced by proteolysis to cinnamate, thereby minimizing the intake of phenylalanine. A considerable amount of research effort is now being made to protect the enzyme from proteolytic inactivation within the human gut (Gilbert and Jack 1981).

## PRODUCTS OF PHARMACEUTICAL AND COSMETIC VALUE

### Skin Respiratory Factor

A yeast-derived product with wound-healing properties was first demonstrated by Sperti (1943). This product has been commercially available since then for the production of hemorrhoidal preparations and is often referred to as skin respiratory factor (SRF) or live yeast cell derivative (LYCD).

It is known that wound healing is directly linked to the oxygen consumption by the cells in and around the damaged tissue. Any treatment that could bring about an increase in the rate of respiration in the damaged tissue could, in turn, enhance the rate of wound healing. SRF prepared from yeast is known to exhibit certain biological properties capable of enhancing the oxygen uptake rate, thereby stimulating wound epithelialization and early angiogenesis. Its effectiveness in skin care has made it a commercially attractive ingredient for the cosmetic and pharmaceutical industries.

Although the active component associated with wound healing has not been identified or isolated as yet, a crude preparation is presently available in commercial quantities. The crude product was first prepared by Sperti (1943) by refluxing fresh baker's yeast with 95% ethyl alcohol for 4 hours at 60°-70°C. For maximum recovery of the active component, the residue from the initial extraction was again refluxed with 50% ethyl alcohol under similar conditions. The combined filtrates were then concentrated under reduced pressure with temperatures maintained below 60°C. The commercial preparations exist either as paste or powder.

Goodson et al. (1976) reported the results of the first systematic study of the effect of SRF preparations on wound healing. These studies included tests on the skins at excissional surgery from three patients: on human fibroblasts, on human polymorphonuclear leucocytes, and on the skin of white rats and rabbits. The change in collagen production as a result of SRF activity was evaluated by use of $^{14}C$-labeled proline. Results of these studies clearly indicated the benefical effect of SRF on wound healing.

Despite extensive usage of yeast extract rich in SRF activity in many over-the-counter (OTC) preparations, the active principle associated with wound healing needs to be identified and characterized. Recent studies in our laboratory and certain other laboratories have shown that a low-molecular-weight compound that passes through a dialysis membrane is perhaps the active component that is capable of enhancing respiratory activity in damaged tissues. This active product is, however, known to be stable at autoclaving temperatures.

A unit of SRF activity is expressed as the amount of SRF in milligrams required to increase the oxygen uptake of 1 mg of dry weight of rat abdominal skin by 1% following 1 hour of treatment. The polargraphic method is presently being used to determine the SRF activity of test materials. Generally, a gram of SRF contains 8000-12,000 units of activity. Commercial ointment made using this product claims 2000 units of activity per ounce as an effective level for promoting wound healing. In 1972, the Food and Drug Administration (FDA) published their findings in the *Federal Register*, vol. 43, no. 1511, after reveiwing the data submitted to a panel on the efficacy of SRF. The panel's conclusion was that the available data were insufficent for the FDA to permit final classification of its effectiveness as a skin protectant for OTC use.

## Glycan

The term *glycan* is used broadly to refer to commercial preparations of the insoluble yeast cell wall fraction. It is generally recovered as a by-product during the production of soluble extracts from baker's or brewer's yeast. Sometimes, the insoluble cell wall material is comminuted and washed adequately to improve certain functional properties of the product. Most often, it is commercially available as a spray-dried powder.

Yeast cell wall material is made up predominantly of glucan, mannan, and chitin polymers that are nutritionally nonfunctional in the human digestive tract. These three components account for more than 80% of the dry matter of the glycan fraction. Nonetheless, $\beta$-glucan, which is a homopolymer of glucose linked through $\beta(1\rightarrow3)$, $\beta(1\rightarrow4)$ or $\beta(1\rightarrow6)$-D glycosidic bonds, is the predominant polysaccharide present in the yeast cell wall. It often amounts to 12-14% of the yeast cell dry matter. Two glucan types have recently been identified (Phaff 1971) on the basis of their solubility in alkali. The alkali-soluble glucan accounting for 15-20% w/w (weight/weight) of total glucan does not impart any structural function to the cell wall. The alkali-insoluble fraction serves as the structure and framework for the cell wall. It provides the characteristic shape and rigidity to the cell (Fig. 8-8). A proximate analysis of commercial glycan contains less than 20% noncarbohydrate material that includes protein, nucleic acid, lipids, and ash.

PRODUCTS OF PHARMACEUTICAL AND COSMETIC VALUE 401

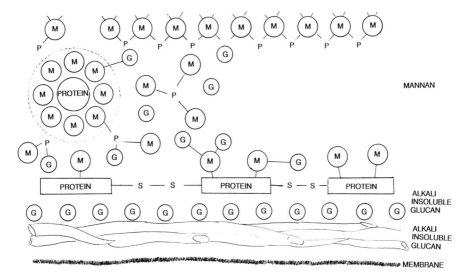

Figure 8-8. Schematic structure of the yeast cell wall.

A desirable property of glycan is its ability to function as a noncaloric food thickener. It does provide a fatlike mouthfeel, thus permitting its use as a fat replacer in the production of low-calorie salad dressings, cheese analogs, and frozen desserts like ice cream (Sidoti, Landgraph, and Khalifa 1973). Its high water-holding capacity has made it a useful ingredient in the formulation of sausages, hot dogs, and certain other meat-based products. More recent research has begun to show certain unique properties in glycan that are likely to permit its use as a remedy for certain human disorders. These specific biological properties of glycan are the result of their physical properties being controlled primarily by their molecular structure. The beneficial effect of glycan in achieving a more uniform fermentation in wine production has been reported by Wahistrom and Fugelsang (1987). Recent studies have also shown the effectiveness of glycan in preventing sticking in wine and other fermentations.

The insoluble cellular debris produced as a by-product in yeast extract production is generally recovered by centrifugation. The particulate material at this stage has an undesirable flavor. Its functional properties are unsatisfactory for use in the food industry. This product is commercially available as feed with a protein content of 35-40%. An attempt is presently being made by the industry to produce value-added products from this waste material.

Although repeated washings could bring about a more bland flavor to baker's-derived yeast cell wall material, additional treatments have to be carried out to eliminate the bitter hop components adhering to cell walls derived from brewer's yeast. Nevertheless, acceptable glycan products of commercial value with the ability to thicken food systems can be achieved by comminuting cell wall material that has been washed with clean water following an alkali treatment. Such a product has potential application in low-calorie, low-fat food formulations. It is the general practice to pass the cell wall fraction through a homogenizer under pressure at about 10,000 psi or to pass it through a bead mill to achieve the desired product viscosity.

Recent studies have demonstrated the beneficial effect of glycan in improving host resistance by enhancing both humoral and cellular immunity to certain malignant tumors as well as to certain bacterial and viral infections. The therapeutic effect of glycan has been primarily due to stimulation of the reticuloendothelial system (RES), which in turn produces increased amounts of macrophages that play a key role in the body's natural immune system. The precise function of the macrophages is to absorb and destroy the invading particles through phagocytosis.

The specific immunological properties imparted by glycan could perhaps be due to its physicochemical properties that are principally controlled by its unique molecular structure and the arrangement of monomeric components within the glycan matrix. In general, the yeast glucan is made up of branched and unbranched chains of glucose units predominantly linked by $(1\rightarrow 3)$, $(1\rightarrow 4)$, and $(1\rightarrow 6)$ glucoside bonds that are either $\alpha$ or $\beta$ types. In recent years, x-ray diffraction studies have been helpful in identifying the molecular structure of glucans. These studies have shown that particulate glucans exist in the triple helical configuration (Deslandes, Marchessault, and Sarko 1960; Chuah et al. 1983; Sarko, Wu, and Chuah 1983). Accordingly, the three polysaccharide chains lie parallel and are wound together in phase and stabilized by interstrand hydrogen bonds (Fig. 8-9).

Early studies included the administration of particulate glucan to animals that have previously been induced with malignant tumors. Although some experimental animals responded favorably to the treatment, some degree of toxicity was apparent, even resulting in a high degree of animal mortality. Attention was then focused on testing the efficacy of soluble glucans that are essentially $\beta(1\rightarrow 3)$ polyglucose fragments derived from the yeast cell wall. Although the new type of glucan preparations exhibited improved activity in animal systems, such preparations were not entirely free of toxic components as evidenced by other disorders detected in such test animals.

A recent patent by DiLuzio (1987) describes the effectiveness of a soluble low-molecular-weight phosphorylated glucan preparation in imparting improved immunological response when administered to animals and

PRODUCTS OF PHARMACEUTICAL AND COSMETIC VALUE 403

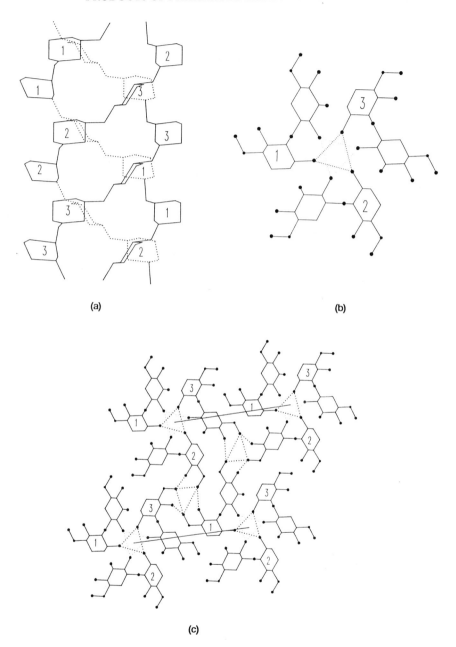

**Figure 8-9.** Triple helical structure of (1→3)-B-D glucan: (a) projection of the triple helix in the XZ plane; (b) projection of the triple helix in the XY plane; (c) probable arrangement of triple helices in the crystalline matrix. Black dots represent the water molecules. (*From Deslandes, Marchessault, and Sarko 1980*)

humans. The special treatment procedure used for the preparation of soluble-phosphorylated glucan eliminated the triple helical arrangement that was characteristic of the particulate glucan. This change in tertiary structure of the glucan could have, perhaps, made it more effective in stimulating macrophage production via the RES. These studies have demonstrated the therapeutic value of phosphorylated glucans in combating a variety of diseases caused by bacteria, fungi, viruses, and also in attempting to change the course of certain neoplastic conditions. Although these findings are of great significance, there is still a need to confirm them, especially with regard to the therapeutic value and safety of such phosphorylated preparations.

The inert nature of yeast glucan has made it an excellent dietary fiber for use in human food formulations. Its hypocholesterolemic effect has been demonstrated by Robbins and Seeley (1981). Although its role in reducing the blood cholesterol is not well understood, there are reports to indicate that the beneficial effect of the dietary fiber is due to its interaction with cholic acid. Nevertheless, the beneficial effect of glycan in the human diet on reversing the incidence of coronary heart disease remains to be confirmed.

## Coenzyme A-Synthesizing Protein Complex (CoA-SPC)

Research is presently underway to develop methods to detect the incidence of cancer at a very early stage, long before any symptoms become apparent in the patient. One such diagnostic technique involves the early detection of $\beta$-protein in the bloodstream, which is an early indication of the body's reaction to cancer.

A biologically active compound referred to as coenzyme A-synthesizing protein complex (CoA-SPC) derived from baker's yeast has found application as one of the components used for detecting $\beta$-protein in human blood. This complex compound is known to first interact with L-cysteine, D-pantothenic acid, and ATP to form an active binding protein that can complex with $\beta$-protein in the serum if the individual is in a very early stage of cancer. In this case, the use of labeled L-cysteine, D-pantothenic acid, or ATP could provide the ability to detect the final complex for positive indentification of the disease.

The biologically active CoA-SPC complex has been identified in the particulate fraction following the lysis of baker's yeast. The particulate fraction also contained large amounts of cell wall material and insoluble proteins. It was thus necessary to free the active component from the cell debris in order to obtain a fraction of CoA-SPC of high purity. This procedure was accomplished by adding a series of recovery steps. A technique was first devised to selectively solubilize certain proteinaceous components adhering to the particulate matter without solubilizing the active component. The procedure required the use of a treatment with salts such as chlorides, nitrates, or acetates of K, Na, Mg, Ca, or Mn. The resulting particulate fraction was then rich in CoA-SPC.

A special procedure was adopted next to solubilize the active CoA-SPC from the CoA-SPC-rich particulate fraction. A low-molecular-weight component referred to as the "t" factor extracted from the baker's yeast during the lysis of the cell at the early stages of the process was found effective in selectively solubilizing the CoA-SPC from the particulate fraction. The molecular weight of the "t" factor was in the 400-1000 range. The purity of the CoA-SPC was largely determined by the purity of the "t" factor used in the final extraction. The procedure for the extraction of CoA-SPC is extensively covered by the patent granted to Bucovaz et al. (1981).

## Genetically Engineered Products from Yeast

The advent of recombinant DNA technology has paved the way for a new era in programming biological functions of living cells to produce novel proteins for food and pharmaceutical value. Over the years, *Escherichia coli* has been the host organism of choice for the production of these valuable proteins, both in the laboratory and in the industry. The widespread use of *E. coli* has resulted from the greater knowledge of its physiology and genetics as compared to other less-studied microorganisms. However, in addition to being a fecal organism, *E. coli* has also been known to produce pyrogenic factors that must be eliminated from any potentially useful pharmaceutical product prior to its use on humans. Furthermore, *E. coli*, like many other procaryotic organisms, is incapable of excreting proteins to the extracellular environment. Excretion of proteins is highly desirable because it simplifies the recovery and purification process. These undesirable traits in *E. coli* have led to an increased interest to explore alternate host/vector systems that would make the process cost-effective and less hazardous under manufacturing conditions.

Among the eucaryotic organisms that can be used for genetic manipulation, perhaps the easiest to manage and the most studied organism is yeast. From a commercial point of view, yeast offers a considerable number of advantages over bacteria. The technology to grow yeast on an industrial scale is well known; it is relatively inexpensive and the studies during the last few years have yielded extensive information on the genetics of yeast at a molecular level. Unlike *E. coli*, yeasts have the capability to secrete the proteins by making use of the signal sequence at their amino terminals. This feature simplifies the product recovering process.

With the technological advances that have already been made in genetic engineering, substantial creative effort has now gone into the development of methodology to manipulate and modify the yeast genome. Considering these unique features, yeast has begun to emerge into the forefront, becoming one of the most frequently exploited microorganisms capable of producing value-added products of commercial importance.

Despite the exceedingly long lag time and high expense of pushing new products through the entire approval process, at least a few biotechnology companies have been able to bring certain genetically engineered yeast-derived products to the final stages of marketing. More genetically engineered products are in human clinical trials and are being evaluated for final approval. Genetically engineered products are likely to gain credibility as more and more products are approved and introduced into the marketplace.

Efforts of research and development (R&D) are accelerating and products are starting to emerge, especially on the food and diagnostic fronts. Although the development of diagnostic devices received only about 10% of the overall R&D expenditure, this segment currently accounts for 55% of the overall sales. Despite substantial R&D support for the development of novel products for human therapy, these pharmaceutical agents represent a minor segment of sales, perhaps due to their more involved procedures for testing and approval.

After considerable publicity over potential cures, one feels disappointed with the number of genetically engineered products that have actually reached the marketplace. At present, the list includes only three well-known products: human insulin, interferon, and hepatitis-B vaccine.

Novo Industries (Denmark) has been producing porcine insulin with one amino acid altered to make it identical to the amino acid sequence found in human insulin. There is, however, considerable interest to replace this form of insulin with a genetically engineered form made from yeast. A single chain that includes both chains of insulin will be produced by the cloned yeast and an enzymatic process cleaves the chain to provide the required double chain for the formation of the active form of the insulin molecule. This gene was cloned early to $E.\ coli$ by researchers at Genentech (U.S). Presently, the Genentech-developed insulin is marketed by Eli Lilly (U.S.) as humulin, one of the many licensing examples between large drug houses and biotechnology companies. The insulin market is obviously large, resulting in stiff competition among genetically engineered insulins from a number of companies.

A therapeutic agent that attracted considerable attention in the past as the new magic bullet against cancer was interferon. This drug actually represents a family of related glycoproteins ($\alpha$-, $\beta$-, and $\gamma$-interferons) that are secreted by fibroblasts and leucocytes. According to reports, the α-interferon as a nasal spray has been found effective in reducing the incidence of the common cold, β-interferon treatment has been effective in reducing the disease systems of multiple sclerosis, and the γ-interferon is presently being testing against rheumatic arthritis.

Although the pioneering work on the production of α-interferon was initiated on *E. coli*, molecular biologists have now begun to consider yeast as better-suited organisms from a commercial standpoint. Interferon Sciences Inc. (U.S.) has been seeking approval from the FDA to market a recombinant α-interferon product made from genetically constructed yeast for use in genital herpes. It is known to relieve pain and shorten the duration of viral shedding from recurrent herpes lesions. Interferons are presently being produced by several companies, shifting the emphasis of the competition from the actual production process to marketing.

Genetically engineered hepatitis-B vaccine is a recent product that has received approval from the FDA for use in the United States and other countries. Hepatitis-B is the most virulent form that affects the human liver, and nearly half a billion people suffer from this disease worldwide. It can be transmitted through blood transfusions, through intimate contact, or from mother to unborn child. The geneticists have been able to induce yeast to produce the protein coating surrounding the hepatitis-B virus so that it can be used in the production of synthetic vaccines against the disease. When injected into humans, the complex protein known as hepatitis-B surface antigen produced by yeast prompts the human body to mount an immune response against the disease. The genetically engineered yeast-derived vaccine can be processed in about 10 days in contrast to approximately 65 weeks to produce the regular plasma-derived hepatitis-B vaccine.

The development of epidermal growth factor (EGF) is more advanced and has received more commercial attention than any other growth factor under development. Many researchers also believe that EGF may have the greatest potential in wound healing. Chiron Corporation (U.S.) is the clear leader in the EGF race and is producing EGF in yeast. This product is presently being used in clinical trials to help heal corneal transplants, nonhealing corneal defects, and surgical epikeratophakia where the donor lens is sewn onto the cornea. Several other companies pursuing this protein make the EGF in *E. coli*.

Other growth factors that include fibroblast growth factor (FGF), platelet-derived growth factor (PDGF), transforming growth factor (TGF), and α- and β-TGF will receive regulatory approval at later dates and will compete with EGF as well as competing for several other medical purposes. Zymo Genetics (U.S.) scientists have been able to produce these growth factors in yeast. In this case, yeast secretes the proteins in properly folded and biologically active form. Insulinlike growth factor (IGF-1), which has the biggest potential in wound healing, is probably for treating bone fractures. The human IGF-1 gene is another of those genes that can be expressed in yeast as has been demonstrated by Chiron Corporation.

Phillips Petroleum (U.S.) has signed five licensees for its *Pichia pastoris* genetic expression system for producing recombinant proteins. This system has been used to produce a number of recombinant proteins intracellularly: streptokinase, $\beta$-galactosidase, tumor necrosis factor (TNF), hepatitis-B surface antigen, and AIDS virus antigen. It has also been used to produce a number of secreted proteins: invertase, bovine lysozyme, human serum albumin, and tissue plasminogen activator (*Genetic Technology News*, August 1989).

Phillips Petroleum has described *Pichia pastoris* as having certain advantages over *E. coli*, *Saccharomyces* species, and mammalian cells in the production of certain foreign proteins. For example, it can produce proteins intracellularly or secrete them into the culture medium. Yields also have been reported as higher than those obtained in other commonly used host cells. Futhermore, the recombinant proteins secreted by *P. pastoris* were determined to be small in size and structure with uniform N-linked oligosaccharides. Similar glycoproteins synthesized by *S. cerevisiae* have been shown to contain larger oligosaccharides.

Scientists at Genentech (U.S.) have been able to clone the gene for human serum albumin in yeast using rDNA technology. Currently, albumin is extracted from human blood and used in treatments to replace blood following surgery, burns, and other trauma. Production of albumin by application of the new technology produces an opportunity to increase the availability of an important product at a potentially lower cost.

Another novel protein of potential therapeutic value is $\alpha$-1-antitrypsin (AAT). This product is being tested on patients suffering from emphysema due to deficiency of AAT in the body. The AAT's function is to destroy excess elastase and the other proteolytic enzymes in the lungs. When the elastase enzyme is present optimally, it helps to clear the lungs and keep passages free of microbes and particulate matter from air pollutants and tobacco smoke. Insufficient AAT causes an elevation of elastase level that would eventually attack the lung tissue causing the disease. Zymogenetics (U.S.) and Synergen Inc. (U.S.) have been able to produce $\alpha$-1-antitrypsin using genetically engineered recombinant yeast. A serious drawback in the initial studies was that the sugar attachment patterns in the yeast-derived AAT were different from those glycosylated proteins observed in human $\alpha$-1-antitrypsin. This difference could result in an immune reaction often harmful to the patients. Cooper Biomedical (U.S.), who had Zymogenetics (U.S.) clone and express the AAT in yeast under a research contract, has demonstrated that AAT does not produce any immune response in animals. The effect of the glycosylated yeast AAT on humans remains to be established.

There are very few genetically engineered products that have applications in food processing. To date, no yeast-derived recombinant enzymes have been cleared for food use by the FDA. A food enzyme that has the FDA's approval process in its final stages is chymosin. It is the principal enzyme found in calf rennet used extensively in the production of different cheeses. Its principal action is coagulation of milk. It also allows casein to bind together and entrap butterfat.

The gene for calf rennet has been inserted successfully into E. coli and yeasts. Collaborative Research's (U.S.) patented process makes use of yeast as the host organism for the production of rennin in fully active, nonglycosylated form. When the protein is produced recombinantly in bacteria, the enzyme is not secreted, the extraction from the cell and subsequent purifications are cumbersome, and the yields are significantly low. Genencor (U.S.) has used *Aspergillus*, a filamentous fungus, as the recombinant host for the production of rennin.

It is thus clear that yeast is continuing to act as humankind's oldest and most reliable ally in the microbial world. With the recent developments in biotechnology and genetic engineering, a second generation of value-added yeast-derived products is beginning to emerge in the marketplace. The fast-developing technologies are adding unprecedented scope and precision to the efforts to modify yeasts to increase their usefulness. It is hoped that these developments will result in revolutionary advances in health care, food, nutrition, and industrial chemistry in the not-too-distant future.

## REFERENCES

Achstetter, T., and D. H. Wolf. 1985, Proteinases, proteolysis and biological control in the yeast *Saccharomyces cerevisiae*. Yeast 1:139-157.

Amoco Food Co. 1974. *Torutein Product Bulletin*. Chicago, Illinois.

Andrews, A. G., H. J. Phaff, and M. P. Starr. 1976. Carotenoids of *Phaffia rhodozyma*, a red pigmented fermenting yeast. Phytochemistry. 15:1003-1007.

Anon. 1970. *Single Cell Proteins*. Protein Advisory Group Guidelines, no. 4. United Nations, New York.

Arnold, W. N. 1971. Heat inactivation kinetics of yeast beta-fructofuranosidase. A polydispersing system. Biochim. Biophys. Acta. 178:347-353.

Bernstein, S. and P. E. Plantz. 1977. Production of yeast from whey. Food Eng. 49(11):74-75.

Biemaan, L., and M. D. Glantz. 1968. Properties of a fungal lactase. Biochim. Biophys. Acta. 167:373-377.

Borglum, G. B., and M. Z. Sternberg. 1972. Isolation and characterization of $\beta$-galactosidase from *Saccharomyces lactis*. J. Food Sci. 37: 619-623.

Bucovaz, E. T., J. C. Morrison, W. D. Whybrew, and S. J. Tarnowski. 1981. Process for the preparation of CoA-SPC from baker's yeast. U.S. Patent 4,284,552.

Chuah, C. T., A. Sarko, Y. Deslandes, and R. H. Marchessault. 1983. Triple helical crystalline structure of curdlan and paranylon hydrates. *Macromolecules* **16:**1375-1382.

Cohn, W. E., and E. Volkin. 1953. On the structure of ribonucleic acids. *J. Biol. Chem.* **203:**319-332.

Corran, H. S. 1975. *A History of Brewing.* David and Charles, Newton Abbot, London.

Daly, W. H., and L. P. Ruiz. 1974. Reduction of RNA in single cell proteins in conjunction with fiber formation. *Biotechnol. Bioeng.* **16:**285-287.

Davies, R. 1964. Lactose utilization and hydrolysis in *Saccharomyces cerevisiae. J. Gen. Microbiol.* **37:**81-98.

Decker, P., and K. Dirr. 1944. Nonprotein nitrogen of yeast. II Comparison of purine fraction and extraction of nucleic acids. *Biochem Z.* **316:**248-254.

Deslandes, Y., R. H. Marchessault, and A. Sarko. 1980. Triple helical structure of (1→3)$\beta$-D-glucan. *Macromolecules* **13:**1466-1471.

DiLuzio, N. R. 1987. Soluble phosphorylated glucan. International Publication No. 87/01037. Publication under the patent cooperation treaty. International Searching Authority, USA.

Gascon, S., P. Neumann, and J. O. Lampen. 1968. Comparative study of the properties of purified internal and external invertases from yeast. *J. Biol. Chem.* **243:**1573-1577.

Gatellier, C., and G. Gilkamans. 1972. Process of improving the food value of microorganisms obtained by culturing on hydrocarbon substrates. U. S. Patent 3,702,283.

Gilbert, H. J., and W. Jack. 1981. The effect of proteinases on phenylalanine ammonia-lyase from the yeast *Rhodotorula glutinis. Biochem J.* **199:**715-723.

Gilliland, R. B. 1956. Maltotriose fermentation in the species differentiation of *Saccharomyces. Compt. Rand. Trav. Lab. Carlsberg. Ser. Physiol.* **26:**139-148.

Goodson, W., D. Hohn, T. K. Hunt, and Y. K. Leung. 1976. Augmentation of some aspects of wound healing by a skin respiratory factor. *J. Surgical Res.* **21:**125-129.

Hata, T., R. Hayashi, and E. Doi. 1967. Purification of yeast proteinases. Part I. Fractionation and some properties of the proteinases. *Agric. Biol. Chem.* **31:**150-159.

Hu, A. S. L., R. G. Wolfe, and F. J. Reichel. 1959. The preparation and purification of $\beta$-galactosidase from *Escherichia coli*, ML 308. *Arch. Biochem. Biophys.* **81:**500-507.

Ikeda, K. 1912. The taste of the salt of glutamic acid. *Orig. Com. 8th Int. Congr. Appl. Chem.* **18:**147.

Johnson, E. A., D. E. Conklin, and M. J. Lewis. 1977. The yeast *Phaffia rhodozyma* as a dietary pigment source for salmonids and crustaceans. *J. Fish. Res. Board of Canada* **34:**2417-2421.

Johnson, E. A., and M. J. Lewis. 1979. Astaxanthin formation by the yeast, *Phaffia rhodozyma. J. Gen. Microbiol.* **115:**173-183.

Johnson, E. A., T. G. Villa, M. J. Lewis, and H. J. Phaff. 1978. Simple method for the isolation of astaxanthin from the basidiomycetous yeast, *Phaffia rhodozyma. Appl. Environ. Microbiol.* **35:**1155-1159.

Katchman, B. J., and W. O. Fetty. 1955. Phosphorus metabolism in growing cultures of *S. cerevisiae. J. Bacteriol.* **69:**607-615.

Kinsella, J. E., and J. K. Shetty. 1982. Recovery of proteinaceous material having reduced nucleic acid levels, U.S. Patent 4,348,479.

Kodama, S. 1913. Isolation of inosinic acid. *Tokyokagaku (J. Chem. Soc. Japan)* **34:**751.

Kuninaka, A. 1986. Nucleic acids, nucleotides and related compounds In Biotechnology. Vol. 4. H. J. Rehm and G. Reed. eds. Verlag Chemie., Florida. 72-86.

Kuninaka, A., M. Fujimoto, K. Uchida, and H. Yoshino. 1980. Extraction of RNA from yeast packed into column without isomerization. *Agric. Biol. Chem.* **44:**1821-1827.

Lenney, J. F., and J. M. Dalbec. 1969. Yeast proteinase B. Identification of the inactive form as an enzyme inhibitor complex. *Arch. Biochem. Biophys.* **129**:407-409.

Lindegren, C. C., S. Speigelman, and G. Lindegren. 1944. Mendelian inheritance of adaptive enzymes in yeast. *Proc. Natl. Acad. Sci. USA* **30**:346-352.

Meister, H. 1965. Yeast invertase: An illusive but useful enzyme. *Wallerstein Lasb. Commun.* **28**:7-15.

Miller, M. W., M. Yoneyama, and M. Soneda. 1976. *Phaffia*: A new yeast genus in the Deuteromycotina (Blastomycetes). *Int. J. Syst. Bacteriol.* **26**:286-291.

Mortimer, R. K., and D. C. Hawthorne. 1969. Yeast genetics. In *The Yeasts*, vol. 1, A. H. Rose and J. S. Harrison (eds.). Academic Press, New York, pp. 385-460.

Nakajima, N., K. Ichikawa, M. Kamada, and E. Fujita. 1961. Food chemical studies on 5' ribonucleotides. I. On the 5' ribonucleotides in foods. (1) Determination of the 5' nucleotides in various stocks by ion exchange chromatography. *J. Agric. Chem. Soc. Japan.* **35**:797.

Nakao, Y. 1979. Microbial production of nucleosides and nuceotides. In *Microbial Technology, Microbial Processes*, vol. 1, H. J. Peppler and D. Perlman (eds.). Academic Press, New York, pp. 311-354.

Newell, J. A., E. A. Robbins, and R. D. Seeley, 1975. Manufacture of yeast protein isolate having reduced nucleic acid content by an alkali process, U.S. Patent 3,867,555.

Newell, J. A., R. D. Seeley, and E. A. Robbins, 1975. Process of making yeast protein isolate having reduced nucleic acid levels, U.S. Patent 4,348,479.

Ohashi, M., and S. Kozutsumi. 1966. Manufacture and utilization of invertase. I. Manufacture of liquid invertase. *Nippon Shokanin Kogyo Gakkaishi* **13**(1):1-7.

Peppler, H. J. 1965. Amino acid composition of yeast grown on different spent sulfite liquors. *J. Agric. Food. Chem.* **13**:34-36.

Phaff, H. J. 1971. Structure and biosynthesis of the yeast cell envelope. In *The Yeasts*, vol. 2, A. H. Rose and J. J. Harrison (eds.). Academic Press, New York, pp. 135-210.

Reed, G. and H. J. Peppler. 1973. Feed and food yeast. In *Yeast Technology*. AVI Publishing Co., Westport, Conn., pp. 328-351.

Robbins, E. A. 1976. Manufacture of yeast protein isolate having a reduced nucleic acid content by a thermal process, U.S. Patent 3,991,215.

Robbins, E. A., and R. D. Seeley. 1981. Process for the prevention and reduction of elevated blood cholesterol and triglyceride levels, U.S. Patent 4,251,519.

Roberts, C., A. T. Ganesan, and W. Haupt. 1959. Genetics of melibiose fermentation in *Saccharomyces italicus* var. *melibiosi*. *Heredity* **13**:499-517.

Saheki, T., and H. Holzer. 1975. Proteolytic activity in yeasts. *Biochim. Biophys. Acta.* **384**:203-214.

Sarko, A., H. C. Wu, C. T. Chuah. 1983. Multiple helical glucans. *Biochem. Soc. Trans.* **11**:139-142.

Schuster, L. 1957. Rye grass nucleases. *J. Biol. Chem.* **229**:289-303.

Sheets, R. M., and R. C. Dickson. 1981. LAC 4 is the structural gene for $\beta$-galactosidase in *Kluyveromyces lactis*. *Genetics* **98**:729-745.

Shimazono, H. 1964. Distribution of 5'-ribonucleatides in foods and their application to foods. *Food Technol.* **18**:294-303.

Sidoti, D. R., G. M. Landgraph, and R. A. Khalifa. 1973. Functional properties of baker's yeast glycan. Presented at the 33rd Annual Meeting of the Institute of Food Technologists, Miami, Florida. Jan. 6-14.

Sotskaya, V. P., V. A. Smirnov, and L. Y. Tikhomirov. 1965. Precipitation of invertase from yeast autolysates with ammonium sulfate. *Izv. Vyssh. Ucheb. Zavedenii. Pischevaya Tekhnol.* **6**:38-42.

Sperti, G. 1943. Toilet preparation, U.S. Patent 2,320,478.
Stimpson, E. G. 1954. Drying of yeast to inactivate zymase and preserve lactase, U.S. Patent 2,693,440.
Takei, S., S. Amao, T. Endo, K. Ishibashi, and T. Ito. 1967. A simplified method for the manufacture of yeast invertase. *Ann. Sankyo. Res. Lab.* **19**:81-85.
Tannenbaum, S. R., A. J. Sinskey, and S. B. Maul. 1973. Process of reducing the nucleic acid content in yeast, U.S. Patent 3,720,583.
Tarantino, A. L., T. H. Plummer, and F. Miley. 1974. The release of intact oligosaccharides from specific glycoproteins by endo-$\beta$-N-acetyl glucosaminidase H. *J. Biol. Chem.* **249**:818-824.
Torii, K., and R. H. Cagan. 1980. Biochemical studies of taste sensation. IX. Enhancement of L-($^3$H) glutamate binding to bovine taste papillae by 5'-ribonucleotides. *Biochim. Biophys. Acta.* **627**:313.
Torii, K., T. Mimura, and Y. Yugari, 1986. Effect of dietary protein on the taste preference for amino acids in rats. In *Interaction of the Chemical Senses with Nutrition*. Academic Press, New York, p. 45.
von Borstel, R. C. 1969. Yeast genetics supplement. *Mol. Genet. Bull.* **31**:1-28.
Wahlstrom, V. L., and K. C. Fugelsang. 1987. *Utilization of Yeast Hulls in Wine Making Observed*. Research Bulletin. California State University, Fresno, pp. 1-5.
Yamaguchi, S. 1987. Fundamental properties of umami in human taste sensation. In *Umami: A Basic Taste*, Y. Kawamura and M. R. Kare (eds.). 1st Edition. Marcel Dekker, Inc., New York, p. 41.
Young, H., and R. P. Healy. 1957. Production of *Saccharomyces fragilis* with an optimum yield of lactase, U.S. Patent 2,776,928.
Ziemba, J. V. 1967. Tailored hydrolysates, how made, how used. *Food Eng.* **19**(1):82-85.

CHAPTER
9

# FOOD AND FEED YEAST

Yeast-fermented foods contribute a significant percentage of nutrients to the human diet. Table 9-1 shows the annual per capita consumption of all fermented foods in the United States. Nutritionally the most important yeast fermentations are baked goods and beer. Foods fermented by bacteria are included in the table to afford a comparison. Nutritionally the most important food fermented by bacteria is cheese.

Table 9-2 shows the percentages of the recommended daily allowance

Table 9-1. Per Capita Consumption of Fermented Foods in the United States

| Yeast Fermentations | | Bacterial Fermentations | |
|---|---|---|---|
| Beer | 32.8 gal | Cheese | 34.8 lb |
| Wine | 3.0 gal | Yogurt | 4.1 lb |
| Distilled | | Sauerkraut | 1.4 lb |
| beverages | 2.8 gal | Olives | 0.5 lb |
| Bread | 47.1 lb | Pickles | 4.0 lb |
| Rolls | 13.3 lb | | |
| Sweet goods* | 3.2 lb | | |
| Soda crackers | 2.7 lb | | |
| Pretzels | 0.8 lb | | |

Source: Based on U.S. Statistical Survey data 1986, 1987, and 1988.
*Includes only yeast-raised sweet baked goods.

Table 9-2. Contribution of Three Major Fermented Foods to Human Nutrients

| | | Weight (g) | Cal. | Prot. (g) | Ca (mg) | P (mg) | Fe (mg) | A (IU) | $B_1$ (mg) | $B_2$ (mg) | Niacin (mg) |
|---|---|---|---|---|---|---|---|---|---|---|---|
| RDA | | | 2400 | 65 | 800 | 800 | 18 | 5000 | 1.4 | 1.7 | 20 |
| Bread and rolls | a | 907 | 2494 | 81.6 | 871 | 925 | 22.7 | 0 | 2.45 | 1.81 | 21.8 |
| | b | 78 | 215 | 7.0 | 75 | 80 | 2.0 | 0 | 0.21 | 0.16 | 1.9 |
| | c | | 9% | 11% | 9% | 10% | 11% | 0 | 15% | 9.4% | 9.5% |
| Beer | d | 100 | 42 | 0.3 | 5 | 30 | 0 | 0 | 0.002 | 0.03 | 0.6 |
| | b | 350 | 147 | 1.05 | 175 | 105 | 0 | 0 | 0.035 | 0.105 | 2.1 |
| | c | | 6.1% | 1.6% | 22% | 13% | 0 | 0 | 2.5% | 6.1% | 10.5% |
| Yeast-fermented foods, total | c | | 15.1% | 12.6% | 31% | 23% | 11% | 0 | 17.5% | 15.5% | 20.0% |
| Cheese | a | 24 | 96 | 6.0 | 180 | 115 | 0.2 | 310 | 0.01 | 0.11 | 0 |
| | b | 43.5 | 173 | 10.9 | 325 | 208 | 0.36 | 561 | 0.018 | 0.2 | 0 |
| | c | | 7.2% | 24.2% | 40% | 26% | 2% | 11% | 1.3% | 12% | 0 |

Note: RDA = recommended daily allowance.
[a] Values for a given weight in grams from USDA tables.
[b] Per capita daily intake (for beer, persons 18 years or older only).
[c] Percentage of RDA contributed.
[d] Values for a 100-gram weight from the literature.

(RDA) of important nutrients. Cheese has again been included to afford a comparison. The figures are not exact because some assumptions had to be made, for instance, all cheeses (except cottage cheese) were assumed to have the composition of cheddar cheese. But the figures accurately portray the important contribution of fermented foods to the American diet.

Yeasts have some importance as foods and as feeds. About 150-180 million pounds of baker's yeasts (dry basis) are produced annually in the United States. Almost all of this yeast is used in the production of baked goods as either compressed or active dry yeast. About 5-10 million pounds are used for direct consumption as dried nutritional yeast. In the brewing industry approximately 0.5-0.8 pounds of yeast (dry basis) become available as excess yeast per barrel of beer. Thus, the total supply is roughly 120 million pounds per year. Most of this yeast is used in feed, usually by codrying it with spent grains, another by-product of the brewing industry. But a considerable percentage, perhaps as much as 30 million pounds, is used in the production of yeast extracts or is sold as debittered brewer's yeast for nutritional purposes.

In some fermentation industries the yeast cannot be recovered as such because the fermented grape must or the fermented grain mash contains insoluble particles that makes separation of the yeast impracticable. In the wine industry the yeast is discarded with the lees. In the production of distilled beverages the yeast is codried with the distiller's solubles or the spent grain and used as feed. The use of active dry yeasts in the feed industry will be discussed at the end of this chapter.

The term *nutritional* yeast may require clarification. Yeast flavor may be pleasant for some people or indifferent or less than desirable to others. It is not used because of its flavor but because of its nutritional contribution of protein, vitamin $B_1$, and mineral nutrition. Yeasts are sold for human consumption as inactive dried yeasts. They are killed by pasteurization or during drum or spray drying. The dried powders or flakes are usually consumed by blending them with fruit juices or by sprinkling over cereals. In the health food industry yeasts are also promoted on the basis of unidentified nutritional factors. Such claims have to be treated with considerable caution because of their promotional aspects. However, it should be remembered that two such previously unidentified nutritional factors were discovered in yeast in the 1950s. They are selenium and chromium, which are now recognized as essential trace elements.

## COMPOSITION

### Gross Composition

The commercially important yeasts *Saccharomyces cerevisiae* and *Candida utilis* (as well as the less important *Kluyveromyces marxianus*) will be dis-

Table 9-3. Gross Composition of Nutritional *S. cerevisiae* Baker's Yeast (as-is basis)

| | | |
|---|---|---|
| Moisture | | 2-5% |
| Protein, crude ($N \times 6.25$) | | 50-52% |
|    Protein | 42-46% | |
|    RNA and DNA | 6-8% | |
| Minerals | | 7-8% |
| Total lipids | | 4-5% |
| Carbohydrates | | 30-37% |
|    Dietary fiber | 17% | |

cussed in some detail. Yeasts grown on petroleum substrates will only be mentioned briefly. At present, they are not produced in significant commercial quantities.

The gross composition of yeasts is quite variable. The percentage of protein and of nucleic acids depends on the substrate and the growth rate. In the following text the composition will be discussed on the basis of *S. cerevisiae* with the understanding that the other two nutritional yeasts have a somewhat similar composition. Table 9-3 shows ranges in composition typical for *S. cerevisiae*.

## Protein

The protein content of yeasts is usually determined by the Kjeldahl nitrogen determination and multiplication of $N \times 6.25$. The total nitrogen includes the nitrogen of the nucleic acids and, therefore, this procedure overestimates the protein content. Protein calculated in this manner should be designated *crude protein*. A factor of $N \times 5.5$ is more appropriate to estimate pure protein. Obviously, the nitrogen determination can yield only an approximate value for pure protein.

General use of the 6.25 factor for calculation of yeast protein has had an unfortunate consequence. It has led to an underestimation of the nutritional value of yeast protein as expressed by the protein efficiency ratio (PER) because animal feeding tests were carried out at an assumed protein value higher than the true value. Much of the literature uses a PER of 1.8 (vs. a casein PER of 2.5).

Table 9-4 shows the amino acid composition of commercial nutritional

**Table 9-4. Amino Acid Composition of Yeast Proteins (in %)**

| | A<br>S. cer. | B<br>S. cer. | C<br>C. util. | D<br>C. util. | E<br>K. marx. | F<br>K. marx. | G<br>C. lipol. |
|---|---|---|---|---|---|---|---|
| Lysine | 9.4 | 8.1 | 7.7 | 7.2 | 6.9 | 11.1 | 7.4 |
| Methionine | | 1.4 | 1.1 | 1.0 | 1.4 | 1.6 | 1.8 |
| Tryptophan | 1.2 | | | 0.04 | 1.8 | | 1.4 |
| Valine | 7.4 | 5.5 | 3.5 | 5.3 | 4.8 | 5.8 | 5.9 |
| Threonine | 5.8 | 4.1 | 7.6 | 4.7 | 5.2 | 5.6 | 4.9 |
| Leucine | 9.0 | 6.6 | 4.4 | 7.8 | 7.5 | 9.6 | 7.4 |
| Isoleucine | 5.8 | 6.0 | 4.4 | 4.3 | 4.3 | 5.2 | 5.1 |
| Histidine | 3.5 | 2.8 | 2.0 | 2.1 | 2.0 | 4.0 | 2.1 |
| Cystine | 1.8 | | | 0.6 | | | 1.1 |
| Alanine | 9.1 | 5.1 | 7.4 | | | 7.2 | |
| Arginine | | 8.3 | 4.7 | 7.2 | 3.2 | 7.4 | 5.1 |
| Glutamic acid | 21.0 | 20.1 | 19.9 | | 13.1 | 15.4 | |
| Aspartic acid | | 9.6 | 11.6 | | | 10.4 | |
| Asparagine | | | | | | 4.2 | |
| Glycine | 5.8 | 4.9 | 5.6 | | | | |
| Proline | 5.5 | 5.3 | 3.3 | | | 4.3 | |
| Hydroxyproline | 5.0 | | | | | | |
| Serine | 5.6 | 4.4 | 8.1 | | | 5.2 | |
| Tyrosine | 5.4 | 3.7 | 4.3 | 3.3 | 3.2 | 4.6 | 3.6 |
| Phenylalanine | | 4.1 | 4.4 | 3.7 | 3.5 | 5.1 | 4.3 |

A = S. cerevisiae, Univ. Foods assay.
B = S. cerevisiae (Kockova-Kratochvilova 1982).
C = C. utilis (Kockova-Kratochvilova 1982).
D = C. utilis (Ridgeway et al. 1975).
E = K. marxianus Univ. Foods assay.
F = K. marxianus (Wasserman, Hanson, and Alvare 1961).
G = C. lipolytica (Gow et al. 1975).

yeasts. Yeast protein is distinguished by its high lysine content that makes yeast an excellent supplement to cereal proteins generally deficient in lysine. The sulfur-containing amino acids methionine, cysteine, and cystine are the limiting amino acids in yeast. Figure 9-1 shows the PER of *S. cerevisiae* protein. The PER of yeast protein is just slightly above 2.

Seeley (1977) reported a PER of 2.2 for a protein fraction of baker's yeast essentially free from nucleic acids. The amino acid composition of yeasts resembles that of oil seed proteins, particularly that of soy protein. The sulfur-containing amino acids methionine, cysteine, and cystine are limiting. Figure 9-1 also shows the effect of d,1-methionine supplementation of yeast protein in raising the PER above that of casein. It must be remem-

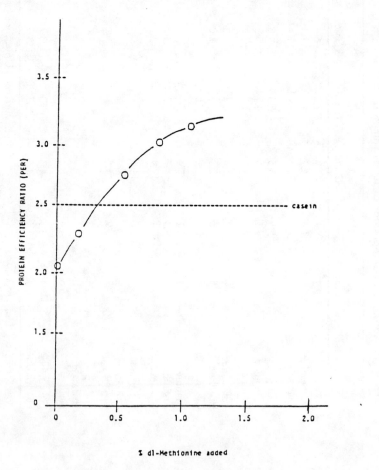

**Figure 9-1.** Protein efficiency ratio of a yeast protein unsupplemented and supplemented with d,1-methionine; in comparison with a casein standard. (*Courtesy Universal Foods Corp.*)

bered that the PER reflects the nutritional value of a protein only if it is fed as the sole protein in the diet. It does not permit an estimate of its value in mixed protein diets.

The listing of amino acids in Table 9-4 requires additional comment. The values can only be accepted as a general guide. A perusal of the extensive literature shows that differences reported for yeasts of the same species are often greater than differences between species.

## Vitamins

Yeasts are good sources of vitamins of the B complex. They lack vitamin C and the oil-soluble vitamins A and D. Several decades ago yeasts were irradiated with ultraviolet light to convert yeast ergosterol to vitamin $D_2$. This process is not economical in view of the low cost of synthetic vitamins $D_2$ and $D_3$. The natural vitamin content of nutritional yeasts has been treated in Chapter 8. It may be increased by growing yeasts on media supplemented with vitamins $B_1$, niacin, and biotin. This case is not true for vitamin $B_2$, riboflavin. For this vitamin one has to rely on the synthetic ability of the yeasts. Riboflavin is produced only by growing yeasts. Its rate of production is proportional to its growth rate in the exponential phase (Tamer, Ozilgen, and Ungan 1988).

## Minerals

The mineral content of a typical baker's yeast is shown in Table 9-5. The predominant anions are phosphates and the predominant cation is potassium.

Table 9-5. Mineral Content of *S. cerevisiae* Baker's Yeast (per gram of yeast dry basis)

| Potassium | 21 mg | Iron | 20 mcg |
|---|---|---|---|
| Phosphorus | 13 mg | Zinc | 170 mcg |
| Sulfur | 3.9 mg | Chromium | 2.2 mcg |
| Magnesium | 1.65 mg | Silicon | 30.0 mcg |
| Calcium | 0.75 mg | Manganese | 8.0 mcg |
| Sodium | 0.12 mg | Copper | 8.0 mcg |
| | | Nickel | 3.0 mcg |
| | | Tin | 3.0 mcg |
| | | Molybdenum | 0.04 mcg |
| | | Lithium | 0.17 mcg |
| | | Vanadium | 0.04 mcg |
| | | Selenium | 5.0 mcg |

The sulfur, magnesium, and calcium contents are not too variable. However, the values shown for trace minerals fluctuate widely since they depend entirely on the composition of the substrate. For instance, the chromium content of brewer's yeast may vary between 0.5 and 2.5 ppm (dry basis) depending on the Cr content of the water used or on the material of the production vessels.

## Chromium

More than three decades ago, Mertz and Schwartz (1955) reported that rats fed with certain nutritionally deficient diets showed impaired tolerance to blood glucose. They were then able to reverse the condition by supplementing the feed with brewer's yeast. They later suggested that the active component responsible for reversing the glucose intolerance was an organic complex rich in chromium.

Diabetes, which is caused by glucose intolerance, is a serious problem among the human population today. This disorder can manifest itself in two ways. Among juvenile diabetic patients, the pancreas fails to secrete insulin normally. This disorder can only be corrected by administering insulin to the bloodstream on a daily basis. Those who are affected by another type of glucose intolerance begin to show symptoms of the disorder in midlife. These patients are able to secrete the hormone insulin into the bloodstream but are incapable of controlling the blood sugar level. It is possible to correct this problem by supplementing the diet of these patients with brewer's yeast. Findings have indicated the importance of another component besides insulin for the proper control of the blood sugar level. This active component, which is a trivalent chromium complex, is now referred to as the glucose tolerance factor (GTF). Patients belonging to the second category have the GTF early in life, but there is a tendency to lose it with age, causing increased vulnerability to late onset of diabetes.

The structure of the GTF is not well established. Nevertheless, Mertz et al. (1974) have been able to identify the components, namely, nicotinic acid, trivalent chromium, and the amino acids glycine, cysteine, and glutamic acid in the active complex. Although the biochemical role of GTF in glucose tolerance is not elucidated, Mertz et al. (1974) have suggested the possibility of GTF functioning as a cofactor for insulin, thereby enhancing the binding of insulin to receptive sites on the membrane of insulin-sensitive tissues. The authors have reported that these bindings are perhaps due to possible formation of bridges between -SH groups of insulin receptors and the -S-S-groups on the A-chain of the insulin molecule. As soon as the insulin is bound to the receptor site, the GTF potentiates the insulin-mediated uptake and utilization of glucose as demonstrated by in vitro studies with adipose and muscle tissue (Mertz 1969).

Likewise, Doisy et al. (1976) observed a reduction in the level of cholesterol and triglyceride in blood of humans by supplementing their diets with brewer's yeast rich in GTF. Although these findings have suggested possible health benefits due to supplementation of diets with organically bound chromium compounds like GTF, more feeding studies are necessary in order to confirm the results before one could suggest brewer's yeast or any source of yeast rich in GTF as possible treatment for lowering the blood sugar or curing lipid disorders.

### Selenium

Selenium has been recognized as an essential trace element for both human and animal nutrition. Recent surveys in some parts of Finland show that people with blood selenium levels less than 0.04 µg/L are three times more prone to heart attacks than their counterparts with normal selenium levels. Also, in China a fatal heart disease in children known as Kaishan's disease is presently treated by supplementing the regular diet with 1 mg $NaSeO_3$ each week.

The importance of Se in animal nutrition is also well documented. A common disease in lambs, white muscle disease, is due to a Se deficiency. The disease can be cured by adding 0.05 ppm Se to the diet. Unfortunately, in all of these cases the role of Se in reversing the disease is not well understood. Inconsistencies in the results of some studies have led to some confusion regarding the importance of Se in animal and human nutrition. It is highly probable that a metabolic step is required to trigger the need for a trace element like Se. Therefore, elucidation of the role of Se in human and animal nutrition may become a difficult task.

A diet containing Se at levels approaching toxicity is effective in reducing some types of cancer in animals. Results of an epidemiological study conducted by Dickson and Tomlinson (1976) showed an inverse relationship between the Se level in the blood and the death rate of the population due to the incidence of some types of cancer. The findings strongly suggest that certain types of cancer can be prevented by maintaining a proper level of Se in the diet.

Early feeding studies were conducted with selenite or selenate as sources of Se. However, the uptake of Se into tissues of internal organs was quite poor. In contrast, organically bound Se, such as selenomethionine found in some natural foods, is incorporated several times faster into the body tissue than inorganic Se. It proved to be more effective in controlling some disorders due to a Se deficiency (Cantor et al. 1975).

The discovery that Se is an integral component of glutathione peroxidase (EC 1.11.1.9) was made by Rotruck et al. (1973). The enzyme catalyzes the reduction of lipid peroxides before they can attack cellular membranes. The enzyme is capable of reducing different peroxides generated within a biological system by using the reducing equivalents derived from the reduced form of glutathione (GSH) according to the following general equation (Combs and Combs 1984):

$$\text{ROOH} + 2\,\text{GSH} \xrightarrow{\text{glutathione peroxidase (Se)}} \text{ROH} + \text{H}_2\text{O} + \text{GS-SG}$$

In this respect the action of Se-dependent glutathione peroxidase appears to be similar to that of vitamin E. Its action within the membrane provides protection to membrane lipids against autooxidation. Se in the organically bound form is now recognized as nutritionally important because of its ability to prevent some vitamin E deficiency disorders, at least in laboratory animals fed with partially purified diets.

Although many parts of the world produce foods and feeds containing adequate Se concentrations to meet human and animal requirements, there are some regions that show inadequate Se levels. The average person's diet in Finland, New Zealand, and parts of China contains less than 60 $\mu$g Se per day. Reports, particularly from Finland, indicate that such Se deficiencies can be corrected by supplementation with Se-rich yeasts that have been propagated to contain high levels of organically bound Se.

There are several factors that restrict Se uptake by yeast under conventional batch propagation conditions primarily because of the toxic effects of Se salts on yeast growth. However, a new procedure for the propagation of food grade Se-rich yeast has been developed by Nagodawithana and Gutmanis (1985). It is based on the concept that, under conditions of sulfur deficiency, sulfur could be replaced by selenium because of similarities between sulfur and selenium metabolism in yeast. In the production of Se-rich yeast the growth medium is fed incrementally so that the Se level never reaches toxic levels. Under these conditions the yeast assimilates selenium as a reaction to sulfur deficiency. The extent to which selenium can replace sulfur in yeast is, however, not well defined. Nutritional yeasts with an intracellular concentration of 1000 ppm of Se have been produced and are sold in the United States as dietary supplements.

There have been a few studies trying to identify the molecular structure of the Se complex present in Se-rich yeast. Korhola, Vainio, and Edelmann (1986) made an extensive study of the distribution of Se in Se-rich yeast with radiolabeled Se ($^{75}$Se). Analysis of the acid-hydrolyzed yeast protein showed that the major Se-containing compound was selenomethionine. The remaining Se was found in compounds such as selenodiglutathione, selenocysteine, and other minor unidentified compounds. The enteric absorption of selenomethionine and selenite was 75% and 45%, respectively. Robinson et al. (1978) showed that the uptake of selenomethionine was significantly faster than that of inorganic selenite by the small intestine, the cecum, and the colon of humans.

## Lipids

The lipid content of yeasts may vary between 4% and 7% (dry basis). About 1% may be extracted directly by fat solvents. The determination of total lipids requires acid hydrolysis prior to solvent extraction. Major constituents are fatty acid glycerides with a predominance of palmitic and oleic acids, sterols, and lipoproteins.

## Carbohydrates

Total carbohydrates account for about 30-35% of the yeast cell (dry basis). They consist mainly of carbohydrate storage compounds such as glycogen and the disaccharride trehalose, and the structural materials of the cell wall: the glucans and mannans. The details of the cell wall structure cannot be discussed here. The reader is referred to a treatise on yeast cell envelopes (Arnold 1981). The complexity of the cell wall structure may be illustrated by the mannan-protein structure shown in Figure 9-2.

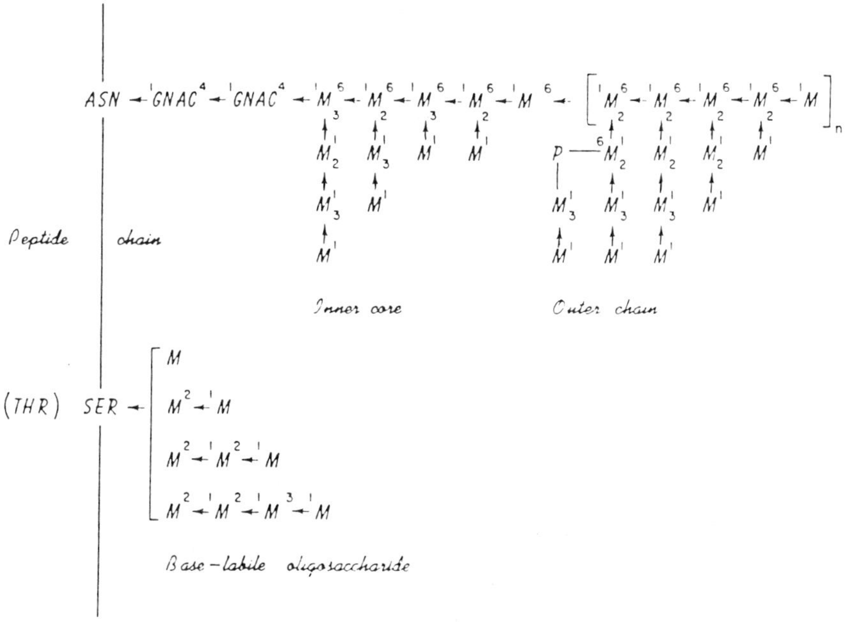

$$ASN \leftarrow {}^1GNAC^4 \leftarrow {}^1GNAC^4 \leftarrow {}^1M^6 \leftarrow {}^1M^6 \leftarrow {}^1M^6 \leftarrow {}^1M^6 \leftarrow {}^1M^6 \leftarrow \left[ {}^1M^6 \leftarrow {}^1M^6 \leftarrow {}^1M^6 \leftarrow {}^1M^6 \leftarrow {}^1M \right]_n$$

*Peptide chain*

*Inner core*   *Outer chain*

$$(THR) \; SER \leftarrow \begin{bmatrix} M \\ M^2 \leftarrow {}^1M \\ M^2 \leftarrow {}^1M^2 \leftarrow {}^1M \\ M^2 \leftarrow {}^1M^2 \leftarrow {}^1M^3 \leftarrow {}^1M \end{bmatrix}$$

*Base-labile oligosaccharide*

**Figure 9-2.** Mannan-protein structure of the yeast cell wall: M, mannose; GNAC, N-acetylglucosamine; ASN, asparagine; SER, serine; THR, threonine; P, phosphodiester group. (*After Bacon 1981*)

Interest in the fiber content of the human diet has led to a determination of nonnutritive fiber in yeast cells. It can be accomplished by sequential digestion with pepsin and trypsin and accounting for fat and minerals in the fiber residue. Whole cells of baker's yeast assayed by this method showed a fiber content of about 18% (dry weight basis). Washed cell walls resulting from the autolysis of yeasts had a fiber content of about 48%. Similar results have been reported by Sarwar et al. (1985). Isolated cell walls of S. cerevisiae have been used in the wine industry to minimize inhibition of the alcoholic fermentation by octanoic and decanoic acids.

## USE OF YEAST AS A MAJOR PROTEIN SOURCE

Candida utilis yeast had been used as an important supplement to the human diet in Germany during World War II. Its early development has been detailed by Butscheck (1962). Sulfite waste liquor, a by-product of the paper pulp industry, and wood hydrolysates were used as raw materials. Several plants that had been built during the war years were dismantled after the war, for instance, an installation that produced feed yeast on a molasses substrate in Jamaica. New plants for the production of food and feed yeast on sulfite liquor were built in the United States during the 1950s, and the production of feed yeast on wood hydrolysates continued in eastern Europe.

Since the 1960s, studies in this field have multiplied and covered a wider range of substrates and microorganisms. This interest was due to three events that occurred at about the same time, that is, between 1960 and 1970. The first was the recognition of acute shortages of protein both for human consumption and for animal feeding. The second was the pioneering work of A. Champagnat at the Société Française de Petroles BP, who studied the microbial oxidation of petroleum fractions beginning about 1957. Finally, the need to combat pollution by industrial wastes of high biological oxygen demand was recognized in the developed countries of the world. It is not possible to say which of these factors was the most important.

Particularly, the exciting prospect of using gas oils and n-paraffins as raw materials for the production of microbial protein led academic institutions and private industry into an ever-widening inquiry. Following the lead of research workers at the Massachusetts Institute of Technology, the products discussed in this chapter are often called single-cell proteins (SCP). Unfortunately, this designation blurs the distinction between whole yeasts and proteins isolated from yeasts. Therefore, currently produced yeasts— all essentially whole cells—will be called *microbial biomass.*

The large-scale production of yeasts on hydrocarbon substrates that looked so promising in the 1970s has not come to pass. The rapid increase in the cost of oil during this period as well as the spurious opposition of regulatory agencies, particularly in Japan and Italy, led to the closing of plants that had been built or had been in actual operation (O'Sullivan 1978). The various processes have been frequently reviewed (Levi, Shennan, and Ebbon 1979; Einsele 1983).

For use as a major source of protein by humans the presence of nucleic acids is a serious obstacle. From 10-15% of the total nitrogen of yeast cells is nucleic acid nitrogen. The intake of nucleic acids leads to elevated blood plasma levels of uric acid and may cause gout. Sources of nucleic acids in the diet are meats, particularly organ meats such as liver, and secondarily beer and possibly other fermented beverages.

Careful studies by Edozien et al. (1970) and Waslien, Calloway, and Margen (1970) have established the relationship between the feeding of yeast and the elevation of uric acid levels in blood plasma. On the basis of their findings, an upper limit of 2 grams of nucleic acids per adult per day has been accepted. This amount limits the daily intake to about 20 grams of yeast, an intake that will provide 10 grams of yeast protein or about 15% of the RDA for a vegetable protein.

A search for methods is underway to reduce the levels of nucleic acids or to remove them completely. Many such methods are available and some have been described in Chapter 8. However, none are presently economical because of the cost of processing and because of the unavoidable losses in yield. The significance of the nucleic acid content of yeasts for human food has been reviewed by Scrimshaw (1975, 1986), and the effect of nucleic acid metabolites on rat growth has been studied by Bryle et al. (1986). There is no need for the removal of nucleic acids from yeast biomass for use in feed.

Yeast biomass in its inactive dried form is used widely as a feed supplement. It is used in poultry rations and pig starter feeds. The use of brewer's and distiller's by-product yeast in feed has already been mentioned. In countries that lack cheap sources of oil seed meals (mainly soy bean meal), *Candida utilis* biomass is used extensively as a protein supplement. Live yeast cells in the form of active dried yeast or yeast culture are also used in the feed industry.

Yeast culture is produced by combining slurries of baker's yeast with cereal feed grains. The mash is incubated to permit yeast multiplication and fermentation. It is then dried at temperatures that preserve yeast viability. This yeast culture as defined by the Association of American Feed Control Officials (1982) may be added to silage: up to 2% of hay chop. It is also used in livestock and poultry feeds at the following levels: ruminants, 1-1.5%; hogs, 1.5-2%; poultry, 2-3% (Peppler and Stone 1976; Peppler 1983). Active dry yeast is also used as a supplement to ruminant feeds at levels that may supply several million yeast cells per gram. It has been reported to increase feed efficiency in cattle, but well-controlled studies have not been published.

## PRODUCTION OF BIOMASS

### Baker's, Brewer's, and Distiller's Yeasts

The production of baker's yeast has already been discussed in a separate chapter. Inactive dried nutritional yeast is produced from the yeast cream of about 18% solids by pasteurization followed by drum drying or spray drying. The spray-dried product has a very fine particle size. The drum-dried product that exits the drums as a thin sheet is available as small flakes or after grinding as a powder.

Spent excess brewer's yeast is recovered in breweries by sedimentation or centrifugation. It may contain from 12-16% total solids of which approximately two-thirds are yeast solids and one-third are beer solids. The slurry must be kept refrigerated before use because even at 4°C respiration reduces the carbohydrate fraction as shown in Figure 9-3. This yeast may be pasteurized and dried as such to a rather bitter dried brewer's yeast, or the yeast may be separated from the beer by decanter or nozzle centrifuges and partially debittered. It is also pasteurized and dried. Distiller's yeast is not recovered as a separate product because of the presence of other insolubles in the still residues.

### *Candida utilis* (Torula) Yeast

Torula is a common name for dried nutritional and feed yeasts. The name is derived from an earlier designation of the genus as *Torulopsis*. It is neither a trade name nor a scientific designation, but is generally used in the commercial trade. The major substrates for its production in the United States are sulfite waste liquors of the paper pulp process, and in Europe and Eastern bloc countries they are wood hydrolysates or sulfite waste. *C. utilis* is used for growth on wood sugars because it assimilates both pentoses and hexoses. Table 9-6 shows the assimilation of sugars and ethanol by commercially grown biomass yeasts.

In the production of paper pulp, lignin is solubilized by treatment of the wood with acid solutions of sodium, magnesium, or calcium salts of sulfurous acid and it forms lignosulfonic acids During this process the hemicelluloses of the wood are hydrolyzed and yield fermentable sugars. For hard woods, such as beech, aspen, or poplar, these sugars consist

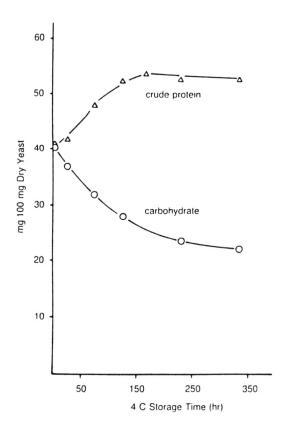

**Figure 9-3.** Changes in the protein and carbohydrate concentration of a brewer's yeast slurry stored at 4°C. (*From Ingledew 1977*)

**Table 9-6.** Assimilation Pattern of Commercial Yeast Species

| Substrate | S. cerevisiae | S. uvarum | K. fragilis | C. utilis | C. tropicalis |
|---|---|---|---|---|---|
| Glucose | + | + | + | + | + |
| Galactose | + | + | + | − | + |
| Maltose | + | + | + | + | + |
| Sucrose | + | + | − | + | + |
| Lactose | − | − | + | − | − |
| Xylose | − | − | − | + | + |
| Ethanol | (+) | − | + | + | − |
| KNO$_3$ | − | − | − | + | − |

*Source:* Based on data from Peppler 1970.
+ = assimilation (growth).
(+) = a few strains assimilate.
− = no assimilation.
S. uvarum, now S. cerevisiae; K. fragilis, now K. marxianus.

mainly of xylose and arabinose. About 20% of the sugars are hexoses, and the sugar concentration in the waste sulfite liquor may be 3% or less. For soft woods, such as spruce, balsam, or pine, the ratio of sugars is reversed. About 75% of the sugars may be hexoses, mostly mannose, and the rest are pentoses. In some instances, it has been possible to ferment the hexoses in a first run with *S. cerevisiae* to produce ethanol. Even baker's yeast has been produced in Canada from waste sulfite liquor. A second fermentation can then be achieved with *C. utilis* growing on the remaining pentoses. At present, *C. utilis* is preferred for obvious reasons.

The major components of waste sulfite liquor are lignosulfonates that may account for more than 50% of the total solids. Wood sugars may be present as 20-30 g/L. Fermentation inhibitors are free sulfur dioxide, acetic acid, and free furfural (Cejka 1985). A considerable fraction of inhibitors must be steam stripped from the liquor before use as a fermentation medium. The growth of *C. utilis* on wood sugars requires the addition of a nitrogen source, phosphate, and potassium to the medium. In contrast to *S. cerevisiae*, the *Candida* yeast does not require the addition of biotin.

The procedures used in the first plants built in the United States in the 1950s have not changed significantly. These are continuous, homogeneous fermentations in single-stirred tanks, often with a capacity of 300 m$^3$. The dilution rate is between 0.25 and 0.3. The pH is about 4.5 and the temperature is 30°C (Peppler 1970). Figure 9-4 is a flow sheet of the early process (Inskeep et al. 1951). It still represents by and large current usage.

The process can be run continuously for months. At a pH of 4.5, bacterial contamination is usually not a problem. The yield of yeast is generally 50% based on fermentable pentoses plus hexoses. A shift in the supply of sulfite liquor from hard wood to soft wood operation (or the reverse) may reduce yield and productivity temporarily until the yeast has adapted to the change. The yeast is recovered by centrifuging. Minimal washing with water leaves a considerable residue of lignosulfonic acids (about 10%) in the product that may be sold as feed yeast. For use as a human food the yeast must be centrifuged at least twice with extensive washing. Drying of the yeast can be done on drum driers or spray driers.

*Candida utilis* has also been produced in the United States on ethanol. A plant built on this principle has produced several million pounds of Torula per year. It is not presently in operation. However, the process is of sufficient interest to warrant a short description (Ridgeway et al. 1975). The continuous single-tank operation is run at a dilution rate between 0.25 and 0.4 per hour. Yield based on ethanol is about 80% and the cell concentration can be maintained at 2.1% based on fermenter volume. Eighty percent of the supernatant may be recycled, that is, it may be used for makeup of the medium. The temperature of the fermentation is

PRODUCTION OF BIOMASS 429

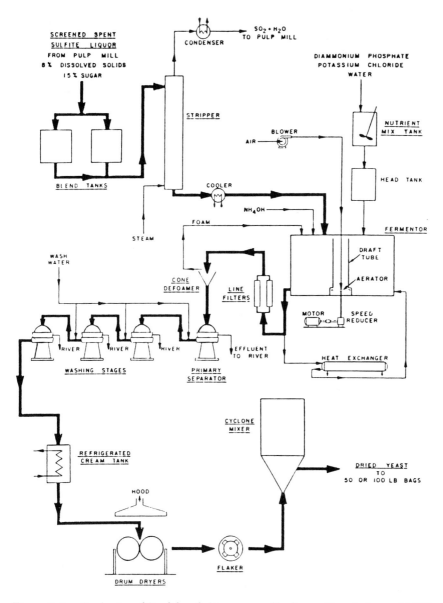

Figure 9-4. Production of *Candida utilis* biomass on sulfite liquor. (*From Inskeep et al. 1951*)

Table 9-7. Inorganic Nutrients for the Growth of *Candida utilis* on Ethanol

|  | Wt. of Element per 100 g Cells | Typical Compound |
|---|---|---|
| Macronutrients | | |
| Phosphorus | 2-4 g | $H_3PO_4$ |
| Potassium | 2-3 g | KCl |
| Magnesium | 0.3-0.6 g | $MgSO_4$, $MgCl_4$ |
| Calcium | 0.001-0.2 g | $CaCl_2$ |
| Sodium | 0.01-0.2 g | $Na_2CO_3$, NaCl |
| Micronutrients | | |
| Iron | 6-13 mg | $Fe[C_3H_4(OH)(COO)_2]$ |
| Manganese | 4-8 mg | $MnSO_4$ |
| Zinc | 2-6 mg | $ZnSO_4$ |
| Molybdenum | 1-2 mg | $Na_2Mo_4$ |
| Iodine | 1-3 mg | KI |
| Copper | 0.5-1 mg | $CuSO_4$ |

Source: Based on data from Ridgeway et al. 1975.

32°-35°C and the pH is kept at 4.0. Ethanol concentration in the fermenter is kept at about 200 ppm.

Table 9-7 shows the inorganic elements in the medium. This older table is presented because it provides more insight into the requirements for mineral elements than can normally be found in the literature. The requirements are expressed on the basis of the amount of yeast grown, and the carbon and energy source, ethanol, does not contribute additional minerals.

In 1988, a plant went into operation that produces a Torula type yeast in high cell density (HCD) fermentations (Anon. 1989). Baker's yeast is produced by the fed-batch process at a final concentration of 40-60 grams of cell solids per liter. HCD fermentations are continuous single-tank processes with cell densities of 120-140 g/L. Substrates such as carbohydrates, methanol, and ethanol and other carbon sources can be used. Species of *Candida, Hansenula, Pichia,* and *Saccharomyces* may be used in the process (Wegner 1983).

The results that can be achieved in 2-liter benchtop fermenters are shown in Table 9-8. In a 1500-liter pilot plant productivities of more than 25 grams yeast solids per L/hr could be produced. The required oxygen transfer rate (OTR) was 800 $O_2$/L/hr and the heat removal rate was 90 kCal/L/hr (Shay and Wegner 1985). Finally, Mermelstein (1989) describes commercial production of *Hansenula jadinii* (the perfect, sporulating form of *C. utilis*) in a 25,000-liter fermenter. The carbon and energy source is a

Table 9-8. Growth of Torula Yeast in High Cell Density Fermentations

|  | Torula Y-1084 | Torula Strain | Torula Strain |
|---|---|---|---|
| Substrate | EtOH | sucrose | molasses |
| Retention time (hr) | 3.7 | 4.7 | 5.9 |
| Cell mass (g/L) | 122 | 160 | 163 |
| Yield (%)* | 78 | 52 | 51 |
| Productivity (g/L/hr) | 33 | 34 | 27.6 |

Source: From Shay and Wegner 1985.
*Yield based on ethanol, sucrose, and fermentable sugars, respectively.

sugar. Maintenance of cell concentrations of 120-150 g/L requires an OTR of 800-1000.

The very high cell densities achieved in these fermentations permit direct drying of the fermentor liquor without separation and washing of the cells. This process also eliminates the requirement for disposal of a high biological oxygen demand (BOD) supernate. The process seems to work well with pure media such as ethanol, methanol, and glucose or sucrose. If molasses or whey are used as carbon and energy sources the final product contains all of the media constituents that have not been assimilated. The process requires substantial aeration. The OTR of 800 mM/L/hr is four to five times higher than that required for baker's yeast production. The requirements for heat removal are also higher per L/hr, but of course they are not higher per gram of yeast produced.

**Wood Sugars as Substrates**

The use of sulfite waste liquor for the production of biomass with C. utilis and the use of wood hydrolysates have already been mentioned. But the use of the most abundant renewable carbohydrate, cellulose or cellulose lignin complexes, for the production of biomass by various microorganisms requires additional comment. No comment is required on the growth of biomass on wood-derived hexoses or hexose pentose mixtures. Any number of yeast species including S. cerevisiae can readily be grown on hexoses and C. utilis can be grown on mixed monosaccharides.

The abundant raw materials from renewable resources are molasses (or cane juice), corn starch, and cellulose. Molasses requires no saccharification. Corn starch can be readily saccharified by available microbial amylases. The saccharification of cellulose is technically much more difficult and costly, thus preventing extensive use of wood sugars for either the production of biomass or ethanol in spite of the abundance and low cost of the raw material. The available raw materials are saw dust, other forest industry by-products, corn stover, and many other cellulosic residues. They consist of hemicellulose, cellulose, and lignin. Cellulose is present at least in part in crystalline form.

The extent of hydrolysis is also hindered by lignin that surrounds the cellulose fibers. In nature, wood is metabolized by several species of wood rot fungi, for instance, by a white-rot fungus *Dichomitus squalens* that can degrade all polymers of the plant cell wall, cellulose, hemicellulose, and lignin (Rouau and Odier 1986). Isolated cellulases react rather slowly. The activity of a pure cellulolytic enzyme on the natural substrate is 1 $\mu$M of glycoside bond split per min/mg enzyme protein (Reese and Mandels 1984). It is one-hundredth of the corresponding attack of amylases on starch. For cellulose hydrolysis the concentration of the reactants is generally not the rate-determining factor. The insolubility of the substrate, its morphology, and its structure are rate-limiting (Becker and Emert 1983).

The development has occurred of various processes for pretreatment of cellulosic substrates by mechanical and/or chemical means. All of these processes are costly. This subject cannot be treated here. Ladish and Tsao (1986) have contributed an excellent review of the engineering and economics of cellulose saccharification systems.

Enzyme hydrolysis of cellulose requires the action of three enzymes for hydrolysis to glucose: endo-$\beta$-1,4-glucanases that split cellulose at random positions of the chain; exo-$\beta$-1,4-glucanases that split the disaccharide cellobiose from the ends of the chain; and $\beta$-glucosidase that splits cellobiose into its two constituent glucose molecules. Complete hydrolysis of cellulose is difficult to achieve because of feedback inhibition by the end products. Production of cellulases by *Trichoderma reesei* and other fungi has been described by Frost and Moss (1987).

Ultimately the conversion of cellulose to monosaccharides may be used in the following three processes. In *ethanol production* ethanol may be formed during enzymatic hydrolysis, thus preventing feedback inhibition. The disadvantage is that efficient organisms for converting pentoses to alcohol are not yet available. In *biomass production* with an organism such as C. *utilis*, which assimilates hexoses and pentoses, the disadvantage is that feedback inhibition makes complete conversion difficult. *Glucose production* for food use has rarely been considered in view of the unsolved problems that have been mentioned.

## Dairy Yeasts

Large amounts of cheese whey are available as a by-product of the production of all cheeses. The whey may be dried and sold as feed or as a food ingredient, but supply by far surpasses demand. Thus, there exists a problem of waste disposal and there have been numerous attempts to use whey as a fermentation substrate for the production of ethanol and for biomass production. Liquid sweet whey contains about 7% solids with the following approximate composition: lactose, 74%; lactic acid, 2%; protein, 12% (lactalbumin and lactoglobulin), minerals, 8%; fat, 4%.

Whey may be concentrated and stored as 30-40% solids whey concentrate. Whey is also commercially ultrafiltered. The retentate is dried and sold as whey protein concentrate. The permeate can also be concentrated and stored as whey permeate concentrate. Concentrated whey and concentrated whey permeate cannot be stored as well as molasses. At lower storage temperatures lactose crystallizes out. At higher temperatures browning reactions occur. Thus, the use of whey as a carbohydrate substrate for fermentations has been severely restricted. Successful operations require that fermentations be carried out at high-capacity cheese plants or in their immediate vicinity.

Dairy yeasts are lactose-assimilating and lactose-fermenting yeasts. The preferred species for the production of biomass is *Kluyveromyces marxianus*. This yeast was previously classified as *Kluyveromyces fragilis* or *Saccharomyces fragilis*. The large and possibly the largest commercial process for the production of food grade dairy yeast is known as the Fromagerie Bel process that has been in operation for approximately 20 years (Blanchet and Biju-Duval 1969). It is carried out in continuous fashion with a residence time of about 4 hours. It shares many features with other commercial processes. The fermentation is highly aerobic; the temperature is 30°C and often higher (up to 38°C) and the pH is kept at 4.5 or lower. Whey protein causes excessive foaming during the aerated fermentation. Therefore, a major portion of the protein, the lactalbumin, may be removed by acid/heat coagulation. Preferably, all of the protein is removed by ultrafiltration. The heat-precipitated lactalbumin or the whey protein concentrates from the ultrafiltration process are valuable by-products.

A U.S. process has been described by Bernstein, Tzeng, and Sisson (1977). Figure 9-5 shows the scheme of this continuous process. Dairy yeast biomass may also be produced in fed-batch fermentations in a process that resembles that of the production of baker's yeast. In all of the commercial processes yeast is recovered from the fermentor liquor by centrifuging. It is pasteurized and drum-dried or roller-dried.

A modification of these processes by a two-step fermentation has been described by Goulet (1986). It consists of the conversion of lactose to lactic acid and galactose by *Streptococcus thermophilus* followed by the growth of baker's yeast. Earlier descriptions of two-step fermentations (using *Lactobacillus bulgaricus* and *Candida krusei*) as well of other alternative processes can be found in an extensive review by Mayrath and Bayer (1979). The high cell density process that has been described above may also be used

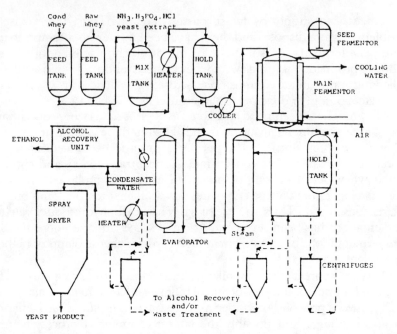

**Figure 9-5.** Continuous production of a dairy yeast (*K. marxianus*) on cheese whey (flow sheet). (*From Bernstein, Tzeng, and Sisson 1977*)

with whey as substrate (Shay and Wegner 1986; Shay, Hunt, and Wegner 1987). With a dilution rate of 0.20, a density of 112 grams of cells per liter could be maintained at a yield of 45% based on lactose and a productivity of 22 grams of cell solids per L/hr. The product resulting from the direct drying of the fermentor liquor had an ash content of almost 19% because of the contribution of the ash of the whey.

Strains of *S. cerevisiae* may be genetically modified so that they will be able to assimilate lactose (Sreekrishna and Dickson 1985). So far these attempts have not led to commercial developments.

## Oleaginous Yeasts

Several genera of yeasts have species that produce considerable concentrations of cell lipids, generally in nitrogen-limited media, for instance, in media with a C to N ratio of 30:1. Ratledge (1982, 1986) reported such fermentations with the following yeasts: *Rhodotorula glutinis, R. gracilis, Trichosporon cutaneum, Candida curvata,* and *Lipomyces lipofer,* among others.

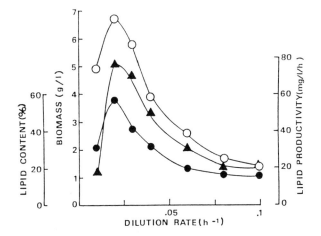

**Figure 9-6.** Lipid concentration (●), biomass production (○), and lipid productivity (▲) of R. glutinis as a function of the dilution rate. (*After Yoon and Rhee 1983*)

Lipid content of the cells varied from 40-70% on a dry weight basis. Oleaginous organisms are assumed to be those with more than 20% lipids.

In continuous fermentations the concentration of biomass as well as the concentration of lipids in the cell depends greatly on dilution rate as shown in Figure 9-6 (Yoon and Rhee 1983). Generally research workers have expressed the yields of lipids based on the weight of substrate used. Yields of 20-30 grams per 100 grams of substrate have been reported. Actually the biomass yields of oleaginous yeasts are lower than those of other microorganisms, but energy yields are higher because of the increased content of intracellular lipids (Pan and Rhee 1986). Lipids produced by oleaginous yeasts can be readily extracted. Crude oil extracted from these yeasts with hexane consists of more than 95% of triglycerides and 2-3% phospholipids (Anon. 1989). The material is frequently called single-cell oil (SCO) in analogy to single-cell protein (SCP), although it is often not clear whether the term refers to the oil-containing biomass or to the oil.

### n-Alkane-Assimilating Yeasts

n-Alkanes are straight-chain saturated hydrocarbons. Crude gas oil may be used as a substrate, but purified n-paraffin fractions with chain lengths from $C_{10}$ to $C_{20}$ are preferred (Champagnat et al. 1963). These hydrocarbons can be completely oxidized and assimilated by many yeast species. *Candida lipolytica* (*Endomycopsis lipolytica*) has been used frequently for

Table 9-9. Growth Parameters of *Candida tropicalis* Growing on Glucose or Hexadecane in Continuous Fermentations (substrate 1% w/v)

| | Dimension | Glucose | Hexane |
|---|---|---|---|
| Yield | g/g substrate | 0.51 | 0.98 |
| Max. specific growth rate | $\mu$/hr | 0.68 | 0.28 |
| Oxygen uptake | mM $O_2$/hr/g | 11.2 | 14.0 |
| Respiratory quotient (RQ) | | 1.0 | 0.46 |

*Source:* Based on data from Einsele 1983.

commercial production in the 1970s. Other Fungi Imperfecti are *Candida tropicalis* and species of *Rhodotorula* and *Trichosporon*. Genera of Ascomycetes-containing species of hydrocarbon-assimilating yeasts include *Debaromyces, Endomyces, Metschnikovia,* and *Pichia*.

Hydrocarbon fermentations are almost always continuous operations. The fermentation temperature of 30°C and the pH of 4.5 are similar to the other yeast fermentations. However, there are some fundamental differences. The liquid n-paraffins are not miscible with water. Therefore, the size of the hydrocarbon droplets is critical since growth of the yeast cells takes place through direct contact of the cell walls with the hydrocarbon droplets (for tridecane and higher alkanes). Usually 2 grams of $O_2$ are required to grow 1 gram of yeast cells.

Faust (1987) gives the following material balance for *C. lipolytica* for the process developed by British Petroleum: 1.12 g paraffin + 2.56 g oxygen + 0.13 g ammonia = 1 g biomass + 1.76 g carbon dioxide + 1.08 g water + 33.5 kJ. The oxygen requirements and the heat evolved are very much higher than for growth of yeasts on carbohydrate substrates. Comparative data for growth of *C. tropicalis* on glucose and hexadecane are shown in Table 9-9 (Einsele 1983).

Alkane-grown yeasts can be recovered by centrifuging. The cells are washed with water and may be spray-dried. The composition does not greatly differ from carbohydrate-grown yeasts although the lipid content may be higher. The following texts deal with the subject in more detail: Levi, Shennan, and Ebbon (1979); Einsele (1983); and Faust (1987). The earlier literature has been reviewed by Humphrey (1970) and by Laine, Snell, and Peet (1976).

## Methanol-Assimilating Yeasts

Methane-assimilating yeasts are rare and have not been investigated to any extent. The following species assimilate methanol: *Hansenula capsulata,*

*Hansenula polymorpha, Pichia lindnerii, Candida methanolica,* and *Torulopsis labrata* among others. But the only processes that have been practiced on the pilot or commercial level use bacteria of the genus *Methylomonas* (Gow et al. 1975). The bacteria grow at a pH of 6.8 and at 37°-40°C. Their crude protein levels are between 60% and 80% on a dry weight basis with a favorable protein composition. Reviews of methanol fermentations have been published by Faust and Praeve (1983) and Faust (1987).

## FURTHER READINGS

The following is an abbreviated list of general references to the field of biomass production for food and feed: Butscheck (1962); Peppler (1970); Rose (1979); Reed (1981); Roltz and Humphrey (1981); Reed (1982); Einsele (1983); Oura (1983); Dziezak (1987); and Faust (1987).

## REFERENCES

Anon. 1989. Single cell oil: A technology looking for applications. *Food Engineering*, Sept., 108-109.
Arnold, W. N. 1981. *Yeast Cell Envelopes: Biochemistry, Biophysics, and Ultrastructure.* CRC Press, Boca Raton, Fla.
Association of American Feed Control Officials. 1982. *Official Publication*, D. H. James (ed.). Department of Agriculture, State Capital Building, Charleston, W.V.
Bacon, J. S. D. 1981. Nature and disposition of polysaccharides within the cell envelope. In *Yeast Cell Envelopes: Biochemistry, Biophysics and Ultrastructure*, vol. 1., W. N. Arnold (ed.). CRC Press, Boca Raton, Fla.
Becker, D. A., and G. H. Emert. 1983. Evaluation of enzyme cost for simultaneous saccharification and fermentation of cellulose to ethanol. *Dev. Industr. Microbiol.* **24:**123-129.
Bernstein, S., C. H. Tzeng, and D. Sisson. 1977. The commercial fermentation of cheese whey for the production of protein or alcohol. In *Single Cell Protein from Renewable and Nonrenewable Resources*, A. H. Humphrey and E. L. Gaden (eds.). Biotechnol. Bioeng. Symp. 7, Wiley, New York.
Blanchet, M., and F. Biju-Duval. 1969. *International Dairy Federation Seminar on Whey Processing and Utilization at Weihenstephan*, German Fed. Rep.
Bryle, D., G. Sarwar, R. W. Peace, H. G. Bottin, and L. Saboie. 1986. Effect of feeding food yeast and nucleic acid metabolites on rat growth. *Proc. Int. Symp. Food and Biotechnology*, Quebec, Aug., p. 275.
Butscheck, G. 1962. Nutritional and feed yeasts (in German). In *Die Hefen*, vol. 2, F. Reiff et al. (eds.). Verlag Hans Carl, Nurnberg.
Cantor, A. H., M. L. Langevin, T. Noguchi, and M. L. Scott. 1975. Efficacy of selenium in selenium compounds and feed for prevention of pancreatic fibrosis in chicks. *J. Nutr.* **105:**106-111.
Cejka, A. 1985. Preparation of media. In *Biotechnology*, vol. 2, H. J. Rehm and G. Reed (eds.). VCH Publishing Co., Weinheim, West Germany.

Champagnat, A., C. Vernet, B. Laine, and J. Filosa. 1963. Microbial de-waxing with production of protein-vitamin concentrates. *6th World Protein Congress*, June, Frankfurt, German Fed. Rep.

Combs, G. F., and S. B. Combs. 1984. The nutritional biochemistry of selenium. *Ann. Rev. Nutr.* **4:**257-280.

Dickson, R. C., and R. H. Tomlinson. 1976. Selenium in blood and human tissue. *Chim. Acta* **16:**311-321.

Doisy, R. J., D. H. P. Streeter, J. M. Freiber, and A. J. Schneider. 1976. Chromium metabolism in man and biochemical effects. In *Trace Elements in Human Health and Disease*, vol. 2, A. S. Prasad (ed.). Academic Press, New York.

Dziezak, J. D. 1987. Yeasts and yeast derivatives. Definitions, characteristics and processing. *Food Technol.* **41:**104-121, 122-125.

Edozien, J. C., U. U. Udo, V. R. Young, and N. S. Scrimshaw. 1970. Effects of high levels of yeast feeding on uric acid metabolism of young men. *Nature* **228:**180.

Einsele, A. 1983. Biomass from higher n-alkanes. In *Biotechnology*, vol. 5, H. Dellweg (ed.). VCH Publishing Co., Weinheim, West Germany.

Faust, U. 1987. Production of microbial biomass. In *Fundamentals of Biotechnology*, P. Praeve et al. (eds.). VCH Publishing Co., New York.

Faust, U., and P. Praeve. 1983. Biomass from methane and methanol. In *Biotechnology*, vol. 3, H. Dellweg (ed.). VCH Publishing Company, Weinheim, West Germany.

Frost, G. M. and D. A. Moss. 1987. Production of enzymes by fermentation. In *Biotechnology*, vol. 7a. H. J. Rehm and G. Reed (eds.). VCH Publishing Co., Weinheim, West Germany.

Goulet, J. 1986. Upgrading of cheese whey: Some initiatives. In *Food and Biotechnology*. Proc. Int. Symp., Quebec, Aug. 19-22.

Gow, J. S., J. D. Littlehailes, S. R. L. Smith, and R. B. Walter. 1975. SCP production from methanol: Bacteria. In *Single Cell Protein II*, S. R. Tannenbaum and D. I. C. Wang (eds.). MIT Press, Cambridge, Mass.

Humphrey, A. E. 1970. Microbial protein from petroleum. *Proc. Biochem.* **5**(6):19-22.

Ingledew, W. M. 1977. Spent brewer's yeast—Analysis, improvement, and heat processing. *Tech. Quart. Master Brew. Assoc. Am.* **14**(4):231-237.

Inskeep, G. C., A. J. Wiley, J. M. Holderby, and L. P. Hughes. 1951. Food yeast from sulfite liquor. *Ind. Eng. Chem.* **43:**1702-1711.

Kockova-Kratochvilova, A. 1982. *Kvasinky a Kvasinkovite Microorganizmy (Yeast and Yeastlike Organisms;* in Slovak). ALFA, Bratislava, Czechoslovakia.

Korhola, M., A. Vainio, and K. Edelmann. 1986. Selenium yeast. *Ann. Clin. Res.* **18:**65-68.

Ladish, M. R., and G. T. Tsao. 1986. Engineering and economics of cellulose saccharification. *Enzyme Microbiol. Technol.* **8**(2):66-69.

Laine, B. M., R. C. Snell, and W. A. Peet. 1976. Production of single cell proteins from n-paraffins. *Chem. Eng. (London)* **310:**440-443, 446.

Levi, J. D., J. L. Shennan, and G. P. Ebbon. 1979. Biomass from liquid n-alkanes. In *Economic Microbiology*, vol. 4, A. H. Rose (ed.). Academic Press, London.

Mayrath, J., and K. Bayer. 1979. Biomass from whey. In *Economic Microbiology*, vol. 4, A. H. Rose (ed.). Academic Press, London.

Mermelstein, N. H. 1989. Continuous fermenter produces natural flavor enhancers for foods and pet foods. *Food Technol.* **43**(7):50-53.

Mertz, W. 1969. Chromium occurrence and function in biological systems. *Physiol. Rev.* **49:**168-239.

Mertz, W., and K. Schwartz. 1955. Impaired glucose tolerance as an early sign of dietary necrotic overdegradation. *Arch. Biochem. Biophys.* **58**:504-506.
Mertz, W., E. W. Woepfer, E. E. Roginski, and M. M. Polansky. 1974. Present knowledge of the role of chromium. *Federation Proc.* **33**:2275-2280.
Nagodawithana, T. W., and F. Gutmanis. 1985. Method for the production of selenium yeast, U.S. Patent 4,530,846.
O'Sullivan, D. A. 1978. BP contests petroprotein plant health issue. *Chem. Eng. News* **56**(12):12.
Oura, E. 1983. Biomass from carbohydrates. In *Biotechnology*, vol. 3, H. Dellweg (ed.). VCH Publishing Co., Weinheim, West Germany.
Pan, J. G., and J. S. Rhee. 1986. Biomass yield and energetic yields of oleaginous yeasts in batch culture. *Biotechnol. Bioeng.* **28**:112-114.
Peppler, H. J. 1970. Food yeasts. In *The Yeasts*, vol. 3, A. H. Rose (ed.). Academic Press, London.
Peppler, H. J. 1983. Fermented feeds and feed supplements. In *Biotechnology* vol. 5, G. Reed (ed.). VCH Publishing Co., Weinheim, West Germany.
Peppler, H. J., and C. W. Stone. 1976. Feed yeast products. *Feed Management* **27**(8):17-18.
Ratledge, C. 1982. Microbial oils and fats: An assessment of their commercial potential. *Prog. Industr. Microbiol.* **16**:119-206.
Ratledge, C. 1985. Lipids. In *Biotechnology*, vol. 4, H. Pape and H. J. Rehm (eds.). VCH Publishing Co., Weinheim, West Germany.
Reed, G. 1981. Use of microbial cultures: Yeast products. *Food Technol.* **35**(1):89-94.
Reed, G. 1982. Microbial biomass. In *Prescott and Dunn's Industrial Microbiology*, 4th ed., G. Reed (ed.). AVI Publishing Co., Westport, Conn.
Reese, E. T. and M. Mandels. 1984. Rolling with the times: Production and application of *Trichoderma reesei* cellulase. In *Annual Reports on Fermentation Processes*. vol. 7, G. T. Tsao (ed.). Academic Press, New York.
Ridgeway, J. A., T. A. Lappin, B. M. Benjamin, J. B. Corns, and C. Akin. 1975. Single cell protein materials from ethanol, U.S. Patent 3,865,691.
Robinson, M. F., H. M. Rea, G. M. Friend, R. D. H. Stewart, P. C. Scow, and C. D. Thomson. 1978. On supplementing the selenium intake of New Zealanders. 2. Prolonged metabolic experiments with daily supplements of selenomethionine, selenite and fish. *Brit. J. Nutr.* **39**:589-600.
Roltz, C., and A. Humphrey. 1981. Microbial biomass from renewables: Review of alternatives. *Adv. Biochem. Eng.* **21**:1-53.
Rose, A. H. 1979. History and scientific basis of large-scale production of microbial biomass. In *Economic Microbiology*, vol. 4, A. H. Rose (ed.). Academic Press, London.
Rotruck, J. T., A. L. Pope, H. E. Ganther, A. B. Ivanson, D. E. Hafeman, and W. G. Hoekstra. 1973. Selenium: Biochemical role as a component of glutathione peroxidase. *Science* **176**:588-590.
Rouau, X., and E. Odier. 1986. Production of exocellular enzymes by the white-rot fungus *Dichomitus squalens* in cellulose containing liquid culture. *Enzyme Microbiol. Technol.* **8**:22-26.
Sarwar, G., B. G. Shah, R. Mongeau, and K. Hoppner. 1985. Nucleic acid, fiber and nutrient composition of inactive dried food yeast products. *J. Food Sci.* **50**:353-367.
Scrimshaw, N. S. 1975. Single cell protein for human consumption—an overview. In *Single Cell Protein II*, S. R. Tannenbaum and D. I. C. Wang (eds.). MIT Press, Cambridge, Mass.

Scrimshaw, N. S. 1986. Nutritional and tolerance considerations in the feeding of single cell protein. In *Food and Biotechnology*, Proc. Int. Symp., Quebec, August.

Seeley, R. D. 1977. Fractionation and utilization of baker's yeast. *Tech. Quarterly Master Brewer's Assoc. Am.* **14**(1):35-39.

Shay, L. K., H. R. Hunt, and G. H. Wegner. 1987. High productivity process for cultivating industrial microorganisms. *J. Industr. Microbiol.* **2**:79-85.

Shay, L. K., and G. H Wegner. 1985. Improved fermentation process for producing Torula yeast. *Food Technol.* **39**(10):61-66, 70.

Shay, L. K., and G. H. Wegner. 1986. Nonpolluting conversion of whey permeate to food yeast protein. *J. Dairy Sci.* **59**:676-683.

Sreekrishna, K., and R. C. Dickson. 1985. Construction of strains of *S. cerevisiae* that grow on lactose. *Proc. Natl. Acad. Sci. USA* **82**(23):7909-7913.

Tamer, I. M., M. Ozilgen, and S. Ungan. 1988. Kinetics of riboflavin production by brewer's yeasts. *Enz. Microbiol. Technol.* **10**(12):754-756.

Waslien, C. I., D. H. Calloway, and S. Margen. 1970. Uric acid levels in men fed algae and yeast as protein sources. *J. Food Sci.* **35**:294-298.

Wasserman, A. E., J. E. Hanson, and N. F. Alvare. 1961. Large-scale production of yeast in whey. *J. Water Pollution Control Fed.* **33**:1090-1094.

Wegner, G. H. 1983. Biochemical conversions by yeast fermentation at high cell densities, U.S. Patent 4,414,329.

Yoon, S. H., and J. A. Rhee. 1983. Quantitative physiology of *Rhodotorula glutinis* for microbial lipid production. *Proc. Biochem.* **18**(5):2-4.

CHAPTER
10

# USE OF YEASTS IN THE DAIRY INDUSTRY

The occurrence and use of yeasts in the dairy industry is not as fundamental as it is for industries treated in the preceding chapters. Nevertheless, there are some interesting and challenging applications that will be discussed briefly.

## CHEESE

In the production of cheese, milk is inoculated with lactic acid bacteria that are responsible for acid formation prior to enzymatic precipitation of the curd, and subsequently for flavor formation during the ripening period. In some cheeses the typical flavor is produced by inoculation with mold spores, for example, with *Penicillium roquefortii*. In some mold-ripened cheeses a considerable population of yeasts develops. These yeasts generally belong to the species *Kluyveromyces lactis*, *Saccharomyces cerevisiae*, and *Debaromyces hansensi*. In a Roquefort cheese the yeast population may reach $10^9$ cells per gram after 30 days of ripening (Gripon 1987). As a secondary flora, yeasts play a part in the development of the flavor of such cheeses.

## LACTOSE-FREE MILK

There is considerable interest in the removal of lactose from some dairy products, such as milk, for use by lactose-intolerant individuals. Lactose removal can be achieved by enzymatic conversion of the disaccharide to its constituent monosaccharides, glucose and galactose. Rao, Godbole, and D'Souza (1988) have suggested the removal of lactose by fermentation with immobilized cells of *Saccharomyces fragilis* (now *Kluyveromyces marxianus*). The yeasts were immobilized in calcioum alginate by well-known methods. Use of the immobilized cells, which can then be removed from the milk, avoids the yeasty flavor that would be contributed by free cells. This work was carried out with buffalo milk, but it is equally applicable to the milk of other dairy cattle.

## ACIDOPHILUS-YEAST MILK

A commensal interaction between *Lactobacillus acidophilus* and lactose-fermenting yeasts has been established by Subramanian and Shankar (1985). *L. acidophilus* growth was stimulated by the presence of the yeast, but growth of the yeast was not affected by the bacterial culture. The mixed culture could be easily maintained since the relative proportions of the two organisms remained fairly constant even after six transfers. The yeasts used were *S. fragilis* and *C. pseudotropicalis*. The number of colony-forming units of yeasts was usually in the neighborhood of $50 \times 10^6$ cells per millimeter and for lactobacilli $10^9$ cells per milliliter. These amounts suggest a considerably larger mass of yeasts than of bacilli because of the larger size of the yeast cells.

## KEFIR

Some fluid or semifluid dairy products are produced by fermentation with yeasts and lactobacilli in symbiotic culture. Bacterial and yeast fermentations either jointly or in sequence have been treated in preceding chapters: the malolactic fermentation of wines, the sour mash whisky fermentation, or the rye sour fermentation. With the possible exception of the San Francisco sour dough fermentation, these fermentations do not depend on truly symbiotic relationships.

In contrast, the production of kefir and koumiss is based on the symbiosis of yeast and bacterial species. The source of the microbes are kefir grains, generally small, white or beige, elastic granules (Duitschaever, Kemp, and Smith 1988). The matrix consists of a polysaccharide/protein/lipid complex. Approximately one-half of the mass of the matrix is a polysaccharide, kefiran, that is excreted by *L. brevis* (LaRiviere, Koolman, and Schmidt 1967). The population imbedded in this matrix consists of yeasts (*Saccharomyces delbrueckii* and *S. cerevisiae*), mesophilic bacilli (*Lacto-*

bacillus brevis and L. casei), and mesophilic streptococci (Leuconostoc mesenteroides and Streptococcus durans), as well as a few acetic acid bacteria (Bottazzi 1983).

The kefir grains are separated after the fermentation. They can be used over and over again. Kefir grains may be suspended in cold water and stored at 4°C for a few days. For longer storage they are dried at a low temperature, and they can then be stored at a cool temperature for a year. The yeast and bacterial cells apparently are not washed out of the matrix during continued use (Chandan 1982). The kefir grains are nature's prototype of immobilized microbial cells.

Kefir is made from cow's, sheep's, or goat's milk. Figure 10-1 shows the

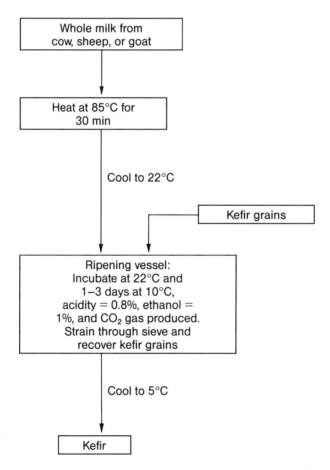

**Figure 10-1.** Scheme of the manufacturing process for kefir. (*After Kosikowski 1977*)

scheme of the manufacturing procedure (Kosikowski 1977). The product is acid in taste, slightly alcoholic, and somewhat fizzy due to $CO^2$ formation during the alcoholic fermentation. As with other natural fermentations, the composition of kefir is quite variable. The fermented milk is consumed in considerable volume in the USSR (4.5 kg per capita in the 1970s). In spite of some promotional efforts (Kemp 1984; Liu and Moon 1983), it could not be successfully introduced in the United States. A kefir drink available in Central Europe appears to be essentially the end product of a bacterial fermentation.

## KOUMISS

This product is made from mare's milk. The alcohol concentration may vary from 1-2.5% and the lactic acid concentration from 0.7-1.8%. The product does not curdle as cow's milk does because of the lower protein content of mare's milk, and hence it forms a smooth, viscous liquid. Koumiss is produced in the USSR and Mongolia where mare's milk is available. Guan and Brunner (1987) suggest the use of a blend of cow's milk with clarified sweet whey (1:1) and fortification with 2.5% sucrose to imitate the gross composition of cow's milk. They could produce a koumiss-like drink with cocultures of *Streptococcus lactis* and *Kluyveromyces lactis*.

## DESUGARING OF EGGS

Dried egg products such as dried albumin (egg white), dried egg yolk, or dried whole eggs deteriorate upon storage. The products brown, they develop a stale flavor, and in the case of egg albumin the whipping characteristics suffer. The deterioration is caused by reaction of glucose with egg proteins (Maillard reaction) as well as with a phospholipid, especially cephalin. The functional properties are mainly affected by the Maillard reaction. The off-flavor develops as a result of the glucose-cephalin reaction (Hill and Sebring 1977). Glucose concentrations of 0.4-1.0% have been reported in fresh eggs (solid basis), and the sugar must be removed before drying.

Glucose can be removed by a bacterial fermentation, by a yeast fermentation, or by conversion with glucose oxidase. Until the 1940s, spontaneous fermentations were permitted to occur, sometimes with disastrous results due to the development of contaminants such as *Salmonella*. Thereafter, selected cultures of lactic acid bacteria were used for the fermentation. The use of glucose oxidase, which is quite common in the industry, need not be described here.

Yeast fermentations are carried out with baker's yeast (*S. cerevisiae*) at a level of 0.34% at 30°C. The process has the advantage that it does not produce any acids, it is a short process that minimizes the danger of contamination, and the yeast is readily available (Hill and Sebring 1977). The use of immobilized cells of baker's yeast was reported by D'Souza and Godbole (1989). The method of immobilization was reported by Rao, Godbole, and D'Souza (1988). Egg melange (mixed whole eggs) could be completely desugared within 30 minutes at the natural pH of the melange and without a change in pH. Washed beads of immobilized cells could be reused repeatedly.

## REFERENCES

Bottazzi, V. 1983. Other fermented dairy products. In *Biotechnology*, vol. 5, H.-J. Rehm and G. Reed (eds.). VCH Publishing Company, Weinheim, West Germany.

Chandan, R. C. 1982. Other fermented dairy products. In *Prescott and Dunn's Industrial Microbiology*, 4th ed., G. Reed (ed.). AVI Publishing Co., Westport, Conn.

D'Souza, S. F., and S. S. Godbole. 1989. Removal of glucose from egg prior to spray drying by fermentation with immobilized yeast cells. *Biotechnol. Lett.* **11**(3):211-212.

Duitschaever, C. L., N. Kemp, and A. K. Smith. 1988. Microscopic studies of the microflora of kefir grains and of kefir made by different methods (in English). *Milchwissenschaft* **43**(8):479-481.

Gripon, J. C. 1987. Mould ripened cheeses. In *Cheese: Chemistry, Physics, and Microbiology*, vol. 2, P. F. Fox (ed.). Elsevier, New York.

Guan, J., and J. R. Brunner. 1987. Koumiss produced from a skim milk-sweet whey blend. *Cultured Dairy Prod. J.* **22**(1):23.

Hill, W. M., and M. Sebring. 1977. Desugarization. In *Egg Science and Technology*, 2nd ed., W. J. Stadelman and O. J. Cotterill (eds.). AVI Publishing Co., Westport, Conn.

Kemp, N. 1984. Kefir, the champagne of cultured dairy products. *Cultured Dairy Prod. J.* **19**(3):29-30.

Kosikowski, F. V. 1977. *Cheese and Fermented Milk Foods*, 2nd ed. Edwards Bros., Ann Arbor, Mich.

LaRiviere, J. W. M., P. Koolman, and K. Schmidt. 1967. Kefiran, a novel polysaccharide produced in the kefir grain by *Lactobacillus brevis*. *Arch. Mikrobiol.* **59**:269-278.

Liu, J. A. P., and N. J. Moon. 1983. Kefir, a "new" fermented milk product. *Culuured Dairy Prod. J.* **18**(3):11-12.

Rao, B. Y. K., S. S. Godbole, and S. F. D'Souza. 1988. Preparation of lactose free milk by fermentation using immobilized *Saccharomyces fragilis*. *Biotechnol. Lett.* **10**(6):427-430.

Subramanian, P., and P. A. Shankar. 1985. Commensalistic interaction between *L. acidophilus* and lactose fermenting yeasts in the preparation of acidophilus-yeast milk. *Cultured Dairy Prod. J.* **20**(4):17, 24, 26.

# INDEX

Acetaldehyde
  beer flavor, 119
  in sherry, 209
Acetic acid, as yeast growth inhibitor, 184
Acetic acid bacteria
  as brewery contaminants, 125
  as wine contaminants, 200
Acetoin, formation by wine yeasts, 191
Acidophilus-yeast milk, 442
Acrolein, in distilled beverages, 231
Active amyl alcohol, formation by wine yeast, 189
Active dry baker's yeast
  drying process, 301
  equilibrium moisture content, 308
  production of, 297-308
  protected, 361
  rehydration of, 361
  storage stability, 307
  use of, 359-363
Active dry distiller's yeast, 244
Active dry wine yeast, 159-162
  in champagne production, 213
  contaminants in, 162
  properties, 150-161
  rate of inoculation, 164
Active dry yeast, as feed supplement, 425-426
5'-Adenylic acid, 385
Adjunct mashing, brewing process, 109-110
Aeration, baker's yeast production, 285
Aerobic growth, of baker's yeast, 264-267
Alcohol dehydrogenase, 398
Alcoholic fermentation, by-products of, 185-193
Alcohols. *See* Fusel alcohols
Aldehydes
  in beer flavors, 127
  in bread flavors, 335
  formation in sherry, 209
n-Alkanes, as yeast growth substrates, 435-436
Alkylating agents, mutagenesis, 46

Amino acids
  concentration in yeasts, 387, 417
  uptake by brewer's yeasts, 96-98
  in wine fermentations, 179
$\alpha$-Amylase, use in baked goods, 319
$\alpha$-1,4-Amyloglucosidase, use in baked goods, 321
$\alpha$-1-Antitrypsin, expression in yeast, 408
Armagnac, 239
Ascorbyl-6-decanoate, as wine preservative, 182
Ascospores, shapes of, 14
Ascosporogenous yeasts, characteristics, 15
Autolysates, 380
Autolysis of yeast, 371-377
  in champagne fermentations, 180, 213

Baked goods, recycling of yeast, 315
Baker's yeast. *See also* Active dry baker's yeast
  aerobic growth, 264-267
  commercial strains, 263
  contamination of, 296
  cryoresistance of, 353
  elemental composition, 266
  fermentation activity, determination of, 331-334
  freezing of, 350
  genetic improvement, 56-71
  growth in doughs, 348-349
  heat inactivation, 325-326
  history, 262
  manufacturing process, 262
  recycling of, 261
  strain selection, 318
Baker's yeast nutrients
  minerals, 275
  nitrogenous, 275
  vitamins, 276
Baker's yeast production
  aeration, 285
  centrifuging, 292
  effluent BOD, 309

INDEX 447

filtration, 292
harvesting, 292
nutrient feed, 284
oxygen concentration, 265
packaging, 294
sequence of fermentations, 288-290
Baker's yeast products
  activity of, 360
  shelf life of, 362
Bakery mixes, complete, 361
Baking. *See* Doughs
Baking powder, 316-317
Barley malt, use in baked goods, 319
Basidiosporogenous yeasts, 27-30
  taxonomic characteristics, 28
Beer. *See also* Brewing; Brewer's yeast
  concentration of flavorants, 127
  dextrin concentration, 139
  low calorie, 138-140
  threshold of flavorants, 127
Berlin short sour dough process, 355
Biogenic amines, in wine, 202-204
Biotin
  as baker's yeast nutrient, 276
  as brewer's yeast nutrient, 99
*Botrytis cineria*
  grape infection, 201
  Sauterne flavor, 202
Bottom fermenting yeasts, 105
Bourbon, 227-233
Brandy, 284-240
Bread. *See also* Doughs
  flavor, 334-336
  formation during baking, 334
  technology, 336-363
*Bretanomyces*, spoilage of wine, 200
Brewer's fermentations, contaminants
  bacteria, 124-126
  wild yeasts, 121-123
Brewer's spent yeast, respiration in storage, 426
Brewer's wort. *See also* Malt wort
  oxygenation, 128
  fermentation of, 111-112
Brewer's yeast
  fermentation of wort sugars, 102-105
  genetic improvement, 72-81
  mating types, 91-92
  metabolic pathways, 128
  ploidy, 91
  vitamin requirement, 99-100

Brewing
  high gravity, 142-145
  history, 89
  yeast growth, 97, 115
  yeast propagation system, 113-116
Brewing process, 105-121
  adjunct mashing, 109-110
  decoction mashing, 108-111
  double mash system, 109-111
  infusion mashing, 108-111
Brimec bread process, 343
Budding, 17
  growth of baker's yeast, 283
Budding cycle, 191
Butanediol, formation by wine yeast, 191
Butylated hydroxyanisole, production of active dry baker's yeast, 305

Cadaverine, in wine, 203
Calcium carbonate, as preferment buffer, 338
Calvados, 240
Canadian whiskey, 236
*Candida*
  high cell density growth, 430-431
  occurrence on grapes, 154
*Candida krusei*
  description of, 159
  in sour doughs, 355
*Candida pseudotropicalis*, lactase production, 396
*Candida stellata*, description of, 159
*Candida utilis*
  biomass production, 426-430
  production on ethanol, 428
  production on sulfite liquor, 428-430
*Candida vini*, description, 159
Candy, soft-centered, use of invertase, 393
Carbohydrate content of yeasts, 423-424
Carbon dioxide
  effect in wine fermentations, 175
  production rate in baker's yeast, 332-333
Carbonyl compounds, bread flavor, 335
Catabolite repression, in baker's yeast production, 270
Cellulose, as fermentation substrate, 431-432
Cell wall
  dietary fiber content, 424
  mannan-protein structure, 423
  removal of, 49

Cell wall (continued)
  rupture of, 387
  structure of, 374-375, 401
Cell wall material, use in wine making, 182
Champagne
  Charmat process, 211
  méthode champenoise, 210-211
Cheese, flavor contribution by yeasts, 441
Chill proofing of beer, 121
Chitin, in yeast cell wall, 377
Chorleywood bread process, 342, 344
Chromium, glucose tolerance factor, 420
Chymosin, expression in microorganisms, 409
Citric acid, formation by wine yeast, 187
Coenzyme A-synthesizing complex, 404
Coenzyme Q, 18
  in yeast taxonomy, 11
Cognac, 239
Compressed yeast, stability of, 295
Consumer baker's yeast, 361
Continuous fermentation
  fuel ethanol production, 255
  of wines, 215
Corn starch, as fermentation substrate, 431
Corn syrup, as baker's yeast substrate, 278
Corn whisky, 235
Crabtree effect, 128
Cryoresistance of baker's yeast, 353
Culture collections, 34
Cysteine, use in short-time doughs, 343
$5'$-Cytidine monophosphate, 385

Damaged starch, 319
*Debaromyces*, taxonomic characteristics, 15
Decanoic acid, inhibition of wine fermentation, 182
Decoction mashing, brewing process, 108-111
Dessert wines, processing, 207
Deuteromycetes. *See* Fungi Imperfecti
DEX genes, 74
Diabetes, yeast glucose tolerance factor, 420
Diacetyl
  beer flavor, 119
  formation in beer, 132-135
  formation by wine yeast, 191
Dimethyl sulfide, flavor threshold in beer, 136
Distilled beverages
  ethanol yields, 230
  mashing, 227
  sour mashes, 231
Distiller's fermentations
  bacterial counts, 232
  effect of glucose concentration, 246
  effect of yeast concentration, 246
  yeast counts, 232
Distiller's spent grains, 233
Distiller's yeasts, 227-255
  productivity, 243
DNA
  base composition in yeast taxonomy, 11
  cloning technology, 51-56
Double mash system, brewing process, 109-111
Dough fermentation
  effect of absorption water, 329
  effect of osmotic pressure, 327-330
  effect of pH, 323-324
  effect of temperature, 324-326
Doughnuts, 347
Doughs
  freezing of, 350
  mechanical development, 342
  rate of carbon dioxide formation, 322

Eggs, desugaring with yeast, 445
Ehrlich pathway, fusel alcohol formation, 131, 118
Epidermal growth factor, expression in yeast, 407
Ergosterol
  effect on wine yeast stability, 176
  in yeast biomass, 390
*Escherichia coli*, cloning system, 52
Esters
  as beer flavors, 127
  concentrations in beer, 135
  flavor thresholds in wine, 189-191
Ethanol
  assimilation by yeast, 427
  determination, as process control in baker's yeast production, 291
  extraction during fermentation, 254
  inhibition, in wine fermentations, 170-171
Ethyl acetate, formation by wine yeasts, 190
Ethyl carbamate, in wine, 203
Ethyl decanoate, in wine, 190
Ethyl hexanoate, in wine, 190
Ethyl octanoate, 190

INDEX    449

Fed-batch fermentation, 265
Feed yeast, 425-426
Fermenters, baker's yeast production, 285
Ferulic acid, 75
Fiber, dietary
  in yeast, 424
  yeast glycan, 404
Fission, 17
Flavor enhancers, 381-385
Flavors, yeast-derived, 370-381
Flocculence
  of brewer's yeast, 79-81, 93-95
  mechanism of, 81
FLO genes, 80
Flor yeast, 208
Food fermentations, history, 369
Foods, fermented
  loss of calories, 5
  nutritional value of, 3-4
  per capita consumption, 413
  preservation of, 2
Fortified wines. *See* Dessert wines
Freezing, of bread doughs, 328
Frozen doughs, 349-353
Frozen dough yeast, genetic improvement, 65-66
Fruit brandy, 240
Fruit wines, processing, 206
Fuel ethanol, production of, 252
Fungi, kingdom of, 12
Fungi Imperfecti, 23-27
  genera, 25
  taxonomic characteristics, 26
Furfural
  in bread flavor, 335
  inhibition of distiller's fermentations, 245
Fusel alcohols
  in bread flavor, 335
  concentrations in beer, 130
  flavor threshold in beer, 130
  flavor thresholds in wine, 189

$\alpha$-Galactosidase, 67
$\beta$-Galactosidase, 396-397. *See also* Lactase
GAL genes, 68
Gas oil, as yeast growth substrate, 435
Generation time, 265
Genetically engineered products, 405-409
Genetic transformation, of wine yeast
  flocculence, 199
  isoamyl alcohol fermentation, 199

Genetics, 37-81
  of baker's yeast strains, 263
  of wine yeast, 199
Gin, 238
Glucan
  of yeast cell wall, 400
  structure, 403
Glucoamylase, use in making low-calorie beer, 140
$\beta$(1-3) Gluconase, yeast autolysis, 373
$\beta$(1-6) Gluconase, lysis of cell wall, 373
Glucono-delta-lactone, 317
Glucose tolerance factor, 420
Glycan
  functionality in foods, 400
  of yeast cell wall, 377, 400
Glycerol
  effect on osmotolerance, 59
  formation by wine yeast, 186
  transport in *Saccharomyces rouxii*, 59
Glycolytic pathway, 129
Glyoxalate pathway, 129
Grain-neutral spirits, enzymatic conversion of substrate, 236
Grape varieties in wine making, 264
5'-Guanylate, 381

Hansen, Emil Christian, 8, 37
*Hanseniaspora*
  occurrence on grapes, 154
  species characteristics, 18
*Hanseniaspora guillermondii*, description, 158
*Hansenula*
  high cell density growth, 430
  taxonomic characteristics, 15
*Hansenula anomala*, description, 158
Heat evolution
  in baker's yeast production, 282
  in wine fermentations, 168
Hepatitis B vaccine, expression in yeast, 406-407
Heterothallism, 39, 91
Hexose monophosphate pathway, 129
High cell density growth, 430
Higher alcohols. *See* Fusel alcohols
High gravity brewing, 142-145
Histamine, in wine, 203
Homothallism, 39
Hybridization, 46-48
Hydrogen sulfide
  effect on beer flavor, 119

Hydrogen sulfide (continued)
　formation in beer, 119
　in wine fermentations, 178
Hydrolysis, acid, production of yeast extracts, 379-380
Hydroxymethylfurfural, inhibition of distiller's, fermentations, 245

Immobilization of malo-lactic bacteria, 197
Immobilized yeast
　champagne production, 215
　fuel ethanol production, 253
Indolyl acetic acid, as yeast growth activator, 279
Industrially important species, 21
Industrial yeasts, nomenclature of, 21
Infusion mashing, brewing process, 108-111
5'-Inosinate, 381
Instant active dry baker's yeast, 299
Insulin, human, expression in yeast, 406
Interferon, expression in yeast, 406
Inulin, tequila fermentation substrate, 240
Invertase, 393-396
　of baker's yeast, 61
　gene dosage effect, 62
　properties, 394-395
Irish whisky, 235
Isoamyl alcohol, formation by wine yeast, 189
Isobutanol, formation by wine yeast, 189

Karl mutation, 78
Kefir production, 442-444
$\alpha$-Ketoglutaric acid, formation by wine yeast, 187
Ketones, in bread flavor, 335
Killer factor
　in brewer's yeast, 76-79
　introduction into wine yeast, 184-185
Killer toxin, 77, 141
Killer yeast, in brewing, 140-141
*Kloeckera*, occurrence on grapes, 154
*Kloeckera apiculata*, description, 159
*Kluyveromyces*, taxonomic characteristics, 15
*Kluyveromyces marxianus*
　biomass production, 433-434
　immobilized for lactose hydrolysis, 442
　lactase production, 396
　yeast extract production, 371
Koumiss production, 444

Laccase, 200
LAC genes, 71, 278, 397
Lactase, 396
Lactic acid bacteria
　as baker's yeast contaminants, 297
　in rye bread production, 354-355
　spoilage of wine, 200
*Lactobacillus*, as brewery contaminant, 124
*Lactobacillus casei*, malo-lactic fermentation, 193
*Lactobacillus hilgardii*, malo-lactic fermentation, 193
*Lactobacillus plantarum*
　malo-lactic fermentation, 193
　soda cracker fermentation, 358
*Lactobacillus sanfrancisco*, in dough fermentation, 357
Lactose fermentation, by baker's yeast, 70-71
Lactose-free milk, 441-442
Lactose intolerance, 396
Lactose permease, 396
Lager beer fermentation
　effect of pH and temperature, 118-119
　primary, 116-117
　secondary, 118
Leached solids, of active dry baker's yeast, 305
Lean dough processes, 343-347
Leavening gases, 316
Leloir pathway, 69
*Leuconostoc mesenteroides*, malo-lactic fermentation, 193
Leu-2-gene, 68
Life cycle, 38-42
Light beer. See Beer, low calorie
Linneaus, Carolus, 9
Lipids
　concentration in oleoginous yeasts, 435
　concentration in yeasts, 423

Maillard reaction during bread, baking, 335
MAL genes, 57, 61, 345
Malic acid, degradation by yeasts, 197-198
Malo-lactic bacteria
　immobilization, 195
　inhibition of, 196
　nutritional requirements, 195
Malo-lactic fermentation, 193-197
　effect of pH and temperature, 195
　metabolic pathways, 193-194

Malt
  gibberellin treated, 227
  peated, 233
Malting, 105-107
Malt kilning, 107
Maltose
  adaptation of yeasts, 56-58
  fermentation in doughs, 344-347
  utilization by brewer's yeast, 72-73
Maltose permease, 346
Malt rootlets, 5′-phosphodiesterase, 383
Malt wort. See also Brewer's wort
  mineral content, 101-102
  sugar concentration, 102-105
  vitamin content, 100
Mass mating technique, 48
Mating type, 39
  interconversion, 41
MEL genes, 67, 397
Melibiase, 397
Melibiose
  fermentation of, 104
  utilization by baker's yeast, 66-69
Mercaptans, in wine fermentations, 178
Methanol
  as carbon source, 23
  concentration in wine, 187
  as substrate for yeast growth, 436-437
*Metschnikowia*, occurrence on grapes, 154
*Metschnikowia pulcherrima*, description, 158
Minerals
  nutritional contribution of yeasts, 419
  required by baker's yeast, 273
  required by brewer's yeast, 101-102
  required by *Candida utilis*, 430
Mixing, of bread doughs, 342
Molasses
  composition, 271
  high-test, use of invertase, 393
  storage, 274
  yeast growth inhibitors, 274
Mold inhibitors, use in doughs, 331
Mutagenesis, 44-46
*Mycoderma vini*, description, 159

*Nadsonioidea*, taxonomic characteristics, 16
Natural fermentations, 1
Natural yeasts
  occurrence on grapes, 154
  in wine fermentations, 152

Nitrogen metabolism, of brewer's yeast, 95-98
Noble rot, 202
Nucleic acids, in the human diet, 386
Nutrition, role of yeast fermented foods, 414
Nutritional requirements
  of baker's yeast, 275-278
  of brewer's yeast, 95-102
  of *Candida utilis*, 429-430
Nutritional yeast, definition, 415

*Obesumbacterium proteus*, as brewery contaminant, 125
*Oideum lactis*, baker's yeast contaminant, 297
Oleaginous yeasts, 434-435
Osmotic pressure
  effect on dough fermentations, 327-330
  effect on wine fermentations, 175
Osmotolerance of baker's yeast, 58
Osmotolerant hybrids, baker's yeasts, 60
Oven spring, 317, 324
Oxygen, as leavening gas, 316
Oxygen transfer coefficient, 266
Oxygen transfer rate, 268, 430

Pannetone, 357
Pasteur, Louis, 37
Pasteur effect, 128
Pathogenic yeasts, 34
*Pediococcus*, as brewery contaminant, 124
*Pediococcus damnosus*, malo-lactic fermentation, 193
*Penicillium*, 5′-phosphodiesterase, 383
2,3-Pentanedione, in beer, 132-135
Pentosans, in rye flour, 354
Pentoses, assimilation by *Candida utilis*, 427
Periodicity, growth of baker's yeast, 283
Pesticides, inhibition of wine fermentations, 181
pH
  in baker's liquid ferments, 339
  effect in wine fermentations, 174-175
  of yeast protoplast, 323
*Phaffia rhodozyma*, astaxanthin production, 391-392
2-Phenethyl acetate, formation by wine yeasts, 190

# 452  INDEX

Phenethyl alcohol, formation by wine yeasts, 189
Phenylketonuria, 398
5'-Phosphodiesterase, 383
Pichia
  high cell density growth, 430
  taxonomic characteristics, 15
*Pichia membranefaciens*, description, 158
*Pichia pastoris*, genetic expression system, 408
*Pichia saitoi*, in sour doughs, 355
Pigments, in yeast species, 390-393
Pita bread, 359
Pitching yeast, brewing process, 112-116
Pizza doughs, 347-348
Plasmid cloning vectors, 51
Plasmolysis, production of yeast extracts, 377-378
Ploidy, 38
Polyphenolic compounds, effect in wine fermentations, 181
Preferment dough process, 336, 338-342
Pretzels, 348
n-Propanol, formation by wine yeasts, 189
Proteases of yeasts, 373, 375
Protein, concentration in yeasts, 416-419
Protein efficiency ratio, of yeast protein, 386, 418
Protoplast fusion
  intrageneric and intergeneric, 71
  techniques, 48
Pulque, 240
Putrescine, in wine, 203

Raffinose, utilization by baker's yeast, 275
Rapid fermentation kinetics, baker's yeast, 62-65
Recombinant plasmids, introduction of, 54
Respiratory quotient
  in baker's yeast production, 270-271
  process control, 291
  for yeast growth on hexane, 436
*Rhodotorula*, pigment formation, 391
Rhodotorula glutinis
  lipid fermentation, 434-435
  phenylalanine ammonia-lysase production, 398
RNA
  excretion by yeast, 382
  extraction from yeast, 382-383

phosphodiesterase hydrolysis, 384
  removal from yeast biomass, 387-388, 425
Rope spores, baker's yeast contaminants, 297
Rum, 240-242
Rye flours, 354
Rye sour doughs, 353-357
Rye whisky, 235

*Saccharomyces*
  high cell density growth, 430
  properties, 20
  taxonomic character, 15
*Saccharomyces bayanus*
  description, 158
  sherry fermentation, 210
*Saccharomyces cerevisiae*
  brewer's yeast, 113
  description, 157
  in sour doughs, 355
  spheroplast fusion, 50
*Saccharomyces diastaticus*, distiller's fermentations, 244
*Saccharomyces mellis, Saccharomyces cerevisiae* hybrid, 318
*Saccharomyces pombe*
  degradation of malic acid, 198
  description, 158
  rum fermentation, 241
*Saccharomyces prostoserdovii*, sherry fermentation, 210
*Saccharomyces uvarum*
  brewer's yeast, 113
  description, 160
Saccharomycetoidea, genera, 16
*Saccharomycodes*, taxonomic characteristics, 15
*Saccharomycodes ludwigii*
  description, 158
  taxonomic characteristics, 19
*Saccharomycopsis*, taxonomic characteristics, 15
Salt, effect on osmotic pressure in doughs, 328-330
San Francisco French sour dough, 357
Sauterne, flavor of, 202
*Schizosaccharomyces*, taxonomic characteristics, 15
Schizosaccharomytedia, genera, 16
Scotch whisky, 233-235

Selenium compounds, in yeast, 421-422
Selenomethionine, 421
Serology, in yeast taxonomy, 11
Serum albumin gene, cloning in yeast, 408
Sexuality. *See* Mating type
Sherry
  baked, 210
  solera process, 207
  submerged culture, 209
Single-cell oil, 435
Skin respiratory factor, production from baker's yeast, 399-400
Snake venom, 5'-phosphodiesterase, 383
Soda crackers, 358-359
Soda cracker sponges, yeast counts, 359
Spheroplast fusion, 50
Spoilage of wine by microbial growth, 199-200
Spoilage yeasts, 33-34
Sponge dough process, 336-338
Spontaneous fermentations, 1
  in wine production, 155-157
Sporulation, of industrial strains, 47
STA genes, 74
Starch
  damaged, 319
  utilization by brewer's yeast, 73-76
Sterols
  activation of wine fermentations, 171-173
  concentration in wine yeasts, 233
Stills
  Coffey stills, 233
  patent stills, 233
  pot stills, 226
  two-column stills, 226
Straight dough process, 336-337
Stuck wine fermentations, 181-184
SUC genes, 62, 394
Sugars
  assimilated by yeast, 153
  concentration during dough fermentation, 319-321
  effect on osmotic pressure in doughs, 327-330
  fermentable in flour, 319
  fermentable by yeasts, 153
Sulfite waste liquor, 426-428
Sulfur compounds, beer flavor, 127, 136-138
Sulfur dioxide, use in wine making, 176-179
Sweet baked goods, 347

Taxonomy
  history of, 9
  principles of, 8
  revision of, 30
Tequila, 240-241
Titratable acidity of wines, 187
Top fermenting yeasts, 105
Torula, 426
*Torulaspora*, taxonomic characteristics, 15
*Torulaspora delbrueckii*
  as baker's yeast, 264
  description, 158
  as frozen dough yeast, 318
*Torulopsis holmii*
  in soda cracker production, 358
  in sour doughs, 355, 357
Transformation. *See* DNA, cloning technology
Trehalose
  in baker's active dry yeast, 298
  in frozen dough yeast, 353
  stabilizing effect, 60
Tyramine, in wine, 203

Ultraviolet light mutagenesis, 45
Umami, 381
Urethane. *See* Ethyl carbamate
5'-Uridine monophosphate, 385

Vicinal diketones
  in beer, 132-135
  concentration in wines, 192
  flavor threshold in wines, 192
Vinegar, as mold inhibitor in doughs, 331
4-Vinylguaiacol, 75
Vitamin D, in irradiated yeast, 390
Vitamins
  nutritional contribution of yeasts, 419
  in yeast biomass, 388-390
*Vitis rotundifolia*, occurrence of yeasts, 155
Vodka, 238
Volatile acidity of wines, 187

Wheat sour doughs, 357
Whey
  as baker's yeast substrate, 278
  as fermentation substrate, 433
Whisky, 227-236
  maturation in oak barrels, 251
White bread formulations, 336
Wild yeasts, 152

Wine
  spontaneous fermentation, 155-157
  table, processing, 204-206
  taste, effect of yeast strain, 192
Wine active dry yeast
  fermentation activity, 161
  inhibition by sulfur dioxide, 161
  rehydration, 161
Wine fermentation
  alcohol yields, 173-174
  assimilation of N compounds, 179
  biochemistry of, 165-184
  inhibitors, 181-184
  starter cultures, 162-165
  succession of yeast species, 156-157
Wine yeasts
  descriptions, 157-159
  ecology, 151
  fermentation rate, 166
  heat inactivation, 168-169
  rate of bread dough fermentation, 323
  reclassification, 152
  species
    identification, 164
    synonyms, 152

Yeast cream, 294
Yeast culture, as animal feed, 425
Yeast extract, 370-381
  bitterness, 371
  as distiller's fermentation nutrient, 247
  use in wine making, 179, 183
Yeast hulls. *See* Cell wall material
Yeast nutrients
  nitrogenous compounds, 179-181
  use in dough fermentations, 330
Yeasts. *See also specific yeasts*
  cell volume, 176
  classification, 7-35
  gross composition, 415-416
  industrial, nomenclature, 21
  selected pure cultures
    distiller's fermentations, 244
    wine fermentations, 159-162
  spoilage, 33-34

*Zygosaccharomyces bailii*, description, 158
*Zygosaccharomyces rouxii*, description, 158
*Zymomonas*, as brewery contaminant, 124